The 64th Army at Stalingrad 1942 – 1943

A Day-by-Day Account of a Soviet Combined Arms Infantry Army During the Battle for Stalingrad

Dann Falk

2019

Copyright © 2019 Dann Falk

Second Edition

All Rights Reserved

No reproduction without written permission from the author

You may contact the author at falkenbooks.com or on Facebook

Published by Falken Books

ISBN: 978-1-7326074-1-5

To All Those Who Fought

All photographs are noted by source, and where necessary every reasonable effort has been made to trace copyright holders to obtain permissions for their use in this book. The author and publisher apologize for any errors.

Special thanks to Jason Mark of Leaping Horseman Books for supplying photos from his personal collection.

Also special thanks to Sherry Anderson for her editing efforts and making the second edition possible.

No work like this can be completed without putting in years of research, then years of writing and further time gathering together bits and pieces of information. It was a very long journey. But I was not alone in this journey; family and friends helped out along the way, and many new friends from around the world also freely gave valuable support and guidance as I needed it. Finally, I owe a great deal to my wife Sarah for making this book real. Without her unwavering assistance and tolerance of my jabbering about events from 75 years ago, my efforts would have ended in failure.

Table of Contents

Maps, Figures, Photos, Abbreviations — v

Preface — 10

Introduction — 11

Chapter 1	December 1941	19
Chapter 2	January 1942	22
Chapter 3	February 1942	25
Chapter 4	March 1942	26
Chapter 5	April 1942	29
Chapter 6	May 1942	32
Chapter 7	June 1942	36
Chapter 8	July 1942	40
Chapter 9	August 1942	75
Chapter 10	September 1942	129
Chapter 11	October 1942	150
Chapter 12	November 1942	167
Chapter 13	December 1942	192
Chapter 14	January 1943	209
Chapter 15	February 1943	235

Conclusion	256
Appendix	258
Notes	267
Selected Bibliography & Websites	280
Index	285

Maps, Figures, Photos, Abbreviations

All maps and figures were produced by the author, and the author takes full responsibility for any errors or omissions. They are not intended to be used as navigation charts, but as reference aids to assist in understanding the flow of battle and the movements of armies and units.

Maps

1. Operation Barbarossa – June to September 1941 — 15
2. Operation Typhoon – 30 September to 15 December 1941 — 17
3. Stalingrad Front Deployment Plan – Mid to late July 1942 — 44
4. 1st Reserve Army Deployment and Training Areas – Late June 1942 — 48
5. 64th Army Entraining Locations – Early July 1942 — 51
6. 64th Army Journey to the Stalingrad Area – Mid July 1942 — 53
7. 64th Army Detraining Locations – July 1942 — 59
8. 64th Army Deployment – 22-24 July 1942 — 61
9. 64th Army Deployment – 25-29 July 1942 — 69
10. 64th Army Deployment – 30-31 July 1942 — 73
11. 64th Army Deployment – 1-5 August 1942 — 76
12. Stalingrad Defensive Positions — 78
13. Map Set – 64th Army Deployment – 4-7 August 1942 — 85
14. Map Set – 64th Army Deployment – 8-11 August 1942 — 91
15. 64th Army Deployment – 17-19 August 1942 — 95
16. Stalingrad Area Situation – 20-27 August 1942 — 98
17. 64th Army Withdraws – 26-30 August 1942 — 105
18. Rush to Stalingrad – 31 August to 3 September 1942 — 130
19. Overall Situation – 4 September 1942 — 134
20. Into Stalingrad – 4-12 September 1942 — 141
21. 64th Army Deployment – End of September 1942 — 149
22. 64th Army Attacks – 25-31 October 1942 — 164
23. Operation Uranus Plan – November 1942 — 173
24. 64th Army Deployment and Attacks – 20 November 1942 — 181
25. Stalingrad Pocket – End of Operation Uranus – 30 November 1942 — 191
26. 64th Army Deployment – 5 December 1942 — 196
27. The End of Operation Winter Tempest – 23 December 1942 — 204
28. Start of Operation Ring – 10 January 1943 — 216
29. Operation Ring – 10-22 January 1943 — 218
30. Final Attacks into Southern Stalingrad – 25-31 January 1943 — 225

31.	64th Army Deployment – 20 February 1943	250
32.	64th Army Journey to Valuiki – Belgorod Region	257

Map Key

Rail Lines

Villages, Towns, Cities, Rail Stations ●

Rivers

Front Line Russian

Movements, Attacks Russian

Front Line German

Movements, Attacks German

Dates 23-8   23 August

Bridges ⟨B⟩

Ferries ⟨F⟩

Major Combat Units 29 RD

Army or Front Boundries ▬ ▪ ▪ ▪ ▪ ▪ ▪ ▪

Armies Russian   62 A   64 A   57 A

Armies German or Axis   6 Army   4 Pz Army

Prominent Peak   145.5 ▲

Figures

1. Geographic Layout Stalingrad Area — 70
2. Don Volga Cross Section — 71

Photos

All photographs are noted by source.   Page

1. Joseph Stalin (author's collection) — 108
2. Adolf Hitler (Bundesarchiv – Bild 121-0720) — 109
3. General Eremenko, SE & SG Front Commander & Nikita Khrushchev, Politburo Member (author's collection) — 110
4. Colonel General von Weichs, Commander of Army Group B (Jason Mark / Leaping Horseman Books) — 111
5. Field Marshal von Manstein 1942, Commander of Army Group Don (author's collection) — 112
6. Colonel General Richthofen and Field Marshal von Manstein (Bundesarchiv – Bild 183-B18912) — 113
7. Major General Timofey T. Khryukin, Commander of the 8th Air Army (Russian State Film and Photo Archives at Krasnogorsk (RGAKFD) and RussianArchives.com) — 114
8. General of Armored Troops Friedrich Paulus, Commander of 6th Army (Bundesarchiv – Bild 101I-021-2081-31A) — 115
9. Lieutenant General Vasilii Chuikov, Commander of 62nd Army, at his HQ in Stalingrad (RGAKFD and RussianArchives.com) — 116
10. General von Weichs, General Paulus, and General of Artillery Walter von Seydlitz, Commander of 51 Corps (Bundesarchiv – Bild 00216-033) — 117
11. Colonel General Richthofen, Commander of *Luftflotte 4 (Air Fleet 4)*; Lieutenant General Warner Kempf, Commander of 48 Pz Corps; and Colonel General Hermann Hoth, Commander of 4th Army group Panzer Army (Bundesarchiv – Bild 101I-216-0410-08) — 118
12. Lieutenant General Warner Kempf, Commander of 48 Pz Corps (Bundesarchiv – Bild 101I-218-0515-11) — 119
13. Field Marshal Paulus, General Schmidt, and Colonel Adam arriving at 64th Army HQ in Beketovka (Jason Mark / Leaping Horseman Books) — 120
14. Field Marshal Paulus in captivity (Jason Mark / Leaping Horseman Books) — 121
15. Lieutenant General Konstantin Rokossovsky, Commander of the Don Front, with Field Marshal Paulus (RGAKFD and RussianArchives.com) — 122
16. General Chuikov, General Shumilov, and Nikita Khrushchev in Stalingrad. (RGAKFD and RussianArchives.com) — 123
17. Editorial Staff for the 64th Army Paper *"For the Motherland"* (RGAKFD and RussianArchives.com). — 124
18. 64th Army checkpoint behind the front (RGAKFD and RussianArchives.com). — 125
19. 64th Army Soldiers during training (RGAKFD and RussianArchives.com). — 126

20. 64th Army Troops after the Battle for Stalingrad (RGAKFD and RussianArchives.com). 127
21. General Shumilov and his driver in a Jeep (RGAKFD and RussianArchives.com). 128

Abbreviations - Soviet

A – Army – 57th Army – 57 A
RC – Rifle Corps – 18 Rifle Corps – 18 RC
RD – Rifle Division – 422 Rifle Division - 422 RD
RB – Rifle Brigade
MRB – Motorized Rifle Brigade
NRB – Naval Rifle Brigade
RR – Rifle Regiment
MRR – Motorized Rifle Regiment
MR – Mortar Regiment
CRR – Cadet Rifle Regiment
GRD – Guards Rifle Division
GRB – Guards Rifle Brigade
GRR – Guards Rifle Regiment
FR – Fortified Region
NKVD RD – NKVD Rifle Division (Secret Police)

TA – Tank Army – 5th Tank Army – 5 TA
TC – Tank Corps – 13 Tank Corps – 13 TC
TB – Tank Brigade
TR – Tank Regiment
GTB – Guards Tank Brigade
GTR – Guards Tank Regiment
CC – Cavalry Corps
CD – Cavalry Division
CR – Cavalry Regiment

AA – Anti-Aircraft or Air Army – 8th AA
AT – Anti-Tank
ART – Artillery
GMR - Guards Mortar Regiment - Katusha rocket launcher
Stavka – Headquarters of the Supreme High Command
NKO – People's Commissariat for Defense
GKO – State Defense Committee
VGK – Supreme High Command - General Staff of the Red Army
RVGK – Reserve of the Supreme High Command
VVS – (*Voenno-Vozdushnye Sily*) Red Army Air Force

German - Axis

OKW – Armed Forces High Command
OKH – Army High Command (Command Authority for Eastern Front)
AG – Army Group - AG South
A – Army – 6th Army
Pz A – Panzer Army – 4th Pz A
AC – Army Corps – XXV AC (25 Corps)
Pz C – Panzer Corps – XXXXVIII (48 Pz C)
ID – Infantry Division – 5 ID
IR – Infantry Regiment – 215 IR
IB – Infantry Brigade – 23 IB

Pz D – Panzer Division – 24 Pz D
MD – Motorized Division – 29 MD

Ru A – Romanian Army - 4th Romanian Army – 4 Ru A
Ru ID – Romanian Infantry Division – 20 Ru ID
Note: Russian / German usage (*Румынский* = Ru) (*Rumänisch* = Ru) on period maps.

It A – Italian Army – 8th Italian Army – 8 It A
It ID – Italian Infantry Division

Hu A – Hungarian Army – 3rd Hungarian Army – 3 Hu A
Hu ID – Hungarian Infantry Division

Note: To help distinguish among the different armies and types of troops, all Soviet/Russian infantry units are referred to as Rifle units and armored units are called Tanks units. German/Axis infantry units are referred to as Infantry and armored units as Panzer units.

Note: For BSSA and BSA definitions see Appendix 1.

Preface

This volume was made possible due to a long struggle by many authors and researchers to reveal the true history of the Soviet/German war during World War II.

In the beginning, historians were limited to few sources of information and produced early works based mostly upon personal accounts and officially approved views of the fighting. German and Russian language titles dominated, with a growing number of titles being translated into English.

Authors like Alexander Werth – 1946, F.W. von Mellenthin – 1956, Heinz Schroter – 1958, Paul Carell – 1963, V.I. Chuikov – 1963, Alexander Werth – 1964, Alan Clark – 1965, Geoffrey Jukes – 1968, and a slew of English language titles coming from the Soviet Union by Progress Publishers in the 1960s, all set the stage for further development of the subject.

Then gradually things began to change. More intense research at various archives brought to light insights and critical pieces of information never before seen, and the stories began to change. Soon the number of volumes increased and the quality and accuracy improved. A new group of authors appeared, Albert Seaton – 1971, William Craig – 1973, Manfred Kehrig – 1974, John Erickson – 1975, and Walter Kerr – 1978, and others. With these new titles building upon earlier efforts, a firm base had been established to move forward into the unknown.

Adding to this growing body of work was the 12 volume Official Soviet *History of the Second World War 1939-1945* edited by Marshal G.A. Grechko, 1973-1982. Then Germany started their groundbreaking 13-part official series, *Germany and the Second World War* – 1979, which was followed by the two volume *US Army Historical Series* of the German-Soviet Conflict, by Earl Ziemke – 1987. These added further details and clarity to the field along with careful research. With the idea of *glasnost* (openness) from the Soviet Union beginning in 1985, new avenues of research opened within the Soviet Union. A series of Soviet General Staff Studies and other official WW II documents began to surface from behind the Iron Curtain with works by Louis Rotundo -1989, V.E. Tarrant -1992, Ewin Hoyt – 1993, Antony Beevor - 1998, and others.

All this hard work set the stage for the use of the newly developed Internet which began a worldwide revolutionary exchange of information and ideas that will eventually allow growing access to previously restricted Soviet and current Russian World War II archives. Leading the way in this expanding field of study were authors like David Glantz and Jonathan House, with their prodigious pioneering series of books, atlases and papers covering the Russo/German war in unpresented detail. Others soon followed, R.L Dinardo – 1991, Jason Mark – 2002, Walter Dunn – 2006, Jonathan Bastable – 2006, Michael Jones – 2007, Geoffrey Jukes – 2011, Anton Joly – 2011, Jochen Hellbeck – 2012, Richard Harrison – 2016, Valeriy Zamulin – 2017, Hugh Davie 2017, and so many more. These last few decades have truly turned out to be a golden age of research and publishing regarding the Soviet/German war. With growing confidence, we can once again step into the unknown and push back the darkness of hidden truths a bit further.

Without the unending efforts of these authors, researchers, historians and numerous other dedicated individuals, my efforts would have utterly failed to produce anything of value. Finally, I am pleased to add my small piece of the historical puzzle to this distinguished body of work, and in so doing, maybe reveal a bit more of the truth.

The 64th Army at Stalingrad 1942 – 1943

Introduction
Blitzkrieg in the East
The Beginning
1941

"You only have to kick in the door
and the whole rotten structure will come crashing down."

Comment by Adolf Hitler about invading the Soviet Union

"The war against Russia will be such that it cannot be conducted in a knightly fashion. This struggle is one of ideologies and racial differences and will have to be conducted with unprecedented, unmerciful and unrelenting harshness."

Adolf Hitler (1941)

"Nowadays wars are not declared. They simply start."

"The Red Army and Navy and the whole Soviet people must fight for every inch of Soviet soil, fight to the last drop of blood for our towns and villages...onward, to victory!"

"Die, but do not retreat."

Joseph Stalin

These comments give us some idea of Hitler's impression of what an attack upon the Soviet Union would be like. In turn, Stalin makes his own thoughts clear: fight to win or die trying.

Early on the morning of 22 June 1941, war came to Russia. Hitler had finally put his words into practice with Operation Barbarossa; the Nazi invasion of the Soviet Union had started.

Along a 930-mile front, the German army waited to go over to the offensive. Three Army Groups, North, Center, and South, containing three million troops, were ready. Hundreds of aircraft from three Luftwaffe Air Fleets were already airborne on their way toward Russian targets. Joining in on the attack were also 300,000 Finns; 358,000 Romanians; and 44,000 Hungarians which pushed the total attacking force to well over the 3,800,000 mark. At 03:15 hours, thousands of artillery pieces suddenly opened fire. The war between Nazi Germany and Soviet Russia was on, transforming a once peaceful border into the Russian Front.

On the Soviet side of the border, this attack came as a complete surprise to the approximately two and a half million defending troops. They had not been told about the German buildup, and they had not been put on alert. Confusion was widespread at all levels of command as the German bombardment chewed up the forward Russian positions and broke vital phone lines. There would be little help or guidance from higher headquarters during the first days of the assault. The shock of the attack extended all the way up the chain of command to even Joseph Stalin. His shock appears to have been even greater because during the preceding months, the British government had warned Stalin of the impending attack, but Stalin did not believe them. Stalin also was receiving reports from spies and even his own troops that Germans forces were moving into assault positions near the border, but he would not believe these either. In the days and weeks before the attack, no action was allowed to take place that might provoke German aggression. Nothing was done that might give the Germans cause to attack. At the last minute, just after midnight on 22 June to be exact, an alert order was sent from

Moscow to the army HQs along the border. This order arrived too late to be effective. For the troops in the field, the surprise was total.

However, behind closed doors, other steps were being taken to prepare the Red Army for the coming battle. The Red Army was quietly being mobilized deep in the interior of the country; a secret general mobilization was already in progress. Reserves were being transferred to assembly locations; new divisions and even armies were being formed. The Red Army was gathering its strength.

For many years, Stalin had been playing a dangerous game with Hitler for control of Europe. The other main players were of course France, England, Italy and the far-off United States. Both Stalin and Hitler were skilled politicians capable of playing one nation against another to gain an advantage. They used force to achieve their aims in some cases; in others, treaties were worked out. Both leaders had matching ambitions for land and power that would eventually lead to a clash of wills. Nevertheless, from about 1939 on, Stalin also had been playing for time. After the German conquest of Poland, Denmark, Norway, Luxemburg, Belgium, Netherlands, and France, the battle-tested German forces appeared to be unstoppable. Stalin knew the Red Army was not ready to fight the more advanced and experienced German military. The Red Army was large but not up to date in tactics or equipment. The Soviet arsenal was full of old and outdated weaponry. New modern weapons like the T-34 and KV-1 tanks were just starting to be produced in large quantities. Secret weapons like the Katyusha rocket launchers also were being developed. New fighters and bombers for the air force, along with ships and submarines for the navy were being planned. All these new weapons had not yet been produced in sufficient quantities to reequip the Soviet military.

By making Russia useful to the German state, Stalin hoped to delay any conflict with Nazi Germany until he was ready to fight. Trading crude oil, food, and strategic raw materials to Germany was one method of appeasing Hitler. Another method was agreeing upon spheres of influence and establishing understandings about future military or political efforts. In the end, appeasement did not pay off. In many ways, Soviet aggression against neighboring states fueled the ill feelings between the two dictators. Soviet compliance in the Nazi attack on Poland went a long way toward isolating the Soviet Union from other European powers in 1939 and 1940. The blatant Soviet attack upon Finland was soon followed by the occupation and annexation of the three Baltic States: Estonia, Latvia and Lithuania. These aggressive moves pushed Hitler closer to his final decision about his eastern neighbor. The occupation and annexation of Bessarabia and Northern Bukovina along with expansion of Soviet influence into the Balkans made it clear to Hitler that only war could resolve his concerns about Soviet Russia. Hitler had been talking publicly about conquest of the East ever since his book *Mein Kampf* was first published in 1925. Ultimately, he would unleash the largest military campaign in history, Operation Barbarossa. Hitler would pit Nazi Germany's might against the vast resources, both human and material, of the Soviet state. Nazi vs. Communist ideology, guided by two ruthless dictators, would ensure the resulting war would be savage.

Many books have been written about Operation Barbarossa and the war on the Russia Front. I do not intend to cover the same information yet again, but a review of the main points during the first few months of combat is appropriate.

The massive 22 June surprise attack proceeded as planned all along the front. In the air, surprise also was achieved as the initial German Luftwaffe attacks struck hundreds of Soviet airfields. By the end of the first day, more than 1,200 Soviet Air Force (VVS) aircraft had been destroyed. Over the next few weeks, thousands more would fall victim to the superior Luftwaffe aircraft and pilots. The Luftwaffe's dominance of the airspace over the battlefield had begun. (see Map #1)

On the ground, German Panzer forces swiftly advanced to the east. Army Group North was thrusting into the Baltic States of Estonia, Latvia and Lithuania with an ultimate objective of Leningrad. Army Group Center was plunging toward Minsk, Smolensk and ultimately Moscow. Army Group South was headed deep into the Ukraine toward Kiev and Rostov at the mouth of the Don River.

Large encirclements of Russian forces were taking place as the Soviet High Command lost control of the battlefield. German Panzer spearheads were closing behind huge groupings of Red Army troops. These envelopments were yielding vast numbers of prisoners and mountains of equipment. The Red Army was being systematically destroyed. These encirclement battles, both large and small, were being fought all along the front, leading to the total devastation of the Red Army. In June 1941, the encirclement around Minsk destroyed 10 rifle divisions; in July, 12 more were lost during the Smolensk battle. In August, 15 more rifle divisions were lost at Uman and then 34 rifle divisions were destroyed during the great Kiev encirclement in September. Likewise, tank and cavalry divisions and brigades also were being surrounded and destroyed by the score. Millions of men and huge amounts of equipment were being lost. Eventually the Red Army would lose 140 rifle divisions due to encirclement battles by July 1942 and a further 42 rifle divisions by the end of 1942. At that point the Red Army and Stavka had learned how to avoid or defeat these large German encirclement tactics.[1]

The rapid German advance pleasantly surprised the German High Command and shocked the Soviet command structure. By 28 June, Minsk had fallen and large parts of the Ukraine in the South and Latvia and Lithuania in the North had been taken. Smolensk was reached on 16 July, and despite much hard fighting by Soviet troops, a large pocket was soon formed. With some delay, the Germans were able to clear the Smolensk pocket and continue the advance. In the north, Leningrad had been largely isolated by 30 August when the last rail link into the city was cut. The battle for Kiev ended on 17 September after the huge encirclement of Red Army troops in that area was eliminated. By 30 September, Leningrad had been isolated and only Lake Ladoga offered an opening in the German ring. In the center, German forces were just over 200 miles from Moscow. In the South, the Crimea had been cut off from the mainland and the front lay along the Dnieper River. But the Germans were not finished; they were preparing to launch a final assault upon Moscow, Operation Typhoon.

Up to this point, the Red Army had been inflicting heavy losses on German forces, but it also had been losing vast amounts of territory and entire armies in the process. To keep fighting, the Red Army needed reinforcements to first halt and then drive back the invasion. Ultimately, Stalin and the Soviet High Command (Stavka) planned well for the rapid expansion and the effective reinforcement of the Red Army. Between June and December 1941, the Red Army received 3,544,000 new recruits. This flood of manpower was translated into 260 new divisions.[2] A number of these new divisions had already been destroyed by the end of September, but most had been used to build new armies or rebuild battered ones. Some of these armies were fighting at the front and others had not yet been committed to battle. One of these reserve armies, the 24th Army, happens to be of particular interest to this study because its formation and ultimate destruction directly leads us to the creation of the 64th Army and its great achievements during the war.

On 24 June 1941, just two days after the start of the war, the 24th Army was activated in the Siberian military district as a Stavka High Command reserve army. It contained two rifle corps with three rifle divisions each.[3]

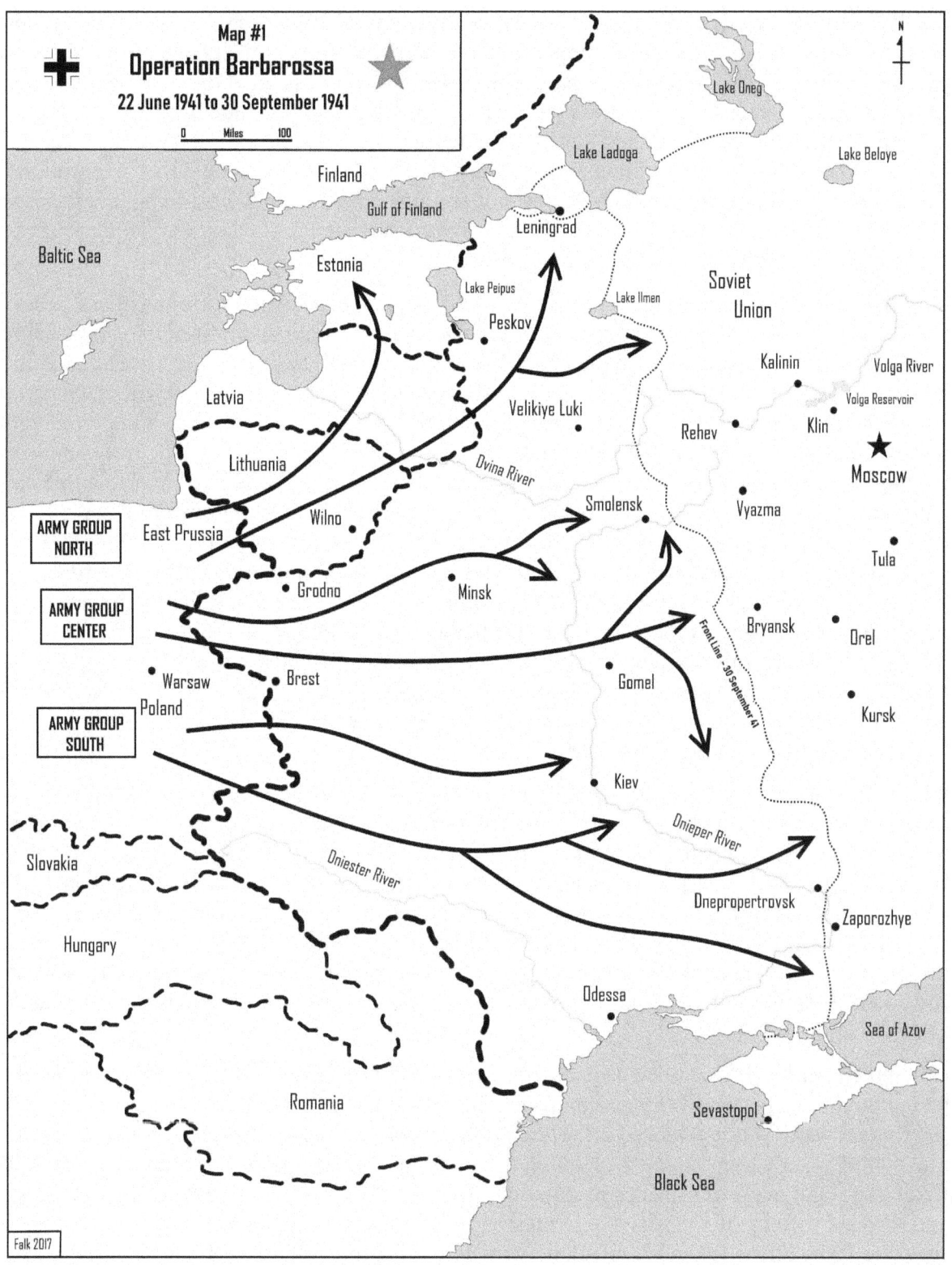

The 24th Army remained in reserve throughout the early part of July when it was moved west of Moscow to strengthen the defense of the capital. This transfer, by rail, must have occurred soon after its formation because by 10 July it was part of the Stavka reserve assigned to prepare defenses in the Vyazma area, halfway between Smolensk and Moscow. What were the thoughts of the officers and men of the 24th Army as they rushed toward the west and the very real war that was fast approaching Moscow? What would become of them? Could they stop the Germans? Would they ever return to their homes and loved ones again? These types of feelings must have been common, but these same Red Army soldiers knew one thing for certain, they were prepared to do their duty to their last breath to defend the Soviet Union from the foreign invaders.

At this point events were moving rapidly at the front. German attacks had surrounded and destroyed the 16th and 20th Armies near Smolensk in July, thus opening a hole in the front line. On 16 July, the German 10 Panzer Division crossed the upper Desna River, south of Smolensk, in an attempt to capture the small but vital town of Yelnya. The 10 Panzer Division found the 19 Rifle Division (RD), of the 24th Army waiting on the other side of the river. The 24th Army had been transferred to the new Reserve Front on 15 July. This new front had been created on 14 July and the front commander moved the 24th Army forward quickly to engage the enemy. The distance from Vyazma to Yelnya is about 60 miles, so units of the 24th Army must have already been marching toward the Yelnya area since 12 - 13 July. The 24th Army would be engaged in heavy fighting on this section of the front from 15 July through September. German accounts compared fighting in this area to Verdun during WWI. Yelnya finally fell on 19 July after the German attack had created a salient into the Russian lines around the city. The Russians were determined to remove this dangerous bulge in their lines. On 30 August, the 24th Army attacked with 13 divisions and over 800 pieces of artillery. After several days of fighting, the Germans finally decided to withdraw from the Yelnya salient, reduce their losses, and straighten the front line. By 8 September, the bulge was gone. The Red Army had finally won a victory, and it would not be their last.

After weeks of intense fighting, the front line in the Yelnya area calmed down. The 24th Army rested and waited. On the other side of the lines, Operation Typhoon, the final German assault on Moscow, was in the final stages of planning. It would be a storm of steel which would sweep away the 24th Army and surge to the very gates of Moscow. On 16 September 1941, the directive for Operation Typhoon was issued with Hitler's full approval despite misgivings from some members of the German High Command. This last offensive of the season was almost ready; victory was within sight. With Moscow so close and the Red Army apparently in a weakened state, the victorious German Army and Luftwaffe were prepared for one final push, one massive assault to end the war and destroy the Soviet Union in 1941.

On 30 September 1941, Operation Typhoon was launched. (see Map #2) At first the attack only came from the South aimed toward Bryansk and Tula, but two days later, on 2 October, the offensive on the direct route toward Moscow kicked off. The 24th Army was directly in the path of this attack. The German 4th Panzer Group rapidly penetrated the front lines and drove deep into the rear areas. The 24th did not stand much of a chance in stopping this powerful armored force. Soon a large pocket was formed around the city of Vyazma by German panzer formations. This pocket contained four Russian armies, the 19th, 20th, 24th and 32nd, along with parts of several others. German infantry units quickly followed to strengthen the ring sealing in the trapped armies. Despite repeated attempts to breakout, very few men were able to make their way out of this tight encirclement.

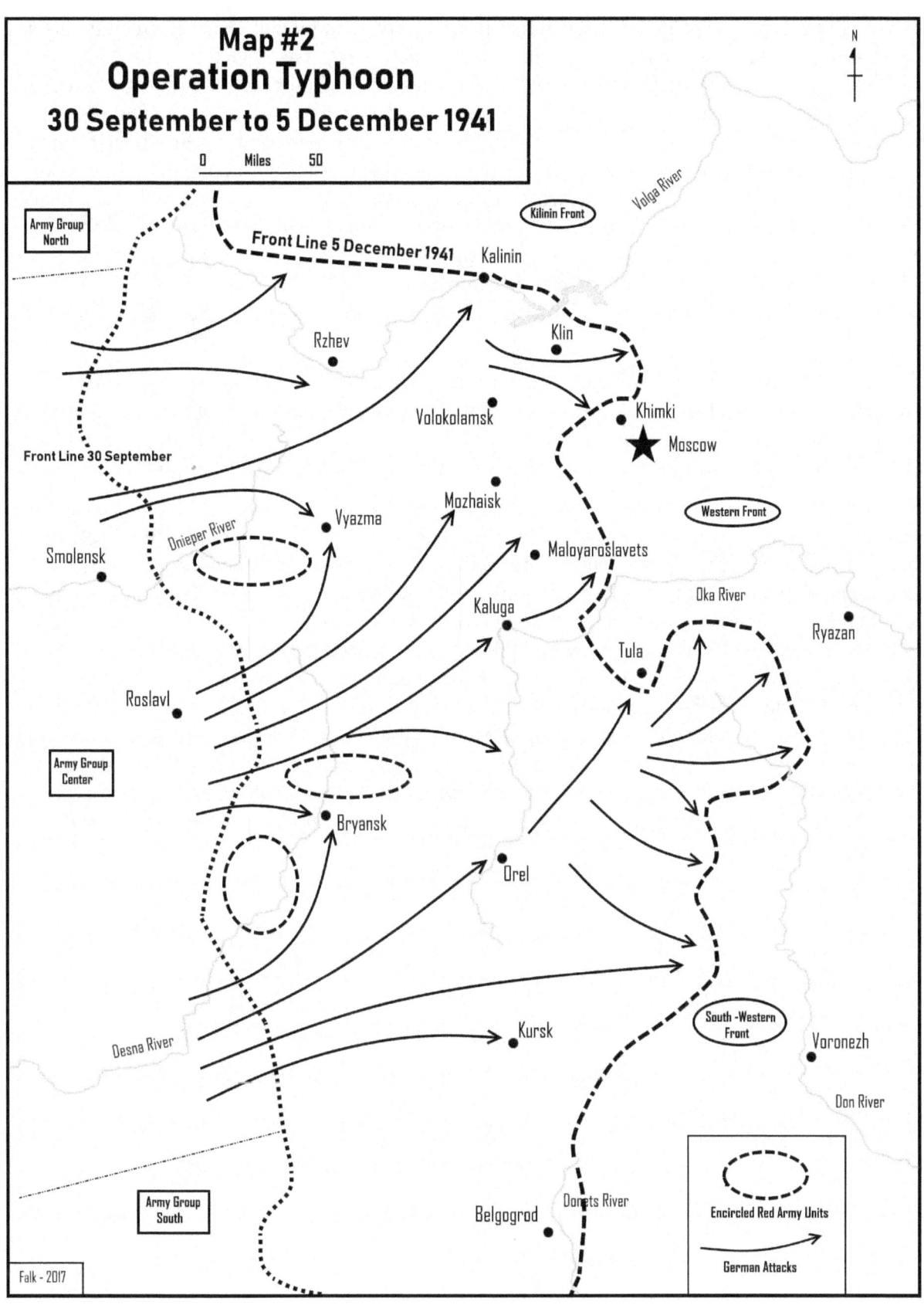

The few who did escape left behind all their weapons and heavy equipment. Once again entire armies were being annihilated by the seemingly unstoppable German Army. By 10 October, the 24th Army had been destroyed and was officially disbanded. Soon after, on 14 October, the Vyazma pocket had itself been completely smashed. By this time, all the surrounded armies and independent units in the Vyazma area had been destroyed. The German Army was now free to continue their advance. Around Vyazma, tens of thousands of officers and men of the Red Army had been killed or captured; among the losses was the commander of the 24th Army, Major General Rakutin, who died within the pocket. So ended the first formation of the 24th Army. The 24th Army would ultimately be formed four different times during the war. During the early part of the war, it was common practice by the Soviet High Command to reconstitute destroyed armies. Almost as fast as the Germans could destroy them, Stavka would rebuild a shattered army and after a few weeks or months of training, the troops would be sent back into combat.

Operation Typhoon would eventually reach the outskirts of Moscow but would advance no further. Fierce resistance by the Red Army caused heavy losses to German spearheads, and a combination of rain, then mud, and finally ice and snow slowed the German attack. In the end, Moscow was saved by the heroic efforts of the defending Red Army. The 24th Army had played their part in this first great victory.

At the end of November, Stalin, Stavka, and General Zhukov, the new commander of the Western front, had finished plans of their own. They were waiting for the moment when German forces had become exhausted by constant battle, lack of supplies and the weather. The Soviet High Command was secretly bringing new units and armies to the Moscow area, building up a strong counterattack force to be used when the German forces were at their weakest. By the start of December, all attacking German armies had been forced to go over to the defensive. Then, starting on 5 December, Red Army units of the Western Front launched the first of a series of powerful counter offensives that would not only drive German troops away from Moscow but also push German forces back all along the front. This general winter counterattack would continue for the next several months as different groupings of Soviet armies, in turn, went over to the offensive. At this point, Operation Barbarossa as well as Operation Typhoon had failed. This failure insured the war would continue into the next year where Soviet strength could only increase.

Formation

1 December 1941	
24th Army	
Assigned – Moscow Defense Zone	
Major General Mikhail M. Ivanov Commanding	
Rifle, Airborne and Cavalry Units	Various units during the month
Artillery Units	-
Armored and Mechanized Units	-
Engineer Units	-
Estimated Combat Strength Estimated Ration Strength (see Appendix 2 for strengths)	Various Amounts About 1,000 men in total supporting the army headquarters[4]

BSSA data - 41, and BSA data - 41 (see Appendix 1)

Even before the launch of the Moscow counter offensive, Stavka and the Soviet Supreme High Command, (VGK), had decided to reconstitute the 24th Army as part of the Reserve of the High Command, (RVGK). On 1 December, the 24th Army HQ appears on a communication scheme within the Moscow Defense Zone. Then sometime during the first three days of December, seven officers gathered in a ministry office on Razin Street in Moscow. These men were to be the core group of officers who would form the new HQs staff of the rebuilt 24th Army.[5] At first, this new HQ had nothing to command and was assigned to the Moscow Military District, but this would soon change.

The first commander of the newly created 24th Army was Major General M.M Ivanov, serving between December 1941 and March 1942. (see Appendix 3) He oversaw the reconstitution of the 24th Army HQ staff even as German forces were approaching to within sight of the capital. At this time, all combat-ready units were either being assigned to front line units or held back as a reserve for a counterattack being planned by General Zhukov. Lacking any real structure or combat units, the 24th Army did not directly participate in the counter attack. Finally, the Moscow counter offensive was launched on 5 December 1941.

As the Red Army forced the Germans away from Moscow, the 24th Army continued the process of reforming. At first it was only composed of a small group of officers. This forming process was most likely accelerated by military necessity and the sound of nearby combat. When the new army HQ was created, Stavka decided to use it as a holding HQ to control recently formed combat units that were on the way to the front. These units would only be assigned to the Army temporarily. During their brief stay, combat units would receive hasty last-minute training and pick up weapons, ammunition and any other missing equipment. Ultimately, the 24th Army could only provide such equipment as was available in the local area around Moscow, and it's doubtful they could fully equip these transient units with everything they needed.

By 8 December, the core of 24th Army HQ had formed and was moved to a new location, south of Moscow along the Parkha River, but still within the Moscow Defense Zone. This new position allowed the Army to fulfil its main mission of receiving temporary units, and at the same time it also represented a provisional defensive line along the Parkha River.

Between 5-10 December 1941, the 24th Army temporarily controlled the following units:

- 21 and 34 Rifle Brigades (RB)

It also had associations with:

- Moscow Defense Zone
- Privolzshk's Military District
- Siberian Military District
- Central Asian Military District.[6]

These different Military Districts were almost certainly related to where newly raised combat units were being formed and which military district would be sending them to the 24th Army. Many of these units were newly raised rifle brigades, which were sometimes called Cadet Brigades because they were formed by using students from officer training schools.

Between 10-16 December 1941, the 24th Army temporarily controlled the following units:

- 213, 334, 358 Rifle Divisions (RD)
- 34, 39, 42, 54, 66, 72 Rifle Brigades (RB)
- 908 Artillery Regiment (AR)

The 213 RD never actually physically joined the army and was reassigned before arriving in the Moscow area. This type of transfer was typical, with the 24th Army only coordinating movements and supply for new units.

Between 20 December 1941 – 22 January 1942, the 24th Army temporarily controlled the following units:

- 334, 358, 385 RD
- 21, 34, 39, 42, 45, 54, 75, 380 RB

The odd numbering of the 380 RB almost certainly was due to the chaotic nature of forming so many new rifle brigades so quickly in different areas of Russia. Between September and December 1941, a total of 159 rifle brigades were formed. This rapid expansion of the Red Army, caused numerous duplications of rifle brigade numbering and other military formations during this time.[7]

If any of these units actually arrived in the Moscow area, they only stayed within the 24th Army a week or two, but the 385 RD arrived on 31 December and stayed with the army much longer. It was originally scheduled to go to the 61th Army. However, since the 385 RD was newly formed, Stavka decided to keep it in the rear for additional training before sending it off to the front. In fact, this is all the 24th Army was going to be doing for the next few months, setting up and training newly formed units that were arriving from the interior of the country. [8] [9] [10]

This rapid turnover of units was supporting not only the Moscow counter offensive, but also the general winter offensive that Stalin unleashed all along the front. The possibility remains that other military formations were briefly assigned to the 24th Army during December, but they fail to appear in the historical record.

1 January 1942	
24th Army	
Assigned – Moscow Defense Zone	
Major General Mikhail M. Ivanov Commanding	
Rifle, Airborne and Cavalry Units	358, 385 Rifle Divisions (RD) 42, 45, 54 Rifle Brigades (RB)
Artillery Units	-
Armored and Mechanized Units	-
Engineer Units	-
Estimated Combat Strength Estimated Ration Strength	32,000 men 33,000 men[11]

BSSA data - 42, and BSA - 42 data (see Appendix 1)

1 January

At the start of the month, the 24th Army Staff issued Operational Summary #1. This summary was reporting on how the Army's connections (units) continued to be loaded and shipped to new destinations. The five listed formations were not with the Army very long. The summary also described having problems with the transfers due to the lack of rail cars which slowed loading rates.[12]

The 358 and 385 RD were at or near full strength, with some 10,000 men each, with the three Rifle Brigades 42, 45, 54 having some 4,000 men each. By this time, the Army HQ also should have controlled several support units like a Signal Battalion, Motor Transport Company, Bakery, Medical Company, HQ Security Company, NKVD company and so on, totaling about 1,000 men.

With the Soviet winter offensive in full swing, the 24th Army was far behind the front lines. Essentially its job was still to receive new combat units moving toward the front from the interior. It also was continuing to provide security along the Parkha River line. Due to the offensive operations of the Red Army, the 24th Army would soon no longer be called upon to defend Moscow.

On 1 January 1942, the Bryansk and Southwest fronts went over to the offensive toward the cities of Orel and Kursk. These new attacks expanded the Soviet winter offensive all along the front from Leningrad in the north to the Crimea in the far south.

3 January

When the 24th Army HQ Staff issued Operational Summary #6, it showed that the newly arrived or assigned 334 Rifle Division was already on the 24th Army's transfer list. This division was ready to be loaded onto trains and moved to the front most likely in just a few days.[13] This is how quickly new units were added and then removed from the 24th Army's control. Note: This rifle division could have been located at a remote location far from the 24th Army itself and was only under its administrative control.

On an unknown day in January, Stavka decided to transfer the 24th Army to the reserve management structure of the Moscow Zone of Defense. This was only an organizational transfer, but it would soon change the Army's situation. After this transfer, the Army continued to perform its basic function of a holding headquarters for several more weeks.

27 January

By late January (date unknown), the 385 RD was finally sent off to the 10th Army at the front. At the end of the month, on 27 January, two Stavka directives were sent to the 24th Army. Directive #151541 and Directive #151549 both related to the movements of six rifle divisions, and the assignment of these rifle divisions to the 24th Army. These directives also ordered the transfer of the 24th Army HQ to the Tula area from Moscow. [14]

The six rifle divisions that were assigned and their final destinations were:

- 42 RD – Tula
- 58 RD – Stalinogorsk
- 397 RD – Volovo
- 152 RD – Venev
- 103 RD – Ryazhsk
- 149 RD – to stay in Ryazan

This general area was a well-known training and assembly location used by the Red Army.

Directive #151541 went on to inform the 24th Army HQ:

Approximate arrival time listed infantry divisions in the new post in the deployment period of 13-18 February 1942.

Accordingly, you should:

1) *During the period of 10-15 Feb 1942 relocate management and all parts and the establishment of the 24th Army to Tula.* [This part of the order almost certainly means the Tula area and not the city of Tula itself.]

2) *Organize a meeting and accommodation of the arriving infantry divisions.*

3) *Organize military preparations according to the provisions of the Main Administration of Organization and Supply.*

The People's Commissariat of Defense (NKO) had the responsibility of ensuring that arriving Rifle Divisions are outfitted with a full complement of weapons and supplies.

These two Stavka directives are very interesting because they convey information about units assigned to the 24th Army that are not found in any other source like BSA or BSSA documents. Since most of these rifle divisions were assigned to the 24th Army for an extremely short period of time, perhaps only a few days and then quickly reassigned to active armies on the frontline, they do not appear in other records.

Only the 42 RD and 58 RD would remain within the 24th Army into February. From this point on, and for the next several months, the 24th Army would occasionally be referred to as a field HQ without troops or only have a few units assigned for a very limited time. The situation at the front and in the rear areas was very fluid as the High Command was trying to coordinate supplies and the reinforcement of its scattered offensive operations.

So, at the end of the month, the 24th Army was certainly planning its transfer to the Tula area some 90 miles south of its current position. Rail transport was not mentioned in the orders, so the army HQ staff and any attached units would have most likely used their own organic transportation to make this move. For most, this meant a nice brisk four or five-day, 90-mile march in the middle of winter. Not a bad accomplishment by itself, but at the same time, there was the need to organize the reception of six rifle divisions that were already on the way. Staff work never ends.

1 February 1942	
24th Army	
Assigned – Moscow Defense Zone	
Major General M.M. Ivanov Commanding	
Rifle, Airborne and Cavalry Units	-
Artillery Units	-
Armored and Mechanized Units	-
Engineer Units	-
Reported Combat Strength Estimated Ration Strength	Without troops HQ Staff only 1,000 men

(BSSA - 42)

At the start of the month, the 24th Army did not have any units assigned to it. The Stavka Directive #151541 from January 1942, stated that the five newly-assigned rifle divisions should be arriving between 13-18 February 1942. This directive also stated the Army HQ itself needed to relocate between 10-15 February 1942 to Tula or the Tula area. So, a great deal of coordination was required to satisfy both parts of this order.

This transfer did not require much effort because once again the Army consisted only of HQs troops and a few support units like a Signal Battalion, Motor Transport Company, Bakery, Medical Company, HQ Security Company, NKVD Company, and so on, totaling about 1,000 men.

During this time (date unknown) the Army also was removed from the Stavka Reserve, but it still was subordinate to the Moscow Defense Zone.

The HQ transfer must have been successful because on 14 February, the 42, 58, 69, 146, and 298 Rifle Divisions were listed as being part of the 24th Army.[15] Note: Only the 42 and 58 RDs from the original January Stavka order actually arrived at the new location, with the 69, 146, and the 298 Rifle Divisions being substituted for the other rifle divisions that never arrived. The missing rifle divisions most likely were redirected en route, which was common.

The 24th Army still was doing its job of training, organizing and preparing combat units, along with quickly giving them a final allotment of weapons, equipment and some basic supplies.

By the end of the month the separate 44 Rifle Regiment and the 118 and 472 Artillery Regiments were added to the Army's control. These additions created a sizeable force of five Rifle Divisions, 42, 58, 69, 146, and 298; one Rifle Regiment 44; and two Artillery Regiments, 118, 472. But this buildup was not to last because ultimately these units were destined for other armies currently fighting at the far-off front. This would be a repeating cycle over the next few months.

Of note, on 23 February the 18 RD began its formation in the city of Ryazan by order of the Commander of the Moscow Military District, through Directive #07391. The 18 RD would eventually be assigned to the 24th Army, which would soon become the 1st Reserve Army.

1 March 1942		
24th Army		
Assigned – Supreme High Command Reserve (RVGK)		
Major General Mikhail M. Ivanov, then Major General Lakov I. Broud Commanding		
Rifle, Airborne and Cavalry Units	42, 58, 69, 146, 298 Rifle Divisions	
Artillery Units	472 Regiment	
Armored and Mechanized Units	-	
Engineer Units	-	
Estimated Combat Strength Estimated Ration Strength	52,500 men[16] 54,500 men (see Appendix 4)	

(BSSA - 42)

At the start of March, the 24th Army still had the 42, 58, 69, 146, 298 Rifle Divisions along with the 472 Artillery Regiment assigned to it.[17] Either these combat units needed more equipment, training and general supplies than earlier units, or Stavka needed fewer reinforcements at the front. At other times the Army was still being listed in BSSA documents as a "Field Control Army, without troops". In other words, the Army was regarded as a mobile field army headquarters temporarily without combat troops assigned to it.

Between 16 – 26 March 1942, the 24th Army temporarily controlled the following units:

- 58, 69, 146, 298 Rifle Divisions
- 459 Separate Rifle Regiment
- 99, 161 Separate Mortar Battalions
- 118, 224, 280, 472, 828 Artillery Regiments
- 109, 211 Separate Anti-Tank Battalions
- 103 Separate Anti-Aircraft Battalion[18]

This illustrates just how fast combat units could be added or subtracted from the 24th Army rolls. The needs of Stavka superseded essential training or equipping efforts, and the 24th Army did their best to produce the required units. To support the larger number of combat units that were arriving, the Army's rear area support formations increased also, maybe up to 2,000 men and women in total.

17 March
Major General Lakov I. Broud takes over command of the 24th Army. He will hold this post until 23 May 1942.

20 March

Stavka Directive #170168 arrived which directed the 24th Army in placing and supporting four more rifle divisions on the way to the front.[19]

The four rifle divisions and their placement locations were:

- 8 RD - the area of Michurinsk
- 29 RD - in the area of Gryazi
- 55 RD - in the area of Tambov
- 280 RD - in Lipetsk region

These locations are all within the Volga Military District and relatively far away from the Tula area, at least 150 miles to the southeast. So, these four rifle divisions never physically joined the 24th Army near Tula. They were all scheduled to arrive at their destinations before 6 April, so the 24th Army had to work quickly to arrange all the preparations that were necessary to support these units at their remote locations.

28 March

Even while the Russian Winter Offensive was continuing, in Germany the OKH (German Army High Command), responding to guidelines from Hitler, produced the final version of the German 1942 summer offensive, Operation Blue. This plan was Hitler's idea of how to cripple the Soviet Union's war effort by attacking and seizing vital Soviet resources. The final plan proposed three major thrusts. The first was by the 2nd Army and 4th Panzer Army toward the city of Voronezh on the upper Don River. The second thrust was with the 4th Panzer Army supporting the 6th Army along the South bank of the Don River to the southeast. This part of the offensive was intended to eventually cross the Don River and to reach the Volga River somewhere in the vicinity of Stalingrad. The third part of the overall summer offensive was by the 1st Panzer Army and 17th Army aimed toward the city of Rostov at the mouth of the Don River. Also, a secondary supporting attack would be launched by part of the 11th Army from the Crimea into the Taman Peninsula.[20] The realistic goals of this plan were the seizure of the Donbass industrial region, the capture of large numbers of Red Army troops by conducting encirclements battles, and the cutting of Soviets transportation supply lines along the Don and Volga Rivers. If carried out to its full extent, this summer offensive would also disrupt Soviet oil shipments from the major Baku oil fields far to the south. Overall, this bold but limited plan was within the capability of German forces; however, Hitler would later modify this plan into a summer offensive with much more ambitious aims. Hitler wanted the outright capture of the oil fields in the Caucasus Mountain region and the capture of a largely obscure industrial city on the Volga River, Stalingrad. David Glantz and Jonathan House explain the situation nicely in their book, *To the Gates of Stalingrad* vol 1, p15-16, "Such an advance would be, to say the least, an operational and logistical challenge greater than any previous German offensive. The straight-line or air distance from Kursk to Groznyi was 760 Kilometers, and the various encirclements of the plan represented a total advance of more than 1,000 Kilometers." It is clear that such an expansion of the original plan could only lead to overextended forces and the dispersal of scarce supplies.

Meanwhile, Stalin and Stavka were working on several different operations of their own for the summer of 1942. Most of these plans were outside the area of the southern part of the front and therefore beyond the scope of this book. But eventually, a Soviet offensive was planned in the south for the recapture of the major city of Kharkov and the surrounding industrial area. Soviet leaders hoped that this offensive also would disrupt any German attacks aimed against Moscow during the summer. This offensive would eventually fail and in so doing would greatly weaken Russian armies along the southern section of the front line. This was, of course, the same area where Operation Blue would ultimately be launched.

All spring long, one plan was pitted against another, one move followed by counter moves. This was all controlled and directed by Stalin and Hitler with their respective High Commands implementing their decisions. But who would make the first move?

1 April 1942	
24th Army into 1st Reserve Army	
Assigned – Supreme High Command Reserve (RVGK)	
Major General Lakov I. Broud Commanding	
Rifle, Airborne and Cavalry Units	29, 58, 278, 280, 298 RDs
Artillery Units	138, 356 Separate AT Battalions 115, 124 Separate AA Battalions 115 Separate Mortar Battalion
Armored and Mechanized Units	-
Engineer Units	-
Estimated Combat Strength Estimated Ration Strength	67,000 men 69,000 men

(BSSA part 2)

Between 28 March and 8 April 1942, the 24th Army temporarily controlled at least the following units:

- 29, 58, 278, 280, 298 RD
- 138, 356 Separate AT Battalions
- 115, 124 Separate AA Battalions
- 115 Separate Mortar Battalion [21]

So, at the start of April, the 24th Army would maintain its fixed HQ supporting strength of around 2,000 men, but with the newly assigned units coming and going, the total combat strength could have been close to 67,000 men for a period of time. The combat units that had been assigned to the Army the previous month either passed through the Army's control quickly, or possibly were still on the way to the Tula area, or were still at their remote locations and not appearing in the historical record at all. Needless to say, trying to track down the exact movements and status of all these different units remained extremely difficult. That is why Stavka had a huge number of staff officers doing just that, tracking the movements, development and current status of thousands of Red Army and VVS units.

By early April, the spring thaw had begun; this was the dreaded *rasputitsa*, "time without roads"[22] or sometimes it was just called the muddy period, where roads turned to mud, ponds into lakes and streams into rivers. These adverse conditions, along with heavy losses in men and equipment, finally put an end to the Russian winter offensive. German forces also had suffered extensive losses during the winter and were using the spring thaw, just like the Russians, to reorganize and to rebuild their depleted armies. The German and Russian leaders also were using this time to develop plans for the summer season…the war would go on. Both sides were intent on delivering a decisive blow to the enemy in the spring; both sides were also struggling to turn plans into reality.

5 April

Hitler issued Directive #41 which covered the expanded plan for the German summer offensive called Operation Blue. This plan outlined major attacks in the southern part of the Eastern Front aiming to destroy or capture all Soviet forces west of the Don River, block water transportation along the Don and Volga Rivers, and occupy the oil fields in the Caucasus Mountain region. Additionally, Stalingrad, a major industrial city and river port on the Volga, would be destroyed or otherwise neutralized during the offensive, but its capture was not considered necessary.

Between 10 - 16 April 1942, the 24th Army temporarily controlled the following units:

- 29, 211, 278, 380 RD
- 840 Artillery Regiment
- 84 Separate AA Battalion
- 280 Divisional Head of Artillery (most likely an artillery division headquarters)
- 368 Separate AT Battalion
- 174 Separate Mortar Battalion[23]

Between 18 - 22 April 1942, the 24th Army temporarily controlled the following units:

- 8, 29, 164, 211, 214, 278 RD
- 829, 847 Artillery Regiments
- 348, 357 Separate AT Battalions
- 110, 173 Separate Mortar Battalions
- 90 Separate AA Battalion

Once again, this series of transfers illustrates the transient nature of units assigned to the Army. It also indicates a small sampling of what was going on behind the front lines. Planning for the future and the formation, supporting, reorganizing and rebuilding of military units never stopped.

26 April

On the Soviet side, Stavka issued Directive #170333 renaming the 24th Army, converting it into the new 1st Reserve Army.[24,] (see Appendix 5) This was an important event in the history of the Army because this renaming signifies a shift in the thinking of Stavka about future combat efforts and requirements. The High Command was anticipating the need for a new, and secret, strategic reserve force that would be ready for use sometime during the summer offensive season. Walter Kerr, in his book, *The Secret of Stalingrad* stated, "[T]his secret strategic reserve was even unknown to some members of the Stavka staff." That was an extraordinary statement. The secret reserve would eventually grow into ten all-arms, infantry field armies numbering approximately 800,000 men. Some members of Stavka were not told about the existence of the reserve force for months. No wonder German intelligence services failed to discover information about this secret reserve until they were committed to battle against them.

The change of title from the 24th Army into the 1st Reserve Army indicated the first step in the creation of this secret strategic reserve. Of course, when and where Stalin and Stavka would use this reserve force during 1942 was unknown. Eventually this secret reserve would grow to ten Combined Arms Infantry Armies, two Tank Armies, and at least three independent Tank Corps.[25] There were also growing indications that Stalin was concerned about a new German offensive directed against Moscow during the summer of 1942, so most of these new secret reserve armies were to be deployed around Moscow or within support range. Stalin wasn't taking any chances with the safety of the capital.

Between 24 - 29 April 1942, the 24th – 1st Reserve Army temporarily controlled the following units:

- 8, 29, 211, 164 RD
- 125 Separate AT Battalion
- 112, 114 Separate Mortar Regiments
- 77 Artillery Regiment
- 159 Separate Mortar Battalion
- 137 Separate Sapper Battalion

Also, during April, the new 1st Reserve Army was assigned its first medical unit, the 2204 PPG or Mobile Field Hospital. This was the first of many hospitals that eventually would support combat operations of the Army. Some records indicated that other medical units might have been assigned to the Army at this time, but they have not been verified. The assignment of the 2204 PPG was a strong indication the 1st Reserve Army was on the way to operational status by being built up into a full-fledged field army.[26] In fact, the 2204 PPG would continue to serve with the Army long after it was committed to battle. Along with the new PPG, the Army also would have begun to receive or organize other medical support units like an evacuation point HQ and horse and motor vehicle ambulance companies. In any case, all these critically needed medical units would be activated before committing the Army to combat.

Toward the end of the month the 29 and 164 Rifle Divisions were assigned to the new 1st Reserve Army as semi-permanent members. This was another strong indication of the changing status of the Army. It was known the 29 Rifle Division had formed in Kazakhstan with some 11,840 men. The 164 Rifle Division likewise was formed in the Urals Military District and would have had a similar number of men assigned. These two additions increased the Army's ration strength by another 23,000 men. With the arrival of these two rifle divisions, the Army started to develop a more stable force structure. Similarly, the reallocation of additional combat units to a reserve army also clearly indicated an end to the Russian winter counter offensive.

Stalin and Stavka were obviously gathering forces for future strategic operations. When these newly assigned combat units finally arrived in the Tula area, the headquarters of the 1st Reserve Army would have a new task to accomplish. Namely to train, equip, and organize an infantry army as a cohesive fighting organization. This was no simple undertaking as the Army would need many additional sub units like logistical, medical, engineering and communication units to function properly. While all this was taking place, General Broud must have been thinking, how much time will I have to train my army and when will the High Command call us away to combat? The war never stops and there was much work to do.

1 May 1942	
1st Reserve Army	
Assigned – Moscow Defense Zone – Stavka Reserve (RVGK)	
Major General L. I. Broud, then Lieutenant General Vasilii I. Chuikov Commanding	
Rifle, Airborne and Cavalry Units	Various Units During the Month
Artillery Units	-
Armored and Mechanized Units	-
Engineer Units	-
Reported/Estimated Combat Strength Estimated Ration Strength	65,500 men 67,500 men

(BSSA part 2)

Between 30 April and 8 May 1942, the 1st Reserve Army temporarily controlled the following units:

- 112, 164, 193, 195*, 211 RD
- 494, 829 Artillery Regiments
- 90, 137 Separate Sapper Battalions
- 110, 130, 157 Separate Mortar Battalions
- 113, 114 Mortar Regiments
- 230 Separate AT Battalion
- 124, 152 Rifle Brigades[27]

* Note: BSSA documents do not show the 195 RD under the control of the 1st Reserve Army at the start of May. This was most likely due to the 195 RD having just been assigned and still was on the way to its new command.

Toward the beginning of May, the 195 Rifle Division was the next combat unit to be assigned to the 1st Reserve Army. This rifle division was raised in the Orenburg Oblast, Volga Military District, and was finally assigned to the 1st Reserve Army in the Ryazan Oblast, Moscow Military District. Just like the previous two Rifle Divisions, 29 and 164 RDs, the 195 Rifle Division was most likely at or near full strength with some 10-11,000 men. With this addition, the 1st Reserve Army controlled five rifle divisions and assorted other battalions and regiments, totaling a combat strength of about 65,500 men.

The early part of May continued to find the 1st Reserve Army deployed in the area just east of Tula as part of the High Command Reserve. As more units were added over the coming months, the Army would eventually be spread out all the way to the city of Ryazan, some 80 miles away. This extra room was needed for training purposes between large size formations, up to divisional size units.

Even as part of the High Command reserve, the 1st Reserve Army was still serving as a training organization for new combat units arriving from the interior. These new units, having been formed throughout the Soviet Union, needed somewhere to go for advanced training and to receive a full set of equipment. Tula was a perfect location for this type of effort. The rail lines converged on this area and the industrial zone around Moscow and Tula provided much of the weapons and equipment these units needed to become combat ready. This effort should not be confused with the struggles of the 24th Army and then the 1st Reserve Army during the December to April period, during the time of the Moscow counter offensive. During its earlier efforts, combat units were quickly brought together, equipped with whatever was needed or was available, and sent off to the front to keep the offensive going. But in May, the army provided real, advanced training to units from the interior.

As the German threat to Moscow passed, and the front moved further west, Stavka began to create other new reserve armies (combined arms armies) during the spring of 1942. Eventually they were numbered 1 to 10. Along with the newly formed 3rd and 5th Tank Armies, these reserve armies were placed in key locations behind the front lines and served as a focus for rebuilding the Red Army. These ten combined arms armies and two tank armies became a significant part of the secret strategic reserve of the Soviet High Command.

The High Command still needed replacement formations for the active army, so combat units of different size sometimes were transferred from these reserve armies and sent off to active armies serving at the front. A continuous cycle formed as combat units would be assigned to these reserve armies, trained and equipped, and then shipped out to the front. If these units were assigned long enough, they were given advanced training up to division level and supplied with a full set of weapons and equipment.

The Russian winter offensive had pushed the Germans back from Moscow, and by the end of April the front had stabilized. Both sides began planning for the summer season of 1942.

Major combat activity had finally slowed due to the *rasputitsa*, or spring thaw, when everything turns to mud. For that reason, the 1st Reserve Army started to acquire and to hold new units much longer. This extra time allowed the units to undertake more advanced and complex training. For example, the 29 RD started forming on 14 December 1941 in the Central Asia Military District. It arrived in the Tula area in March but only was assigned to the 1st Reserve Army in April. The 29 RD was at full strength with 11,840 men.[28] The 164 RD started forming in December 1941 in the Urals Military District. It started moving west and arrived in the 1st Reserve Army by late April 1942. After almost five months assembling, the 164 RD also was most likely at or near full strength of 11,000 men. Then by mid-May, the 164 RD was reassigned to another army.

The two mortar regiments assigned to the 1st Reserve Army at this time were the 113 and 114 MRs. They were independent mortar regiments of the Stavka reserve with about 800 men each. The 113 Mortar Regiment was eventually assigned to the 33rd Army in 1942 for Operation Mars and Jupiter offensives. The 114 Mortar Regiment was eventually assigned to the 21st Army in 1942 where it took part in the Uranus and Saturn offensives.[29]

Between 10 - 22 May 1942, the 1st Reserve Army temporarily controlled the following units:
- 112, 131, 193, 195, 214, 278 RD
- 41, 156 Separate AT Battalions

- 86, 275 Separate Sapper Battalions
- 436, 475 Artillery Regiments
- 104, 112, 113, 114, Separate Mortar Battalions
- 124, 152 Rifle Brigades

19 May

Stavka issued Directive #994012 assigning the 131 and 147 RD to the 1st Reserve Army. They were to arrive by rail with the tentative arrival dates of

- 131 Rifle Division – 26 May, 1942
- 147 Rifle Division – 3 June, 1942

The army HQ staff would, of course, arrange for their quartering within the training area upon arrival. The HQ staff would also arrange to fill any logistical needs and to create a plan to include the two new units into a training schedule, as time would allow.

23 May

As new units were on the way to the Army, Stavka planned a change in command. Lieutenant General Vasilii I. Chuikov was appointed acting commander of the 1st Reserve Army on 23 May 1942.[30] With this change of command, Major General L.I. Broud took over as second in command and as commander of artillery for the 1st Reserve Army.

Since the start of the war, General Chuikov had been serving in the Far East as military attaché and chief Soviet military advisor to the Chinese government of Chiang Kai-shek. He did not like this assignment. Chuikov arrived back in Moscow at the start of March looking for a posting to the front. He wanted a combat command but received instead command of the 1st Reserve Army. Having survived a bad car accident after arriving at his new command (this was his first personal brush with death), Chuikov oversaw what he called "intensive military training" of the units assigned to the 1st Reserve Army during May, June, and early July.[31] As Chuikov was starting to acquaint himself with his new command, events were taking place further south that were to play a great part in the story of the 1st Reserve Army and even of Chuikov himself.

While the 1st RA was slowly forming, Stalin and Stavka were planning to launch an early spring offensive in the south. In early May, the ground had finally dried enough to support operations, and Stalin was eager to attack the Germans somewhere. On 12 May, the Southwestern Front launched an attack toward the city of Kharkov out of the so-called Izyum Bulge. The plan was to capture Kharkov using a two-pronged attack. Kharkov was not only a major population center, but also an important rail and supply center for the Germans. Stavka also hoped this attack would disrupt any German plans for summer operations of their own. Attacking Russian armies initially surprised the Germans and made good progress during the first five days. However, the Soviet command did not know they were attacking into a concentration of German forces. German reserves had been gathering in this very same area preparing for Operation Blue, the planned German summer offensive in the Southern part of the front. Stalin and Stavka did not know it, but their attack on Kharkov was thrusting several Soviet armies into what would become a huge trap.

Taking advantage of favorable conditions and placement of forces, a German counterattack, Operation Fridericus, eventually was launched on 17 May to destroy the attacking Soviet Armies. Operation Fridericus was successfully completed by 29 May, resulting in the encirclement of three Soviet Armies, the 6th Army, 57th Army and 9th Army. The Red Army had suffered yet another defeat, losing 240,000 prisoners, 1,200 tanks, and 2,600 artillery pieces.[32] The dead totaled between 20-

30,000 men. The Kharkov offensive was a major defeat for Stalin's summer plans and did little to disrupt the German buildup for Operation Blue.

At the same time, other German forces were busy further South in the Crimea. The Kerch Peninsula was cleared of Soviet forces by 19 May, with the Red Army losing another 170,000 prisoners. Then the Fortress of Sevastopol finally was reduced by 4 July, resulting in a further loss of more than 90,000 prisoners.[33] In just over two months, the Red Army had lost more than 500,000 troops, just in prisoners alone, with entire armies being destroyed along the way. But these were only the opening moves in yet another pivotal campaigning season. With the depletion of its front-line strength, the Red Army was in no position to defeat the German summer attack in the south. These early victories also convinced Hitler and the German High Command they could start Operation Blue with full confidence.

Back at the 1st Reserve Army, General Chuikov and other staff officers were unaware of these defeats. They were hard at work forming, organizing and training a new army. These men also were destined to play a key part in the upcoming decisive battles of 1942.

1 June 1942	
1st Reserve Army	
Assigned – Supreme High Command Reserve (RVGK)	
Lieutenant General Vasilii I. Chuikov Commanding	
Rifle, Airborne and Cavalry Units	29, 112, 131, 164, 193, 195, 214 Rifle Divisions
Artillery Units	114 Mortar Regiment
Armored and Mechanized Units	-
Engineer Units	1363 Separate Sapper Battalion
Reported/Estimated Combat Strength Estimated Ration Strength	75,500 men 77,500 men[34]

(BSSA part 2)

Between 24 May and 5 June 1942, the 1st Reserve Army temporarily controlled the following units:

- 131, 164, 214, 229 RD
- 124, 152 RB[35]

As expected, the 1st Reserve Army continued to receive new rifle divisions and brigades only to later have them transfer to front line formations. At the start of June, seven rifle divisions were assigned to the Army. This was more than the average number of rifle divisions being allocated to other reserve armies at this time. Of the ten reserve armies that would be ultimately formed by Stavka in early 1942, most went into battle for the first time with six rifle divisions and a few assorted support units. By 1 June, only seven of these reserve armies had been formed and they contained the following: 1st Reserve Army, positioned near Tula, with seven RDs; 2nd Reserve Army, positioned near Vologda, with six RDs; 3rd Reserve Army, positioned near Tambov, with eight RDs; 5th Reserve Army, positioned northwest of Stalingrad, with two RDs; 6th Reserve Army, positioned south of Voronezh, with four RDs; 7th Reserve Army, positioned near Stalingrad, with four RDs; and the 8th Reserve Army, positioned near Saratov, with six RDs.[36] [37]

It must be understood these reserve armies were "infantry armies" in every sense, much like their WW I ancestors. The assigned rifle divisions contained few motorized vehicles and relied mostly on horses for transportation, with most of its soldiers walking into battle. The rifle divisions themselves contained a fair amount of 76mm and 122mm field artillery pieces along with 50mm, 82mm, and 120mm mortars for added firepower, but nothing larger. The AT assets were minimal and were limited to 45mm AT guns and AT rifles. Against enemy Panzer units, an infantry army was extremely vulnerable, with rifle divisions being forced to use 76mm field guns in the AT role. Independent engineering assets also were lacking, so no extensive emplacements could be built which included minefields, anti-tank ditches, barbwire entanglements, bunkers, and so on. Likewise, anti-aircraft units and weapons were also insignificant or entirely absent. Such an army was powerful in itself, but on its own or without supporting tank, artillery or air elements, it was very fragile. If not handled correctly in battle, it would be subject to heavy losses and early destruction.

With the assignment of new rifle divisions and a few sub-units, the 1st Reserve Army was starting to look more like a real fighting force. In addition, it picked up the 114 Mortar Regiment and the 1363 Sapper Battalion (combat engineers). Despite more units being allocated to and staying with the 1st Reserve Army longer, it was still a training organization. Most of these temporarily assigned formations were destined to go into battle with other armies.

Sometime toward the end of the month, the 18 RD located at Ryazan was assigned to the 1st Reserve Army in the Tula area. The 18 RD had three different formations because it was created three different times during the war. The first formation was officially disbanded on 19 September 1941 after being surrounded and destroyed during heavy fighting west of Smolensk. The second formation did so well at the front that it was ultimately transformed into the 11th Guards Rifle Division on 5 January 1942. The third formation was created in February 1942 and despite having a less colorful start was destined to go into battle with the 4th Tank Army at Stalingrad. The 18 RD also had the good luck to survive until the end of the war.

Another new addition to the 1st Reserve Army was the 229 RD which was formed in Siberia and had been filled out with more than two thousand privates after they were granted an early release from their Siberian imprisonment. Naturally, these criminals were given the chance to fight for Mother Russia and to pay for their crimes on the battlefield. Despite this, or maybe because of an allotment of prisoners, the 229 RD was regarded as being in good fighting shape and would later do well in battle.

However, as some units were arriving, others were relocating. At about the same time the 18 RD and 229 RD were arriving, the 93 RD and 195 RD were both reassigned and transferred to the 3rd Reserve Army located near Tambov. As a result, the 1st Reserve Army would end the month still containing seven rifle divisions.

To illustrate what the Army was doing during this hectic time, let's start with its location. The 1st Reserve Army was spread out over a wide area from the industrial city of Tula eastward some 90 miles to the city of Ryazan. A great deal of space was needed to conduct live fire exercises with artillery and small arms for tens of thousands of troops. Room also was needed to conduct long marches for the infantry and to allow for maneuvering during tactical problems. The Tula area was even large enough for separate Battalions, Regiments and even Divisions to conduct combat military exercises, sometimes taking place over the period of several days. By day and by night, the Army and Divisional Headquarters were fully engaged in training, equipping and evaluation the units assigned to them. The experienced officers knew life at the front would be far harder than this playing at war.

The rifle divisions were most likely paired off in groups of two to conduct advanced combat exercises together. It was known that the 112 RD was deployed near Ryazan with the 214 RD to the west near the small town of Maklets.[38] The 1st Reserve Army HQ also was located at Maklets. The 29 RD was located near Volovo south of Stalinogorsk and Maklets. Later on, when the 18 RD arrived, it was located near Ryazan along with the 229 RD. The final rifle division, the 131 RD, was stationed near the small town of Venev, north of Maklets. Supporting sub units of the Army and its rifle divisions would have been spread out in the area between Tula, Ryazan, and Volovo, but currently the exact locations of these small formations are unknown. (see Map #4)

As combat units carried out tactical exercises, army and division HQs staffs were conducting map exercises and actually commanding units during maneuvers. Some of these staff officers were new to their jobs, so planning route marches, making logistical arrangements, and performing other key but routine HQ tasks would have been invaluable preparation before being committed to combat at the front. Training also extended to party-political work with political instructors indoctrinating troops and reinforcing the Soviet dogma about the war and the hated enemy.

After the failure of the Russian attack toward Kharkov at the end of May, combat operations along the southern part of the front calmed down. The surviving Russian armies were depleted of troops and equipment, so they were in no shape to resume offensive actions anytime soon. The lack of activity from the German side was somewhat deceptive. The Russian attack against Kharkov had disrupted German plans to a certain extent, but the lull in activity was mostly due to German commanders finishing preparations for their own summer offensive, code named Operation Blue.

The Soviet High Command knew the Germans were planning some type of summer attack, but the exact starting date and ultimate objectives were unknown. Stalin and several others within Stavka were focused upon Moscow, thinking the capital was the real target of any German offensive. So Stavka issued Directive #170446 which arrived at the 1st Reserve Army HQ on 7 June.[39] This directive outlined how the 1st Reserve Army along with units of the West and Bryansk fronts, would form a defensive line covering the Tula, Stalinogorsk, Ryazan zone. The directive went on to say how, "Simultaneously, the army should be ready to destroy any enemy paratroopers landing in the areas of concentration of the army." Along with laying out a new defensive line and defending against paratroopers, the army should continue with "combat and political training of its soldiers" along with tactical training at "platoon, company, and battalion" levels. This Stavka directive was an indication of just how worried Stalin was about the defense of Moscow. Unexpectedly, on 19 June, a German liaison aircraft crashed behind Russian lines. The pilot was killed and his passenger, Major Joachim Reichel, was shot and killed after the crash. The German Major was carrying documents and maps displaying information about Operation Blue. These important documents were recovered by Russian soldiers and passed up the chain of command. When they finally reached Stalin, he thought they were a cunning deception plan on the part of the Germans, so he took no action. If Stalin had acted upon this information, the Red Army would have been much better prepared for the coming attack.

22-25 June

German forces unleashed Operation Fridericus II. This was a preparatory offensive designed to advance the 1st Panzer Army to the Oskol River and to line it up with the 6th Army which was deployed further to the north. The stage was now set for the main event, Operation Blue.

General Chuikov did not know it at the time, but soon events on the south section of the front would add to the already hectic training pace of his Army. Both the officers and men knew the war would catch up with them eventually, but they were unaware of the current strategic situation. Their training period behind the front lines was quickly running out when the German 1942 summer offensive started on the morning of 28 June.

28 June

The German summer offensive, Operation Blue, opened with the 4th Panzer Army and the 2nd Army attacking due east toward the Don River and the city of Voronezh. Fighting was fierce, and by the evening German tank units had broken through the Russian defenses of the 13th and 40th Armies. In response Stavka committed its reserve 5th Tank Army in an attempt to prevent the German offensive from turning north toward Moscow.

29 June

The battle continued as German panzer and supporting infantry units advanced toward the east. In turn, Luftwaffe aircraft flew supporting combat missions at the point of attack and reconnaissance missions far into the rear of the defending Red Army.

30 June

The German 4th Panzer Army and the 2nd Army continued their push toward Voronezh, while strained Russian forces were unable to prevent their advance. Units of the 5th Tank Army began to arrive from the north, but most of its tanks were out of fuel and their movements and attacks were uncoordinated. The Russian command organization was starting to lose control of the battle. Further south, the German 6th Army joined the offensive by launching its own supporting attack and making a clean breakthrough at the junction between the defending 21st and 28th Armies. Despite fighting bravely, Russian forces started to fall back in disorder all along the line. The Red Army wasn't able to match the skill and concentrated power of these coordinated German attacks.

At this point in the battle, Stalin and Stavka had no real idea what this new German summer offensive was trying to achieve. But Stalin feared that Hitler was aiming to envelop Moscow from the south with the ultimate aim of capturing the capital. Acting on this assumption, Stalin would soon act to prevent this from occurring even though the threat might not be real.

1 July 1942		
1st Reserve Army		
Assigned – Supreme High Command Reserve (RVGK)		
Lieutenant General Vasilii Ivanovich Chuikov Commanding		
Rifle, Airborne and Cavalry Units		18, 29, 112, 131, 214, 229 Rifle Divisions
Artillery Units		-
Armored and Mechanized Units		-
Engineer Units		1363 Separate Sapper Battalion
Estimated Combat Strength Estimated Ration Strength		72,800 men 79,800 men

(BSSA part 2)

With the new German summer offensive underway, Stavka held back most of its reserve armies from the front. Stavka and Stalin were awaiting further developments and clarification of German intentions. Meanwhile, the ten reserve infantry armies continued to train and to process their assigned combat units like normal.

What Hitler and the German High Command planned for the summer campaign season was Operation Blue, a massive attack toward the city of Voronezh and following its capture, further attacks down the lower Don River encircling and destroying Russian armies along the way. This offensive would then be further expanded to capture the oil fields at the foot of the Caucasus Mountains. This was to be Operation Blue's major goal: capture the oil fields and deny them to the enemy. They also planned to reach and cut the Volga River supply line somewhere near Stalingrad and eventually advance down the Volga and capture the city of Astrakhan on the shores of the Caspian Sea. The Volga River was important because it served as a major transportation route for Soviet industry. Everything from oil from Baku, far to the south, to Lend Lease supplies coming up from the Persian Gulf moved along it. Finished goods, spare parts, raw materials, food and even entire military units moved up and down the river. So disrupting shipping along the Volga and nearby rail lines throughout the entire area would be a worthwhile secondary goal for the German attack. Much has been written about Operation Blue, so I will not repeat well-known information. Needless to say, Operation Blue was a bold plan that would demand much from German forces to achieve its objectives. The same could be said about Russian forces. Much would be asked of them to defeat the German plan.

By the start of July, the 193 and the 195 RDs had left the 1st Reserve Army; both were transferred to the 3rd Reserve Army near Tambov. The orders transferring these RDs probably were issued at the very end of June because they do not appear in the 1 July BSSA data. Also, the 114 Mortar Regiment transferred to the Moscow Military District at about the same time. In addition, but also in late June, the 164 RD transferred to the 31st Army, which was at the front near Moscow. To cover these losses, the 1st Reserve Army gained control of two new Rifle Divisions, the 18 and 229 RD. These two rifle divisions transferred in early July or late June. Most of the remaining units within the 1st Reserve had been filled out with troops. The combat strength was listed on 10 July as being 72,800 people.[40] (see Appendix 6) These troops were the largest force that would be assigned to the Army for a long time.

1 July

Far to the south of Moscow, the frontline was under steady attack with Russian armies falling back toward the safety of the Don River.

For the 1st Reserve Army, the new month brought the same intensive training schedule for all units assigned to the Army. There was little concern about the newly-launched German offensive because most of the solders did not hear about it. Only after several days did information slowly reach the command staff. General Chuikov stated that, "The South Western front…was rapidly moving eastward under pressure from attacking German armies…and coming close to Voronezh."[41] In any event, the German attack reached the Voronezh area within a few days and pushed the new front line to within about 130 miles from where the 1st Reserve Army was training.

2 July

The German 4th Panzer and 6th Army linked up near Stary Oskol, on the way to Voronezh, but most of the retreating Russian armies had avoided being encircled.

Stalin and some members of Stavka anticipated a German attack upon Moscow during the summer of 1942. Because of this fear, most of the new reserve armies and other substantial forces were deployed to protect the capital. This was not a bad assumption on the part of Stalin, but it proved to be incorrect. With the start of the German summer offensive on 28 June, the first attacks were directed toward the city of Voronezh, on the upper Don River. The 1st Reserve Army just happened to be in a good position to block any German movement to the north toward Moscow from this area. However, the German offensive would ultimately unfold to the southeast and away from Moscow.

Stalin did not know what the German objectives were, so major Russian forces finally started to move into the Voronezh area to prevent any northward turn in the offensive. Stalin activated two reserve armies and committed them directly to the fighting at Voronezh. The 3rd Reserve (the new 60th Army) came from Tambov and the 6th Reserve (the new 6th Army) from the Borisoglebsk area south of Voronezh. Also, coming down from the north, the 5th Tank Army joined the battle at Voronezh.

Anticipating this type of response, German air units bombed the rail junctions of Voronezh, Michurinsk, Svoboda, and Valuiki that same day, trying to block the movement of these new forces.[42] Despite German interdiction efforts, these two reserve armies were successfully activated and transported directly to the battlefields around Voronezh.

3 July

German Panzer units reached the Don River directly west of Voronezh; the city was actually located on the east side of the river. Stalin immediately ordered the 5th Tank Army to counterattack this threatening advance. The loss of Voronezh would allow the German to easily cross the wide Don River, and Stalin felt that must be prevented at all costs.

4 July

During the day, large tank battles were taking place between the German 4th Panzer Army and the 5th Tank Army that was attacking from the north. The German 2nd Army also was screening the northern flank of the offensive with its infantry divisions which allowed most of the 4th Panzer Army to continue on toward Voronezh.

5 July

Supported by the Luftwaffe, German troops fought their way over the Don and engaged defending Russian forces within Voronezh itself. Stalin reacted to this news by forming a new Voronezh Front. This new Voronezh Front was commanded by General Nikolai Fyodorovich Vatutin and controlled the 3rd, 6th and 40th Armies.

6 – 8 July

Fighting was heavy in and around Voronezh for several days, but the Germans finally captured and cleared the city by 8 July. With its capture, Phase I of Operation Blue was complete, so Hitler ordered major German forces to move toward the southeast following the west side of the Don River. The capture of Voronezh was only intended to be the left flank anchor point for the summer offensive. Unknown to Stalin, the threat to Moscow was starting to diminish as the focus of the German offensive moved further south. Eventually Stalin and Stavka understood the battle had changed, but they also had to deal with this new turn of events. What were the Germans trying to accomplish? It's also interesting to make note that with the capture of Voronezh, the German 4th Panzer Army had reached a point about 100 miles south of the positions held by the 1st Reserve Army. David Glantz and Jonathan House noted in their book, *To the Gates of Stalingrad* vol 1, p 104, that the "4th Panzer Army contained 733 tanks of various types and the German 6th Army fielded another 360 tanks," a most formidable force. These two German armies were destined to have a much closer meeting with the 1st Reserve Army in a few weeks' time.

9 July

By this date Operation Blue was being fully executed by five German and one Hungarian army. The German 2nd and 4th Panzer Army advanced directly toward Voronezh. The German 6th Army and Hungarian 2nd Army attacked toward the middle Don and further south the German 1st Panzer Army and 17th Army attempted a large enveloping attack toward the city of Rostov at the mouth of the Don River. Stalin and Stavka, realizing the danger posed by these powerful German enveloping attacks, ordered their forces to withdraw, give up ground and to fight their way out of any encirclement.

While these events were taking place, the 1st Reserve Army continued to train and to develop the skills of its soldiers. The lands east of Tula were ideal for training different types of units because of its varied terrain. Also the HQ of the 1st Reserve Army was located at the village of Maklets, which was about 30 miles from Chuikov's home town of Serebryanye Prudy (Silver Ponds), so he knew this area well. The landscape was mostly open with rolling hills and small-forested areas dotted about. Depending upon their skill level, small rifle formations practiced maneuvering and assault drills at first in open fields (so that everyone could see what was going on), then moving to high grass, brush, and eventually into forested areas. Dry firing of weapons, to conserve ammunition, was used during all types of weapons training. Units were drilled in marching, entrenching and the use and placement of weapons as well as many other skills needed on the battlefield.[43] The idea was to give all levels of a formation a basic understanding of their jobs, be they riflemen, mortar crews, or machine gunners. Proficiency in these skills only would be achieved after their first battles. This same basic practice was repeated at higher formation levels as the training continued. If time allowed, entire rifle divisions eventually maneuvered against each other in mock combat. By early July, these mock battles actually were taking place.

With the fall of Voronezh to concentrated German attack, Hitler insisted that mobile units be released to continue with Phase II of Operation Blue. As German panzer and motorized units started to move off toward the south, Russian troops in this new area were being shattered by the expanding offensive. Stalin and Stavka planners became so alarmed with the direction and power of this attack that they ordered the southern part of the front to give up ground and conduct a fighting withdraw toward the Don to prevent the formation of any large encirclements. The front was moving to the south and east and opening up. Something needed to be done to counter the German offensive.

At this point a very busy Stavka made some major decisions, issuing Directive #994101 transferring eight independent RD and two RB from the Far Eastern Front into Stavka reserve. These units would at first be heading toward Moscow, via railroad, but along the way they always could be diverted to another destination as needed. Then on 9 July, Stavka added to its reserve by issuing Directive #994102 transferring two more independent RD and one RB from the Trans-Baikal Front. (see Appendix 7) Of these units, the 126 RD (destined to join the 64th Army at Stalingrad) contained 12,553 men, in other words a full-strength rifle division. It took four days to load the division onto 18 trains using two different stations. These Siberian reserves took at least two weeks to make the journey to the active front. Stavka was planning ahead and gathering its forces.

Also, on 9 July, Stavka acted again to plug gaps and to stabilize the front line. On this day, Stavka sent out Directive #994103 renaming three more local reserve armies, thus activating them for front line service. This directive went into effect at 12:00 hours on 10 July. The 7th Reserve Army, deployed near Stalingrad, became the new 62nd Army; the 5th Reserve Army, deployed along the eastern bank of the middle Don River, became the new 63rd Army; and the 1st Reserve Army, deployed in the Tula area, became the new 64th Army.[44] Of these three new armies, the 64th was the most remote from the newly-developing battlefield. Swift action would be needed to ensure the timely arrival of this vital formation. Stalin and Stavka chose to oppose the German offensive toward the south by building an entirely new front line in front of it. The great bend in the Don River was selected as the place to build this new defense line. (see Map #3) In Moscow, a line literally was drawn on a map crossing the open steppe; this was where the new armies would dig in, and this was where the enemy would be stopped. Walter Kerr, in his book *The Secret of Stalingrad*, recounted how General Kolpakchi, the commander of the newly activated 62nd Army was told "the Germans might be in the Don Bend and coming his way, even though Kolpakchi thought the Germans were still some 150 miles further to the west." Kolpakchi was also startled by the thought that his young army would be required to cover about 100 miles of this new frontline by itself, which was a bit much. I agree with this assessment.

Looking at map #3, it's easy to understand the 62nd Army would be dangerously strung out and would badly need additional forces to mount a credible defense, including a supporting army, especially on its left flank. There is where the 64th Army was going to be deployed, but would it arrive in time?

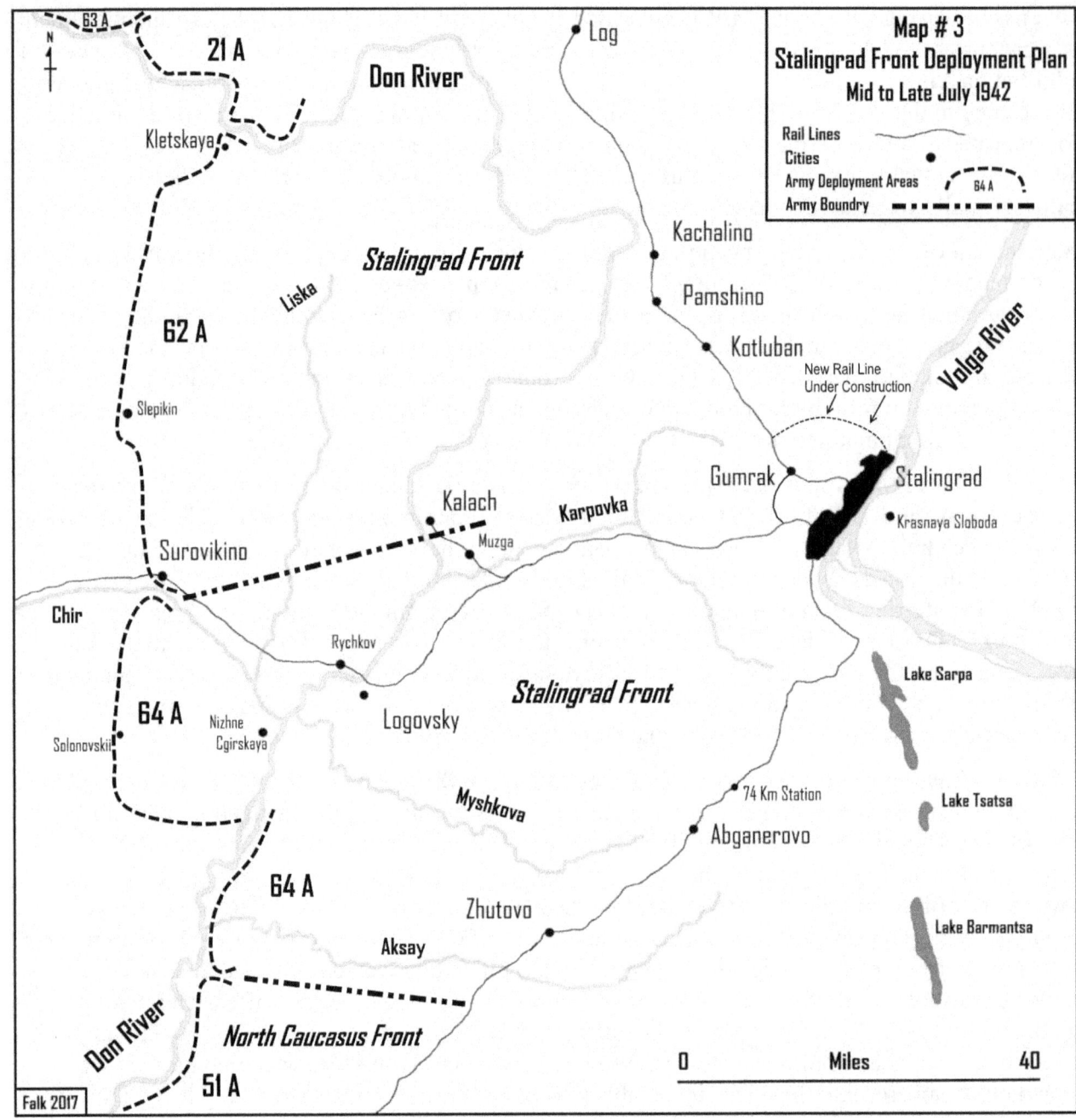

Once Stavka took the decision to activate and then transfer Strategic Reserves to the Stalingrad area, using either local reserve armies or units from Siberia, a complex chain of command and logistical measures were needed to put the plan into action. (Note: Some of these events might have occurred simultaneously or undertaken several different times, or were performed slightly out of the order presented here):

1. Orders and directives from Stavka were transmitted to the General Staff (VGK) to be put into action. The VGK itself was under the direction of the People's Commissariat for Defense (NKO), which had a much broader organization and responsibility. At this point, different sub-divisions of the General Staff, Operations, Transportation, etc., undertook a review of available formations and/or armies. This included an appraisal of the army's location, order of battle, training level, equipment and overall capability. Staff officers within the General Staff most likely performed this task using the most recent strength reports of available formations. Also, Stavka itself might have named specific armies for activation, based upon its own information and appraisal of the situation. (Note: To implement any plan like this, direct and extensive communications between Stavka and the General Staff was required. These exchanges have been omitted from this outline).

2. If suitable armies were found and approved for activation, an alert order was sent out to each individual army headquarters notifying them that further orders would soon follow. In the case of the 1st Reserve Army, the alert order was Directive #994103 which renamed the 1st Reserve Army to the 64th Army. Renaming also implied transfer to the active front.

3. Once suitable forces had been located, staff officers of the Operative Transportation Directorate were notified. They, in turn, would have quickly contacted staff officers within the VOSO (Office of Military Communications) to plan the railroad transfer of the required troops. These two-separate planning/transportation/logistical staffs were all located in Moscow near Stavka HQ, along with the rest of the General Staff and NKO. To deal with the extreme work-load, the NKO was staffed with about 10,000 people.

4. The Operative Transportation Directorate and VOSO staff officers then assessed how many rail assets -- engines and cars -- were available in the local area and matched them with the expected need to move the different units within the different armies.

5. Once sufficient rail assets were found, staff officers undertook a review of stations and loading platforms in close proximity to the units. The idea was to match stations/platforms/loading ramps with the needs of the military formations, light vs. heavy equipment and troops vs. supplies.

6. After adequate trains and suitable loading/entraining stations were found, staff workers/officers produced a scheme (tables, charts and diagrams) to bring these different components together at the proper time and place: how many standard size trains, with the proper number and types of cars, would arrive at which stations, at what time. Staff officers also needed to estimate how long it would take for all the different military units to reach their assigned stations. Then they estimated the entraining time for each unit, at each station, for each train.

7. This detailed plan was then presented back to the controlling Stavka or General Staff officers for approval. Once a plan was produced and approved, most likely only then would movement orders be sent out to the affected military units. Officers within the General Staff most likely issued these orders. Simultaneously, the Operative

Transportation Directorate and VOSO staff issued their own set of orders executing the approved plan from the transportation/logistical point of view.

8. Concurrently to all this, a routing and detraining plan was created for each train along each section of rail line. These plans included times and places, servicing steam engines, replacing train crews, and so on. The staff needed to consider other formations moving along the same sections of tracks, not to mention supply and medical trains. They also needed to review and to plan for enemy interference, accidents, broken equipment, etc. The detraining locations were at first only generally defined. Later on, as the trains neared their destinations, the receiving front HQ provided more detailed detraining information.

9. The General Staff issued another set of orders directly to the receiving front HQ. These orders, stated in simple terms what forces the High Command was sending, and where the detraining location were most likely to be. It also directed the front HQ to take command of these new formations and to provide supplies and support to these troops as they detrained and marched toward the new defensive front designated by Stavka.

10. Finally, the Operative Transportation Directorate and VOSO, working together, oversaw and coordinated the transfer.[45] (see Appendix 8)

Later during 9 July, General Chuikov was reviewing the results of a war game between two of his divisions, the 112 and 214 RD, when he was handed a message from Stavka. It was Directive #994103, renaming the 1st Reserve Army into the 64th Army. General Chuikov knew renaming meant joining the active army and impending action at the front. But he still did not know exactly when or where this would be. As can be imagined, this order would have started a great deal of activity and questions within the army. What was going on? Where were they going and when?

10 July

The next day, Directive #1035042 arrived. This directive ordered the transfer of the Army with its six Rifle Divisions, 18, 29, 112, 131, 214, 229 to a region northwest of Stalingrad. With this order, all rifle divisions immediately stopped training activities, if they had not already done so, and started preparing for movement via railroad to the front somewhere toward Stalingrad and the lower Don River.

Surprisingly at this time, it would appear the six rifle divisions of the 64th Army were in the process of adopting a new organizational *Shtat* or Table of Organization and Equipment (TO&E). All these rifle divisions were most likely formed using the 04/750 *Shtat* that went into effect on 6 December 1941. This 04/750 *Shtat* specified a manning level of 11,628 men for each rifle division, but notably most of these rifle divisions were already above this manning level. This implies the newer 04/200, 18 March 1942 *Shtat* with a manning level of 12,725 men, was being implemented and that process had not yet been completed. To even further confuse the issue, apparently a slight modification of the 04/200 *Shtat* was being applied at this time, one that increased the manning level slightly, by +82 men, to 12,807 men. This addition could have been an AT Rifle Company or an AA Battery; both of these small units were around 82 men in strength. This strength information was taken from a historical document that also lists the recorded strengths of these divisions as of 10 July 1942, as follows:[46]

- 18 RD   12,577 men   Started forming 23 February 42
- 29 RD   11,851   Started forming 18 December 41
- 112 RD   12,768   Started forming 18 January 42
- 131 RD   10,795   Started forming 25 December 41
- 214 RD   12,603   Started forming 25 December 41
- 229 RD   12,229   Started forming 12 December 41[47]

As can be seen from these manpower totals, the divisions of the 64th Army were in good shape, with an overall minor shortfall of some 4,019 men. The slight under strengths of these divisions most likely was due to the shortage of manpower in the Tula area and the lack of time to reach the new *Shtat* totals. Moreover, rifle divisions within the 62nd and 63rd Armies had similar high strength totals, so the 64th Army was not an isolated instance.

This new information goes against the old, but still widely held belief that newly formed Red Army rifle divisions were much weaker during this time period. For instance, Walter Kerr stated in his book, *The Secret of Stalingrad,* "in each of the ten all-arms armies [the secret reserve] … there were six to seven divisions with 5,000 to 8,000 men in a division." This is clearly a substantial understatement. But he was not alone. Albert Seaton in *The Russo-German War* and John Erickson in *The Road to Stalingrad* make similar statements. Understandably these authors had to do research under less than ideal circumstances. With the extreme secrecy of the Soviet Union, they were prevented from gaining access to vital hard data about strengths and losses of the Red Army.

11 July

A barrage of directives soon arrived from the High Command. These directives were sent out almost simultaneously so they should have arrived at the 64th Army HQ at about the same time.

Directive #1035043 ordered the 292 RD to be transferred from the 10th Reserve Army (deployed near Ivanovo, northeast of Moscow) to the 64th Army.

Directive #1035044 ordered the 316 RD to be transferred from the 9th Reserve Army (deployed near Gorki, east of Moscow) to the 64th Army.

Then Directive #1035045 arrived which ordered the 29, 214, and 229 RDs to be transferred from the 64th Army to the 62nd Army, which was then located at Stalingrad.

Finally, Directive #1035046 arrived which canceled Directive #1035042 from the previous day, the one transferring the 64th Army to Stalingrad. So, the new 64th Army (reduced by the three divisions listed above) was staying in the Tula area and not moving south. Directive #1035046 also ordered the 112 and 131 RDs of the 64th Army to relocate to new positions near Tula. The 112 RD was ordered to move by rail from Ryazan to Volovo, a distance of about 203 rail miles. The 131 RD was ordered to move also by rail, from Venev to Stalinogorsk, a distance of about 35 rail miles. (see Map #4) These movements are very telling, as they concentrate these two divisions on the main north-south rail line, replacing the 29 and 214 RD that had been covering this area. It would appear that Stalin was having second thoughts about transferring the entire 64th Army away from the Moscow area. Stalin was still concerned about protecting Moscow from any attack coming from the south.

To sum up the day's directives:

The 64th Army was ordered to stay in the Tula area and not move south.

Three RDs of the Army (29, 214, and 229) were transferred to the 62nd Army at Stalingrad.

Two RDs (292 and 316) from different reserve armies were transferred to the 64th Army.

Two RDs within the 64th Army (112 and 131) were transferred to new positions to cover the main north-south rail line to Moscow.

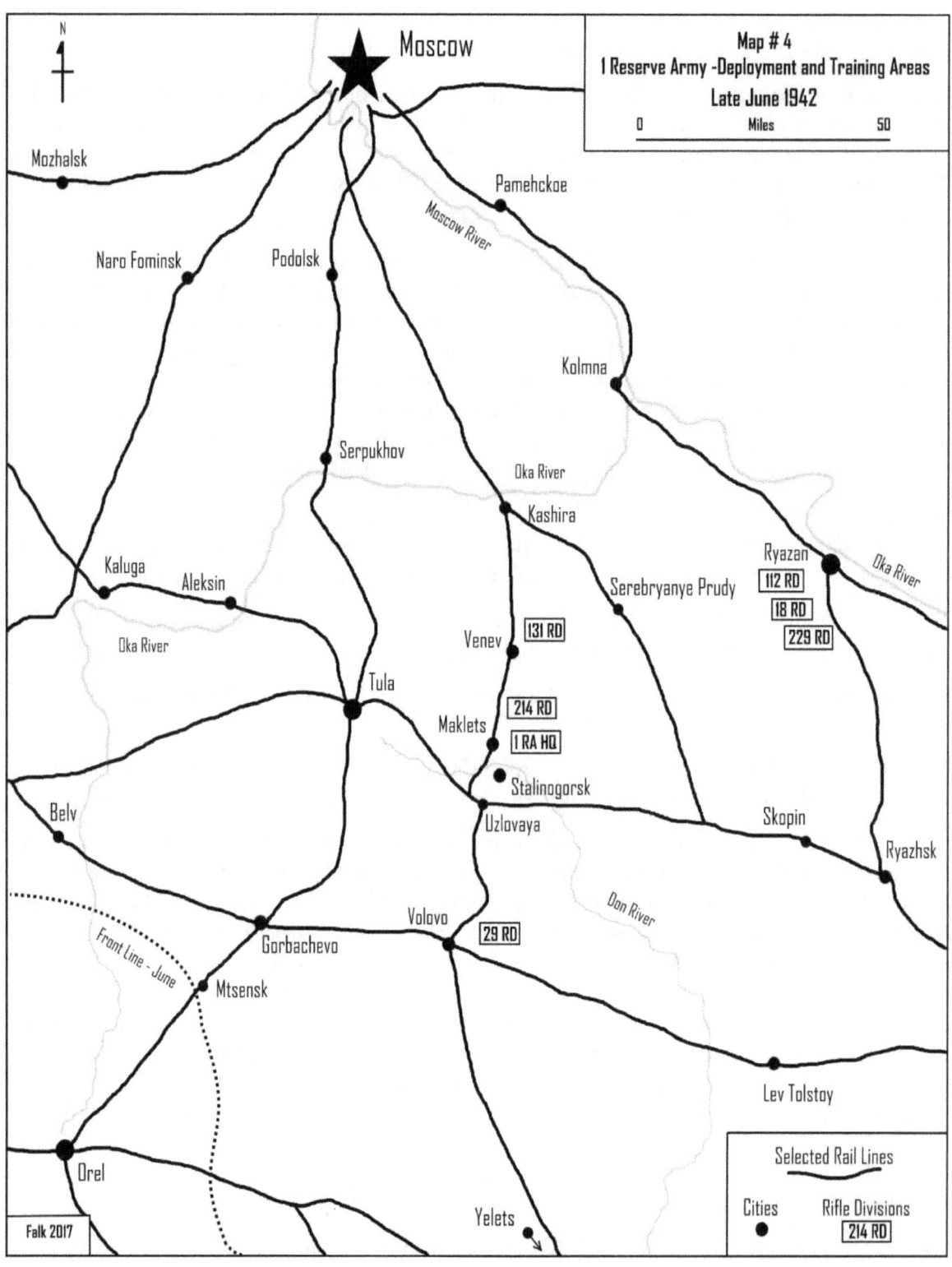

12-13 July

12 July brought more directives. Stavka was thinking about how they were going to control all these new armies. So, Directive #170495 was issued transferring the 9th, 28th, 38th, and 57th Armies from the Southwestern Front to the control of the Southern Front. Then it renamed the Southwestern Front to form the new the Stalingrad Front. The new Stalingrad Front would eventually include the 21st, 62nd, 63rd, and 64th Armies. The Directive also ordered the 62nd and 64th Armies to form a defensive perimeter to the west of the Don River.

Back at the 64th Army HQ, Directive #994111 arrived. This new directive subordinated the 64th Army with its six rifle divisions and other separate units to the new Stalingrad Front upon arrival. There must have been some confusion over this order because the day before, the 64th Army was ordered to stay in the Tula area. We must assume General Chuikov communicated his confusion to Stavka or the VGK because Directive #994111 was confirmed and clarified later that same day by Directive #1035055. This directive specified the order of entraining and movement of the Army HQ and attached units. It also ordered the 29, 214 and 229 RDs to remain under 64th Army control at Stalingrad.

Also, on 12 July, Directive #1035044, the one transferring the 316 RD to the 64th, was cancelled. Most likely, the similar Directive #1035043, the one transferring the 292 RD to the 64th, also was cancelled at the same time. Neither of these rifle divisions were ever under 64th Army control. Eventually, the 316 RD was assigned to the 66th Army and the 292 RD went to the 24th Army.

By issuing all these directives and counter directives, Stavka was clearly trying to quickly improvise a response to the German attack. When the anticipated Moscow attack did not develop, Stalin and Stavka were forced to shuffle major combat formations southward to try to counter the German's breakthrough. One can sense the indecision in these different directives, as Stavka attempted to reinforce the Stalingrad sector and at the same time keep the 64th Army in position covering the approaches to Moscow from the south. In the end Stavka decided to not break up the 64th Army and deployed it intact to the new Stalingrad Front. This was undoubtedly the correct decision on the part of the High Command. The Red Army would need to concentrate all of its forces to defeat the German summer offensive.

Meanwhile, General Chuikov and the 64th Army HQ staff were left to sort out these confusing orders. Troops first were told one thing then another as orders were issued then canceled. Adding to the confusion were the movements of different units themselves. Some formations, the ones that were ready to go, would have already begun marching to their designated loading stations; lucky troops got a ride on a wagon or a truck, but most had to walk. Units that started off nearest their assigned train stations or had been previously loaded were pulling out as fast as possible; the Transportation Directorate and VOSO had a complicated transportation schedule to keep. The 64th Army had a long way to go and was in a hurry, but this confusion at the start of their journey would have a negative impact on the Army later on in battle.

Because of conflicting orders and general confusion, only on or about 12 July did the first trains loaded with formations of the 64th start moving toward their assigned destinations.[48] Some units had orders to start loading on 10 July, but it's doubtful they and/or the trains were ready by that date. Directive #1035055 from the Chief of the Red Army General Staff, A. Vasilevskiy, confirmed the 64th Army was to be moved by rail to the Stalingrad Military District. The order stated the Army should take along one unit of fire (ammunition), one or two units of fuel, and a total of 11 days of rations for the men.[49] Additionally, the Army would have brought along enough fodder and water to feed the hundreds of horses that provided mobility to formations without vehicles. An average rifle division in mid-1942 had about 140 cars and trucks along with 15 tractors to pull artillery, and some 1,800 horses, which pulled around 636 wagons and 93 carts.[50] These numbers could easily vary from unit to unit

As individual units completed their movement preparations, they were sent off to their assigned railroad stations for entraining (loading). Provided trains were at the station, the entraining process would normally proceed quickly. A pure troop train of just infantry and small arms could be loaded or unloaded quite rapidly. But a mixed train of heavy weapons, vehicles, supplies or other equipment could require 2-3 hours or more to load at a station under ideal conditions. Infantry are able to carry their personal weapons, supplies, and equipment with them when they climbed into a boxcar or a passenger car. But it's another thing altogether to load artillery, trucks, tanks, or tractors onto a flatcar. A train station or loading ramp was the best way to load or unload heavy equipment. Additionally, to transport the 1,800 horses allocated to each rifle division would have required approximately 225 boxcars alone (if they were all transported during this initial move), and considerable time to load. It should not be forgotten the added demands of packing boxcars carrying horses with fodder, water, horse handlers, and veterinary staff at the same time.

The 214 RD required 14 trains, of approximately 50 cars each, to carry all of its men, horses and equipment. The 29 RD required 15 trains; the 229 RD, 14 trains.[51] The 18 RD required 13 trains, 112 RD required 14 trains, and the 131 RD required 10 trains. The HQ of the 64th Army along with attached support units would have required another 17 trains.[52] In all, the 64th Army would require at least 97 separate trains and some 4,900 rail cars to move its combat elements to the Stalingrad area. Even with this great effort, most of the army-level supply, medical, and support units would remain in the Tula area for several weeks awaiting transportation. They would not reach the Army until late July or even early August. General Chuikov complained that some "units of the 229 RD and HQ Staff were particularly delayed in route and the last of their troop trains did not arrive in the Stalingrad area until July 23"[53]; some HQ trains did not load until 18 July. Somewhere during the planning stage, the decision was made to give transportation priority to combat elements of the Army, thus leaving supporting units behind until additional trains were assembled. Ultimately this decision would cause great suffering to the troops of the 64th Army. The troops would be engaged in combat for several weeks before the remaining medical and logistical units could arrive to support them. During this early defensive phase of the fighting (July to September) "marked a great shortage of hospital bed capacity. Irretrievable combat losses were enormous," and medical personal of 64th Army were able to transport "not more than 50 percent of the seriously wounded, and the rest remained on the field of battle, dying from hemorrhage and shock."[54] When additional rail transport did become available, it wasn't a fast journey. One medical battalion traveled by rail from Volovo to the area south of Stalingrad on the Aksay River. After taking a roundabout route to the Stalingrad area, they finally arrived after three weeks on the train. That works out to an average of about 30 miles per day.[55]

Related to this lack of transport, the Chief of the Red Army General Staff, A. Vasilevskiy, back in 14 December 1941, issued Directive #575 about overloading trains so that the available rail network would be able to carry the maximum quantity of units with the fewest number of trains. This order commanded, "pack units transferred by railway more densely, and to reduce the number of trains per rifle a division to 16, and per a cavalry division to 9." As a result of this order, loading and unloading times also would have increased.

On or about 12 July, units of the 64th Army started entraining. The 64th Army HQ moved south somewhat and entrained at the Uzlovaya Station. The 112 RD and 18 RD entrained at Ryazan Station and the 131 RD at Venev Station.⁵⁶ The three remaining divisions also had orders to start moving on or about 11-12 July and entrained near their training locations. The 29 RD entrained at or near Volovo Station; the 214 RD also moved south and entrained at or near Uzlovaya Station. The 229 RD entrained at or near the Skopin Station south of Ryazan.⁵⁷ (see Map #5) Some sub units of these rifle divisions also might have entrained at nearby, but unnamed, stations.

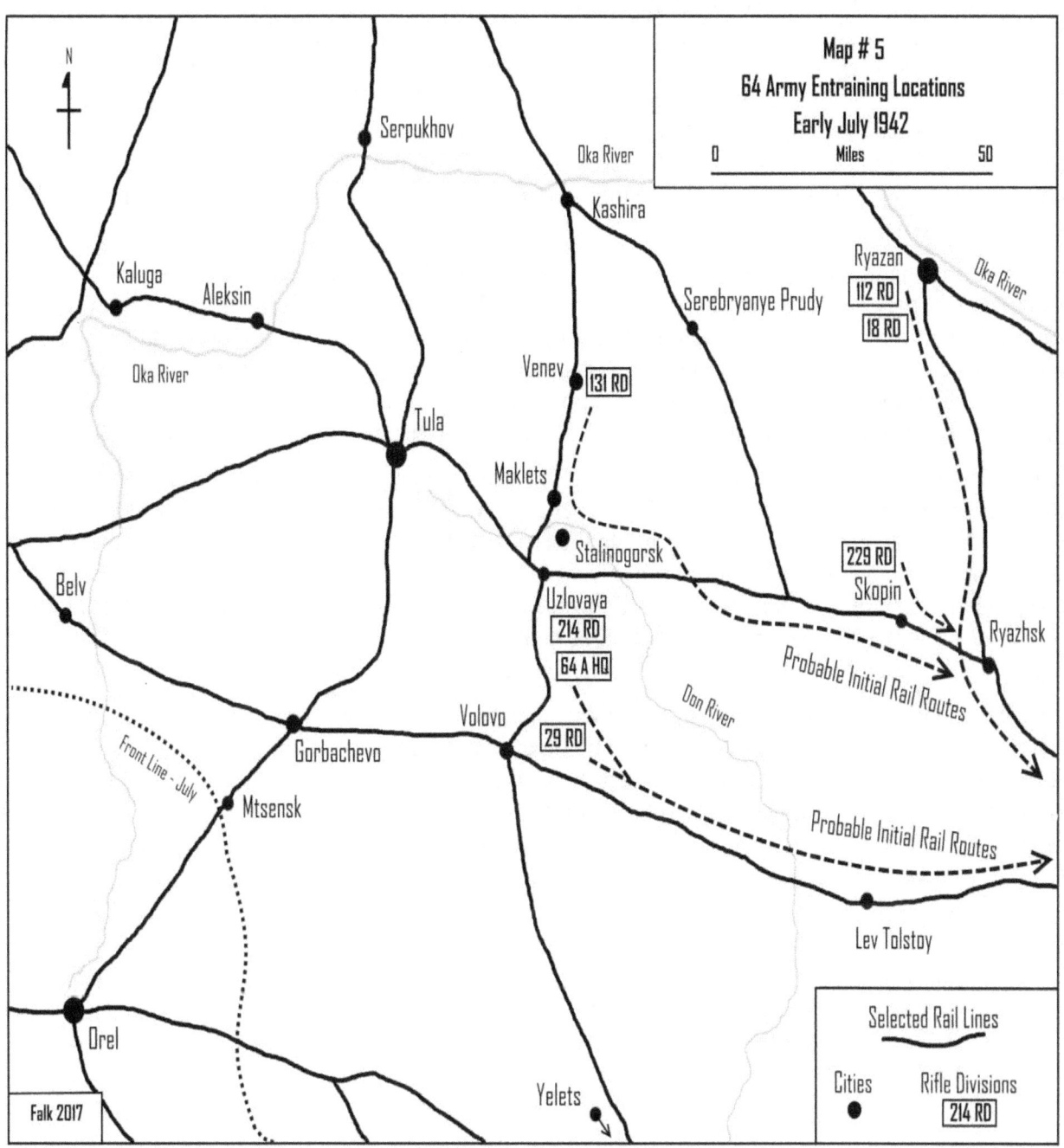

The Army HQ entrained at a rate of three echelons (trains) per day for the first two days and then seven echelons every 24 hours until finished. The 18 RD entrained at five echelons per day. The 112 RD and 131 RD each entrained at seven echelons per day. The 29 RD entrained at seven echelons per day, 214 RD at six echelons per day, and the 229 RD at five echelons per day. So once started, the plan was for the 64th Army to be entraining at a rate of 40-44 trains per day. At this rate, it would still take 2-3 days just to entrain the combat elements of the Army and dispatch them as they were ready. Support and medical units, supplies and spare parts still needed to wait until more trains were made available.[58] The different entraining rates most likely were due to the number of trains that could be delivered each day, and the quality of stations and loading facilities that were available for each unit to use at each location. A certain level of confusion and delay in loading was certainly part of this process and could not be helped. The 64th Army was finally on the move and headed toward the front, and that was the most important development. General Chuikov had at last achieved his wish of a combat command.

Later on during the month, General Chuikov remarked as the 64th Army formations arrived in the Stalingrad area, they were sent off to the front "not in military columns, but in whatever order the troop trains delivered them."[59] There was little time to organize the transfer properly. Stavka wanted to move as many combat units into the Stalingrad area as fast as possible. To help out, the Stalingrad Front was called upon to provide supply and support to the combat elements of the army until their own rear services could arrive. Time would tell if these improvised efforts would be adequate.

Organizational issues were not the only problem. The rail network between the Tula area and Stalingrad was not very well developed. This forced the VGK to use the few available rail lines well beyond their normal capacity in order to move the required forces. Poorly maintained rails and sleepers forced the overloaded trains to move very slowly. On many sections of line, trains were averaging 9-12 mph during the journey. At that speed, a train would be able to cover only about 288 miles in a 24-hour period. Destroyed, missing, or damaged lights, switches and communication equipment also added to control problems, which would further slow the trains' movements.

The transfer route for the 64th Army was as direct as possible given the circumstances. (see Map # 6) The 64th Army HQ, along with the minimum number of support units, and the 112 and 131 RDs were sent from the Tula area via the Tambov, Rtishchevo, Balashov, Povorino, Stalingrad route, for a total of roughly 656 rail miles to their detraining locations. (see Appendix 9)

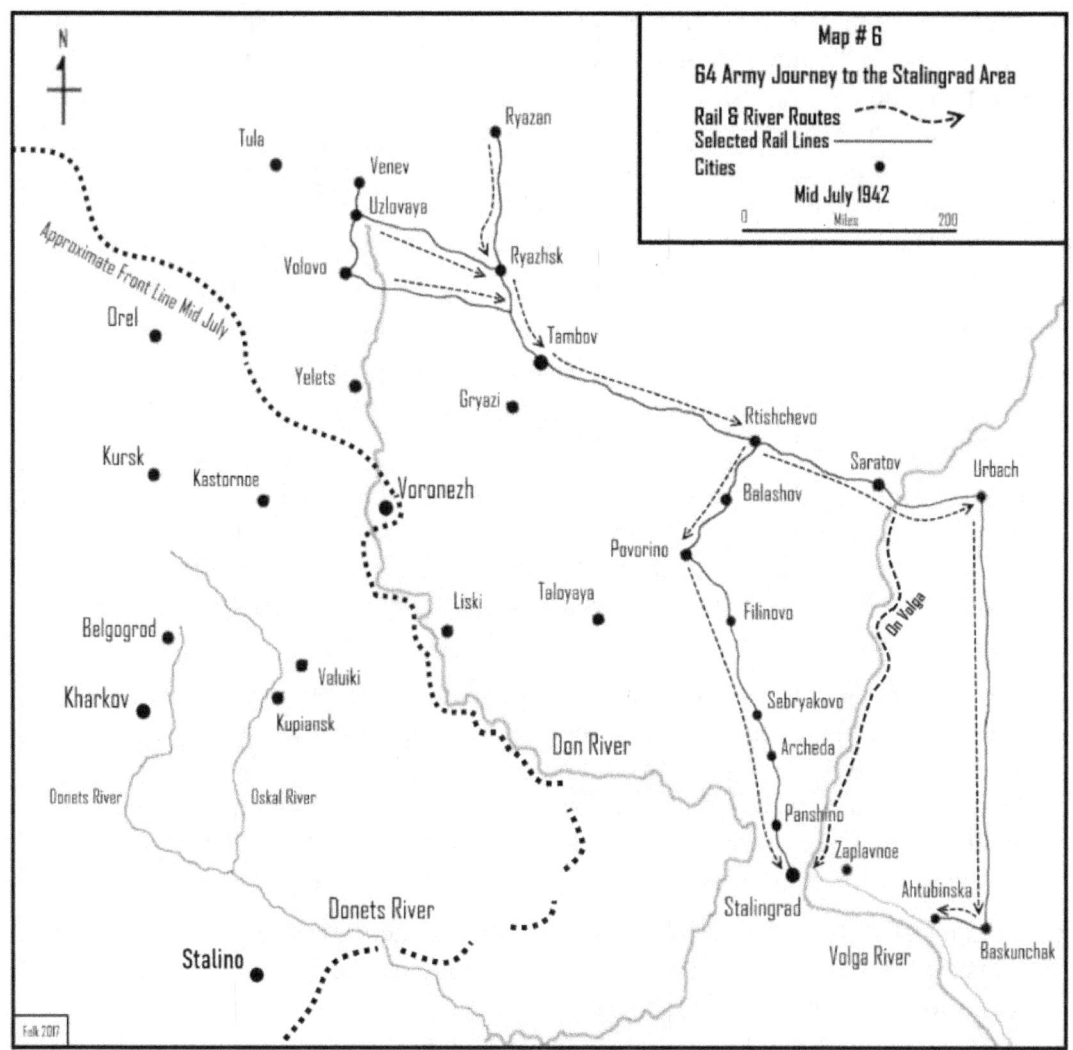

The 18 RD and the 229 RD took a more roundabout and confusing route, Ryazan, Tambov, Rtishchevo, to Saratov.[60] At Saratov, both divisions split up with some trains unloading their units. These troops boarded boats and barges on the Volga, then made their way down river to the Stalingrad area to unload. The remaining units of both divisions, the ones that were still on their trains, continued on to Urbakh. Once on the east side of the Volga, they travelled down to Baskunchak and then to Volodymyriuka Station near the town of Ahtubinska. At this date, this section of rail line did not yet reach the Stalingrad area, so once there, they would have been transferred to barges or some other type of river transport and sailed up the Volga to Stalingrad, or they could have walked to Stalingrad. In any case, the last part from Stalingrad to the front would have been on foot. This long journey totaled roughly 690 rail miles, 300 river miles from Saratov to Stalingrad, 75 river miles (or on foot) to Stalingrad from the Ahtubinska railhead, and a further 93 miles on foot from Stalingrad to the front, for a grand total of 1,258 miles by rail, water and/or foot.[61]

Despite conflicting accounts, it appears the bulk of both divisions, the 18 RD and 229 RDs, moved by rail following the route Ryazan, Tambov, Rtishchevo, Saratov, Urbakh, Baskunchak then Volodymyriuka Station, and up the Volga, by land or water, to Stalingrad.

But something else occurred to the 229 RD along the way. Sometime during transport, the division HQ/administration staff, some rifle battalions and heavy equipment like artillery and AT batteries, were diverted from this path and did not follow the main group of the division to the lower Volga. They were transported by rail, via Povorino, directly to Panshino Station area northwest of Stalingrad and detrained there. Later on, somewhere near Stalingrad, the three separate parts of this division were reunited during the march to their assigned positions at the front. Needless to say, such a fragmentation of the 229 RD only could add to the confusion and delay its deployment at the front.

A quick look at Map #6 explains why these two rifle divisions took a longer, more roundabout route to Stalingrad. Simply put, the Povorino to Stalingrad section of tracks were extremely overloaded. That section of track was a choke point in the rail network and was bound to be attacked by the Luftwaffe more than it already was. In all probability, the Soviets used every means of transporting troops, supplies and equipment to the front.

Due to the earlier conflicting orders about transfers to the 62nd Army, the 29 and 214 RDs most likely were some of the first units to entrain and to arrive in the Stalingrad area. Their actual movement route is unknown, but due to their detraining locations they most likely followed the same Tula area, Tambov, Rtishchevo, Balashov, Povorino, Stalingrad route as did the bulk of the 64th Army. (see Map #6)

The first trains carrying formations of the 64th Army began to arrive in the Stalingrad area on 16 July.[62] Having started to leave the Tula region on 12 July, this initial group of trains took four days to make the 656-mile trip, or about twice as long as would be expected if they had travelled at the 288 miles per day rate. Something had slowed down their progress significantly. Loading problems, train availability, control issues, or maybe something else caused this slowdown. I believe all these issues played their part, but a major factor in slowing their arrival was the Luftwaffe.

Luftwaffe Air Field Manuel #16 (paragraphs 162-177) established principles for air interdiction of enemy troop movements. Seven target categories were outlined with emphasis being placed on attacking troop concentrations and rail interdiction targets.[63] (see Appendix 10) The 64th Army and other units were forced to run this air gauntlet to reach their new area of operations.

Even before July, German Luftwaffe reconnaissance and combat aircraft were flying far behind Russian lines as well as providing close air support to German troops advancing toward Voronezh. The many targets provided by the vigorous Russian defense of Voronezh and the Don River crossings attracted many Luftwaffe attacks. These Luftwaffe efforts greatly helped ease the way for advancing German Army units. In order to block the movement of new Russian forces from arriving into the area, German aircraft also were bombing more distant interdiction targets like rail lines and junctions at Voronezh, Michurinsk, Svoboda, and Valuyki.[64] Also, on 9 July, the Luftwaffe struck the major rail junctions of Yelets, Tambov, and Povorino in attempts to further isolate the Voronezh battlefield.[65]

Of course, when the German ground offensive turned south, the bulk of the Luftwaffe efforts shifted south too. Numerous reconnaissance missions could hardly fail to observe the mass movement of Russian forces toward the middle and lower Don as well as toward the Stalingrad area. The Luftwaffe's General Richthofen commented in early August that, "the enemy attempts to fling troops from every point of the compass into the Stalingrad sector. He's hell-bent on holding the city." [66] Trains moving along key rail lines and major rail junctions were given priority for repeated air attacks.

Many German and Russian sources have stated, at this time, that Russian rail movements were restricted to the night to avoid air attack. But during this period, that might not have been true.

Several additional Russian accounts related how troop movements were taking place during the day, and they suffered losses when the Luftwaffe bombed their trains or the stations they were using.

At the Aleksikovo Station (Povorino to Stalingrad line), V.V. Grinevsky in the book, *Heroic Sixty Fourth* p 7-11, stated that at "[a]pproximately one in the afternoon (I do not remember accurately date), against the station flew 13 the Junkerss [Junkers JU-87 aircraft]. After unrolling, they with the terrible howl began to dive to the military trains. After setting the rail cars on fire …ammunition began to explode. And most unbearable-neigh, the wild neigh of horses in closed wagons. Then the aircraft came back and started to strafe people."

This type of attack, with Ju-87 dive bombers, could have taken place only during the day. Also, at another station, Archeda, a train was approaching the station loaded with tanks when German aircraft attacked. The tankers started their tanks on the rail cars and drove them off "jumping to earth and driving away" from the station. And then the statement, "In the daytime and at night Fascist carrion vultures bombed and fired military trains, [and] railroad junctions." [67] So, it would appear formations did suffer air attacks by both day and night. In any type of mass movement like this, units would have suffered attrition along the way. But some paths to Stalingrad were worse than others.

Clearly, finding and attacking steam engines at night was difficult but not impossible. At night "the high polish of railway lines was…prominent, as were signals [lights]." Rails at night were so visible from the air that they "were prime navigation aids" for aircraft. As for the trains themselves, "At night the steam engines pulling trains sent up billows of sparks from their chimneys, while each time a stocker opened the furnace door to throw on more coal the glow lit up the smoke overhead and was visible for miles in the darkened landscape."[68] Different steps could be taken to reduce nighttime visibility, but trains still would have been in danger of sudden air attack.

One critical section of tracks that received much attention from the Luftwaffe was the Povorino to Stalingrad line. After some repair work, this section could pass 21 pairs of trains every 24 hours. The other critical section was the Urbakh to Baskunchak line which could pass only 19 pairs of trains every 24 hours.[69] These statements are talking about pairs of trains because typically loaded trains head toward a destination and unloaded trains return from that destination using the same rail line.

Povorino itself was bombed each night, disrupting normal railroad operations including servicing locomotives and keeping them supplied with fuel and water. In order to keep priority troop trains moving, the VOSO decided to prepare locomotives at remote locations away from Povorino during the day and to send trains through Povorino at night without stopping. Also, to speed the movement of troops, Russian railroad authorities started one-way movement of trains from Povorino to Stalingrad, keeping the empty cars at stations and sidings in the Stalingrad region until they could be moved elsewhere.

To complicate matters even more, the Luftwaffe destroyed three major district stations along the Povorino to Stalingrad line during this period; they were the Filinovo, Sebryakovo and Archeda Stations. Due to continuous German air attacks, the Russian 13th Railroad Brigade and assigned civilian repair crews could not keep up with the damage and were forced to abandon these three stations. Bypass lines were built around the stations to keep the through lines open. What was left of the rail lines within the stations themselves were used only as short-term parking of trains.[70] The Red Army Airforce, the VVS, was unable to stop or even seriously interfere with these raids against vital rail targets. It was left to ground forces to make speedy repairs of the rail lines to keep the trains running and the armies moving toward the Stalingrad area.

With the 64th Army on the move, General Chuikov stated that he went as far as Balashov with the Army HQ train, but then got off (due to slow progress) with a Member of the Military Council and proceeded by light truck ahead of the train. He went on to state, "We followed the railway line and called in at the big stations, so as to have a clear picture of the movement of our troop trains. The enemy was systematically bombing all these stations in an attempt to stop the movement of our troops."[71]

Lieutenant General I.V. Kovalev, of the Central Transportation Corps, remarked that the enemy "…subjected railroad sections, units, large stations and bridges to the bitter bombings. Attacks were produced by systematically [using] predominantly small (from 2 to 6 aircraft) groups, and with the actions against bridges - by single aircraft." [72]

Details of German tactics used in the 1941 Barbarossa campaign against trains actually supports this account, stating that "most of the attacks were by small units of between two and four aircraft, flying armed reconnaissance at low altitude."[73]

General Kovalev went on to say "…for the guarantee under the conditions for the incessant enemy bombings of the continuous supply of the troops and equipment to Stalingrad, it was necessary…to strengthen the railroad troops in Stalingrad region." [74] In the area of the Voronezh and Stalingrad fronts operated the 5th, 13th, 19th and 27th brigades and some other groups of the railroad troops. These RR Brigades and thousands of conscripted civilian workers kept the threatened rail lines functioning. Eventually the Russians had to resort to positioning repair crews and stocks of equipment along the most often attacked rail lines. After an air raid, these repair crews would emerge from bunkers or trenches and quickly put the damaged tracks back into service.

Despite their best efforts, Luftwaffe attacks upon the Russian transportation system were never strong enough or systematic enough to stop rail operations completely. They did delay the movement of troops and supplies. They also caused some losses, but their greatest effect was to disrupt the tempo of Russian operations. In the end, the Luftwaffe could not achieve much more. The Russians for their part put forth a major effort to keep the rail transportation system from totally failing. Trains were destroyed, stations blasted, rail lines were cut, and still the system functioned, but at a reduced level.

As for the troops of the 64th Army, they waited out the transfer packed into boxcars and laying amongst equipment and vehicles lashed to flatcars. They watched the countryside go by at a slow speed and thought about their future and the war. As they got nearer the front, these troops started to encounter fleeing refugees moving along the tracks going in the opposite direction. The war came upon these raw troops slowly, as if they were entering a strange new world. The landscape gradually filled with signs of death and destruction. Masses of debris filled destroyed railroad stations, then a broken steam engine laid in a ditch, and nearby villages were burned to the ground. These images would occupy the thoughts of the troops for days. As the troops pondered the war, their officers struggled with the logistical nightmare of controlling an army on the move. Despite their efforts, far off events took place that would soon alter the composition of the 64th Army and ultimately decide its fate.

Before even reaching their final destinations, the commander of the Stalingrad Front, Marshal Timoshenko, decided to transfer two rifle divisions out of the 64th Army to serve as Stalingrad Front reserves. This was an operational Directive of the Stalingrad Front # 0023 issued on 17 July 42.[75] When they arrived in the Stalingrad area, the 18 and 131 RDs were assigned defensive positions outside Stalingrad. The 18 RD was sent north of Stalingrad near the Don. Eventually, on 22 July, the 18 RD was reassigned to the newly formed 4th Tank Army that was assembling in this area. The 131

RD was sent southwest of Stalingrad and was eventually assigned to the newly formed 1st Tank Army. With these transfers, the 64th Army lost one third of its strength. The 18 RD had about 11,700 men and the 131 RD had 10,795 men for a total of 22,496 men.

To make up for the loss of two full strength rifle divisions, the Stalingrad Front assigned to the 64th Army some of the flood of units Stavka was sending to the Stalingrad area. After arrival, the 64th Army took control of the 66th and 154th Naval Rifle Brigades with about 5,000 men each, the 40 and 137 Tank Brigades with about 1,341 men each, and four Cadet Rifle Regiments with some 2,500 men each. So, the 64th Army lost 22,495 men due to transfers, and then once it had arrived at Stalingrad it was assigned new units totaling about 22,682 men.[76]

With these transfers, the Army would mostly be brought back up to strength in terms of manpower. The addition of the tank brigades was most welcome, as the 64th did not have any tanks under its control prior to this. However, it was hard to compensate for the removal of these two rifle divisions. The 18 and 131 RD had been training with the Army for weeks, so even before entering combat the 64th Army suffered a major loss in unit cohesion. Meanwhile, the 64th Army HQ had to take control of these newly assigned combat units not knowing their locations or arrival times. This would have been very hard to do on the move, and most likely only occurred after most of the 64th Army units and HQ staff had arrived in the Stalingrad area.

Sometime during the formation of the Stalingrad Front, Stavka developed a plan on how to employ the four armies assigned to this newly formed front. It appears the plan called for the 63rd (former 5th Reserve Army) and 21st Armies to cover the north bank of the Don River. This part of the plan was easily achieved because these armies were close to the Don. Next, the 62nd was to move from the Stalingrad area due west into the Don River bend and form a new defensive line stretching from the Don in the north to the village of Slepikin in the south. This was a change in the original plan that was most likely introduced by the Stalingrad Front command staff themselves. This too was achieved rapidly because the 62nd Army was deployed at Stalingrad, only 90 miles away from the new front. Finally, when the 64th arrived from Tula, it would occupy the line on the left flank of the 62nd Army, with the junction between the armies near the village of Slepikin and then south to the Chir River and beyond.[77] Only part of this initial plan was achieved by the 64th Army before the Germans attacked. (see Map #3)

Using Slepikin as the junction point would only make sense if Stavka or the Stalingrad Front commanders intended the 62nd Army to have two lines of defense, or a defense in depth, with three rifle divisions in the front line and three more rifle divisions in the next line. If this were true, it would have required the 64th to cover the remaining front from Slepikin to Suvorovski, on the Don River, all by itself. To cover this extended line, the 64th Army would have needed to put most if not all its forces in the front line. This would leave only the four cadet regiments in a second line as a reserve. This would not have been as strong a defense as the 62nd Army had, but this deployment was still adequate. On the positive side, this would place the entire 64th Army on the west side of the Don making support and coordinating issues much easier. In this way, the 62nd and 64th would entirely block the great bend in the Don River. If we add in a few tank regiments as local reserves and a few additional rifle divisions as early reinforcements, this defense line started to look relatively strong. Add to this the soon-to-be-formed 1st and 4th Tank Armies, and it made for a robust force blocking the way to Stalingrad. Anyway, that appears to have been the original plan, but when do plans fully work out during war?

14 July

The Presidium of the USSR Supreme Soviet declared martial law in the Stalingrad Oblast, so the military now had total control and total responsibility over Stalingrad and the surrounding area. This act was a clear sign that the Presidium thought Stalingrad itself was in great peril.[78]

15 July

As the 64th Army and other Soviet combat formations made slow progress toward Stalingrad, German units also made sluggish progress, due to supply issues, toward the same destination. Meanwhile, Stavka and Stalin were attempting to arrange for a great battle to take place far to the west of Stalingrad in order to protect the city, and in this they would be partly successful.

16 July

On 16 July, General Chuikov arrived at Stalingrad. The same day the first 64th Army troop trains also arrived in the area. General Chuikov went directly to the Stalingrad Front HQ and was briefed on recent military events. Here he learned about the failed Kharkov offensive earlier in the summer and more recent actions in the local area.

As loaded trains started to arrive in the Stalingrad area, the different units were sent to separate detraining locations. (see Map #7) Hundreds of trains carrying thousands of troops cannot all stop at the same station. What an inviting target that would have been for the Luftwaffe! Instead, trains and stations were matched, as well as could be expected, to the unit's final destination. As it was, there were far too many trains arriving for the limited amount of rail facilities to handle. The large main stations and platforms directly in and around Stalingrad were needed to unload tanks, vehicles and other heavy equipment. Troop trains would be lucky to have an improvised ramp to unload their horses, wagons, and other supplies. Most trains were stopped along some desolate stretch of track far outside the nearest town or village. The Russians had learned the Luftwaffe watched stations more closely and attacked when a target presented itself. Detraining in some isolated spot was far safer for the troops.

On this same day, much further to the south, the German 29 Motorized Division reached the lower Don River near the town of Tsymlyanskaya. At that time, this mobile division was prevented from exploiting this advance further because of the lack of fuel, ammunition and support. Only on 20 July was the 29 MD able to cross the Don and form a bridgehead on the eastern side forcing back units of the 91 RD of the 51st Army. Overall this was a disturbing sign; the Germans had crossed the 1,150-foot-wide Don River and were established on the eastern side.[79] The crossing of the Don by German forces would soon have an impact on the 64th Army further north toward Stalingrad.

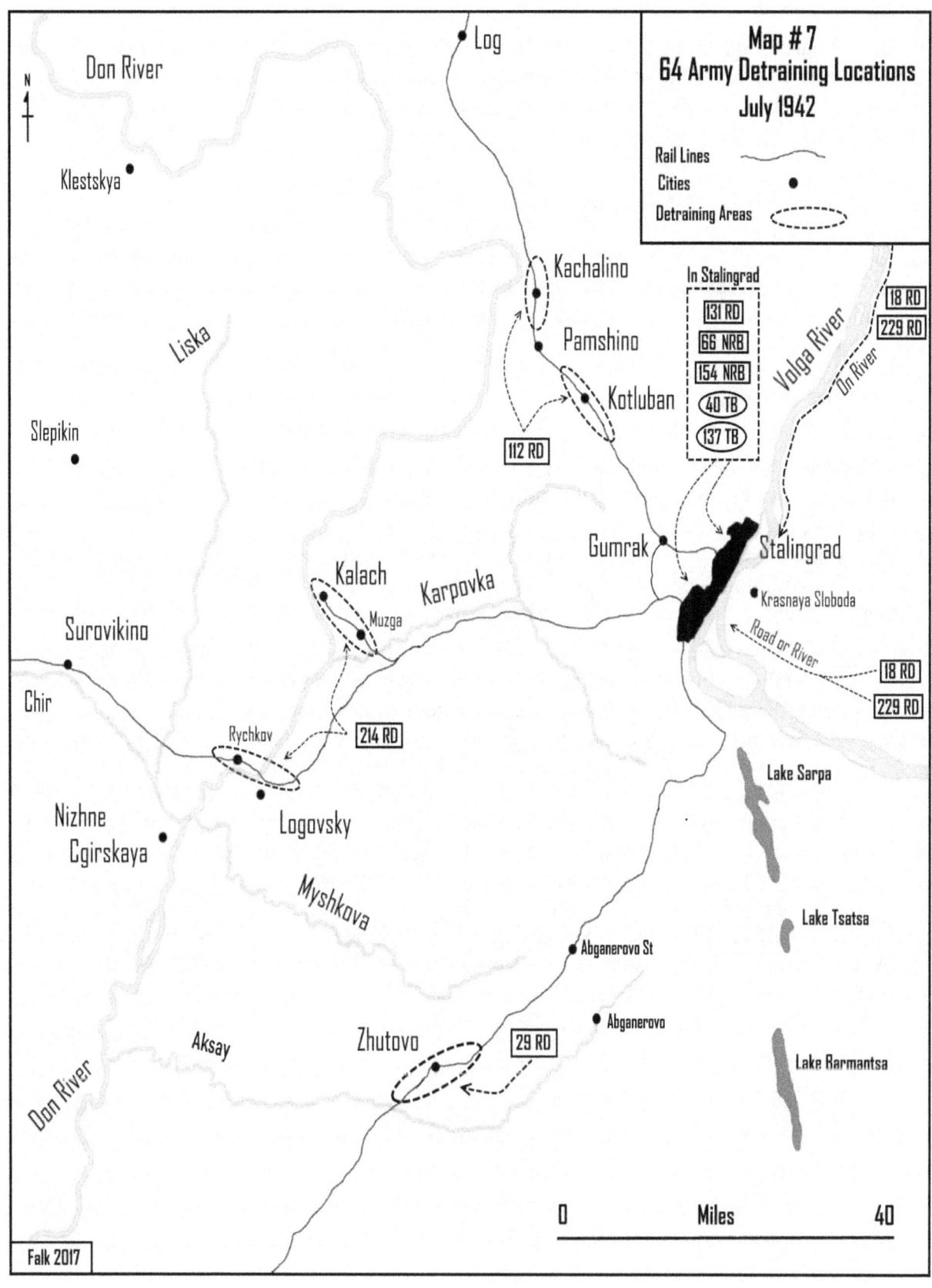

17 July

Even at this early stage of the battle for Stalingrad, Stalin saw need to interfere with how the new Stalingrad Front was handling its forces. Stalin issued a direct order to Marshal Tsimlianskaia, Directive #170520, to send out strong "forward detachments" in front of the newly designated defensive line. Tsimlianskaia was unable to act on this order because most of the required forces had not even arrived in the area yet.[80]

Also, on 17 July, General Chuikov received an order from the Stalingrad Front HQ about where his Army would be deployed. This order stated the Army should be occupying the assigned defensive positions on the west side of the Don River and then in a line south of the Chir River by the night of 19 July. Chuikov was not pleased with these orders because they were "clearly impracticable" as the units had "just left the troop trains…and the units would need to march 125 miles or more to reach their assigned positions."[81] At this point, Chuikov had only a vague idea about the detraining locations and overall disruption of the arriving units. Chuikov needed some time to reorganize his Army and to regain control of dispersed units.

18 July

In the morning, General Chuikov went back to the Stalingrad Front HQ, talked with staff officers, and told them the 64th "could occupy its line of defense not earlier that July 23." At this point, a more junior officer at the HQ changed the date on the orders (to occupy the defensive positions) to 21 July. Chuikov was astounded by this action and went on to say, "Who was in command of the Front?"[82] After this insight into the operations of the Stalingrad Front command staff, Chuikov spent the rest of his time searching for and assembling his detraining units.

Soon many of his disorganized formations started their long, hot trek toward the Don River. In order to construct and then occupy the new defensive line as prescribed by Stavka, most of the Army needed to cross the wide Don. With only a few bridges or ferries available for crossing, forward progress was further slowed by these choke points. Once on the west side of the Don, troops still needed to march to the assigned front-line area. This new defensive line would run from Surovikino on the narrow Chir River south some 40 miles to the middle Don near the village of Suvorovski. Only then could they start the construction of the new front line. A few lucky units did not have that far to go as they remained on the left, east side, of the Don. (see Map #8). This section of the front began at Suvorovski and continued south until reaching the 51st Army on the 64th Army's left flank. This new plan required the 62nd Army to cover the entire area from the Don River in the north to Surovikino on the Chir River to the south. Most likely this deployment change was made because the arrival of the 64th Army was delayed. This section of the front had to be covered, and the 62nd Army was the only one available.

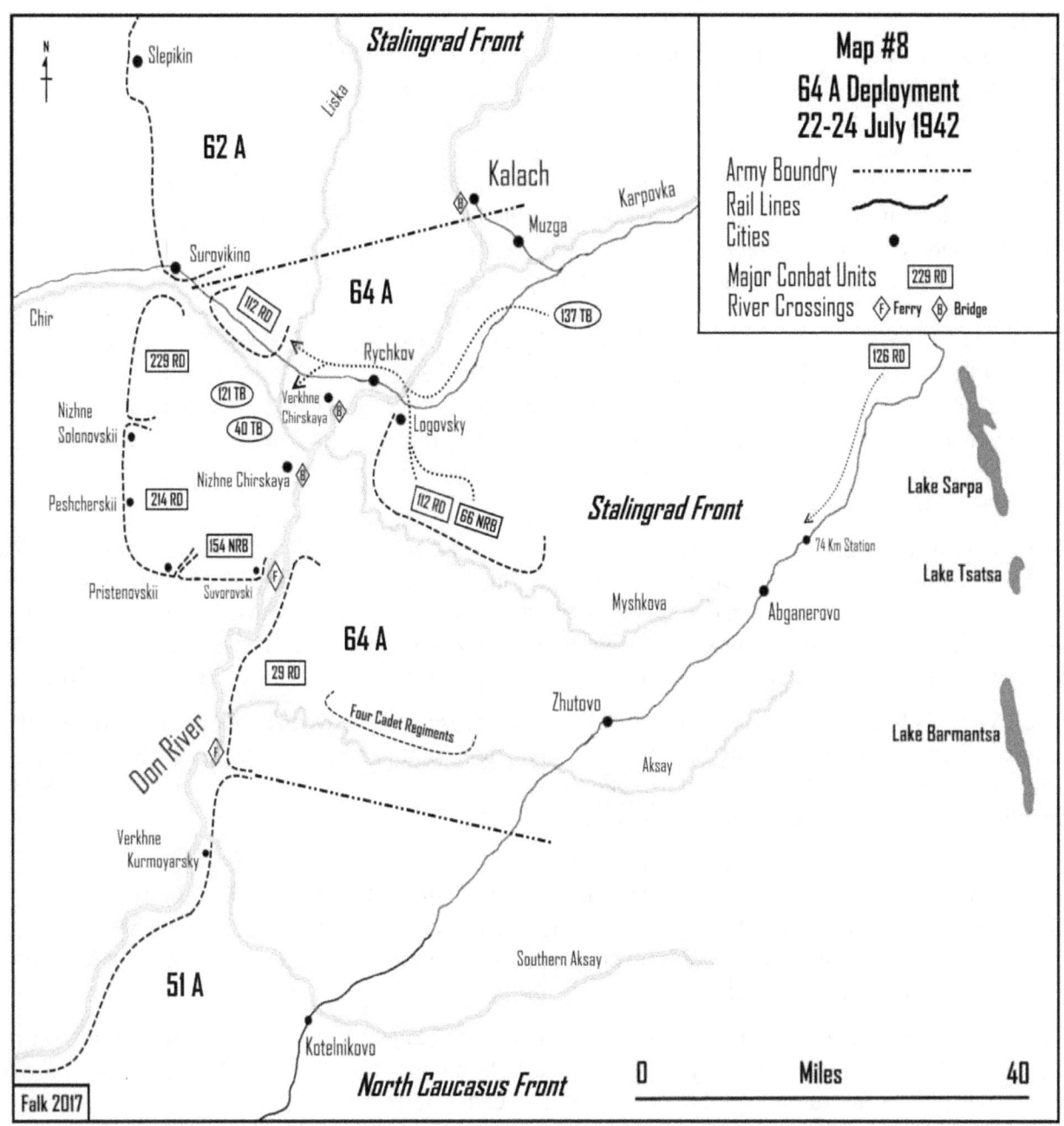

19 July

On 19 July, the Stalingrad Front issued another set of operational orders regarding the deployment of the 64th Army. After laying out what the 64th Army needed to accomplish overall, the orders went on to state that by the night of 21 July ... "229, 214 and 29 Rifle Divisions [and] even [the] 154 Brigade of marines proposed to take urgent measures for the protection of the passages through [over] the Don." So, in addition to establishing a new defensive front line, these formations also were given the task of protecting the several crossing points over the Don, in the rear of the 64th Army.[83] This order is a clear indication the Stalingrad Front HQ was worried about the security of the Don River crossings. The order also indicates these four units were the first ones of the 64th Army to be in position to actively start operations. In any case, by 06:00 hours on 19 July, only 3 echelons out of 14 for the 214 RD had been unloaded. Additionally, only 6 of 15 echelons for the 29 RD and 8 of 14 for the 229 RD had likewise been unloaded. The 112 RD was lagging behind and was still entirely loaded on trains moving toward Stalingrad.[84] The 154 Naval Rifle Brigade (NRB) also must have been in the same general area to have been included in these orders. From what is known about the detraining locations and movements of these units, this order, to cover and to guard the river crossings, was both reasonable and possible.

But the 64th Army had other units assigned to it. Besides the 154 NRB, it also had the 66 Naval Rifle Brigade, the 40 and 137 Tank Brigades (TB), and four Cadet Rifle Regiments. These newly arrived formations were in the Stalingrad area, but their exact locations are unknown. General Chuikov would not have been certain how combat ready these units were, having just arrived himself. Nevertheless, Chuikov ordered them to proceed toward the Don River crossings and the slowly forming front line.

The 154 NRB had recently come down from north of Moscow. The Brigade would have required 6-8 trains, depending upon its strength, to make the transfer and must have arrived in the Stalingrad area sometime in mid-July. The Brigade also needed to have been located somewhere near Stalingrad to be mentioned in the new operations orders issued on 19 July.

The 66 NRB came from the Karelian Front north of Moscow where the Brigade had been part of the 32nd Army. Then in June 1942, it was transferred to the Taman Front. From there it moved to Stavka Reserves and then on to the 64th Army at Stalingrad, showing up under its control sometime after 10 July. It too would have required 6-8 trains to carry its troops and equipment as it joined the flood of units moving toward Stalingrad.

The 40 TB had an interesting history. This TB had fought in the Crimea in May 1942, during the defense of the Kerch Peninsula. By the end of the fighting on 18 May, the 40 TB had been destroyed. At this point it disappeared from the record. But, "a cadre of troops survived" and were sent from the Crimea to Stalingrad to be reequipped with tanks directly from the STZ tank factory (Stalingrad Tractor Factory), which was then producing the formidable T-34 tank. By July, it had been refurbished with men, equipment, and about 53 tanks when it was assigned to the 64th Army.[85]

The 137 TB had been formed during the March-May period of 1942 in the North Caucasus Military District. Then on 10 July it was assigned to the 64th Army.[86] Sometime after this date, the 137 TB made its way north to Stalingrad, almost certainly via rail on 3-4 trains, but it was possible this unit was loaded onto ships or barges and moved up the Volga to Stalingrad. This TB should have been at full strength with some 53 tanks.

Finally, the four Cadet Rifle Regiments (officer cadets) assigned to the 64th Army also came from the south having been formed in the North Caucasus. These regiments came from the Zhitomir and Krasnodar machine-gun and mortar schools, and the 1 and 3 Ordzhonikidze infantry military schools.

All four of these regiments were initially assigned to the 64th Army, but by 1 August one of the Ordzhonikidze regiments was transferred to the 62nd Army. Each regiment would require 2-3 trains to make the move and contained between 2,000-2,500 cadets. One of these school units received its transfer orders on the night of 10 July. School personnel, staff and cadets were called together and the announcement was made; everyone then promptly started preparing to move to the Stalingrad Front by rail, most likely leaving the next day. As some cadets had said, they "hurried to Stalingrad."[87] These officer cadets would ultimately pay a terrible price in blood during the coming battle. One veteran commented, "Only an extreme need forced Stavka to throw these incompletely trained cadets, tomorrow's lieutenants, into battle."

Adding to the confusing situation, during the evening of 19 July, Lieutenant General Gordov arrived at the temporary 64th Army HQ. He had orders to take over command of the 64th Army; consequently, General Chuikov became his deputy commander. Gordov approved of Chuikov's planned dispositions of the Army south of the Chir River but ordered the 112 RD to be deployed not to the Chir but on the eastern side of the Don and to take up positions in the outer defense line of Stalingrad. He also moved the 66 NRB, 137 TB and the four regiments of cadets south to the Aksay River, on the far-left flank of the Army, again on the eastern side of the Don. Chuikov did not like this arrangement because it moved all the 64th Army's reserves to the east side of the Don and took away the second line of defense behind the positions near the Chir River.[88]

20 July

Despite changes in orders and assignments, by 20 July, the first units of the 214, the 229 RDs, and the 154 NRB already had moved over the Don River and started to occupy their assigned defensive positions.[89] These units had arrived just in time because soon German forces would reach them, with the first real combat starting on 23 July.[90] As further formations of the 64th Army came off the trains, they were moved toward the Don River crossings in forced marches. The men knew they were rushing toward an uncertain future, but they marched on despite being tired, hungry, thirsty and afraid.

While the Stalingrad Front Commanders were waiting for the 64th Army to fully deploy, they already had the 63rd and 21st Armies in position along the east side of the Don River, blocking any move by German forces to the north. The 62nd Army also was in position in the bend of the Don. When it was activated, the 62nd Army, formally the 7th Reserve Army, was already in position near Stalingrad. By 20 July, some of its units had been digging in for nearly a week along its 100-mile defensive line. This defensive line started in the north from the village of Kletskaia on the Don River, all the way south to the main rail line linking Stalingrad to Rostov at the village of Surovikino. (see Map #3) The 62nd Army was forced to cover this extended line because the 64th had not yet arrived in strength. The 62nd Army was stretched too thin in trying to cover the entire area all-alone. Besides its own sector, the 196 RD of the 62nd Army was covering an 18-mile section south of the rail line at Surovikino. This section of the front was assigned to the 64th Army, but it was too critical to leave unguarded. So, until the 229 RD of the 64th Army arrived to take over, the 62nd Army had to cover this sector too.

Shortly before this time, both the 62nd and 64th Armies also were required, by order of Stalin, to send an advanced force far out in front of the main defensive line. Chuikov did not like splitting his troops up like this.[91] The end result of this order was that starting on or about 19 July, as the German 6th Army advanced toward the Russian main defense line, it started to overrun and drive back these small Russian so called "advance detachments." These small detachments of troops and weapons had been deployed well forward of their main defense line with no other support. This idea of forward deployed detachments came directly from Stalin himself. Most Russian commanders thought this tactic was a waste of manpower and equipment, but it would have been unwise to argue with Stalin. So, these small detachments of troops were sent out and promptly surrounded, destroyed or forced out of their positions. Most of the men and equipment were lost, with hardly any impact on the German advance. An officer of General Staff of the 62nd Army, a Major Kordovskiy, wrote about these advance detachments in a report to A.M. Vasilevsky. As a result of sending out these advance detachments, the army "…lost a large number of personnel and material part prior to the beginning of battle on the front edge. On their basic task (they) carried out very little."[92] From the German point of view, these advance detachments were hardly even noticed. They thought they were overrunning scattered rearguards or remnants of shattered Russian units.

The overall situation was really a race between armies; which side would be the first to reach and then to cross the Don River near Stalingrad in force? The Russians were moving troops as fast as possible to the west side of the river and building up a new defensive line there. Meanwhile, the Germans were aiming to cross over to the east side of the Don and advance directly upon Stalingrad and the Volga River. Standing in the way of these German plans was the 62nd Army who by itself was not strong enough to defend the direct path to Stalingrad. If the 62nd Army did not get help, the Germans would break through and capture Stalingrad. Would the 64th Army come up in time to help block the approach to Stalingrad, or would the Germans arrive first?

21 July
During this critical time, Stalin summoned General Gordov to Moscow. Gordov had only been in command of the 64th Army for two days, so not knowing why he was called away he hurried off to see Stalin. After arriving back at Army HQ later that night, General Chuikov finally learned that General Gordov had left for Moscow. Certainly, Chuikov was not pleased with this information; the Army had been left without an onsite commander for hours!

22 July
On 22 July, Stalin fired General Timoshenko, the current commander of the Stalingrad Front, and appointed General Gordov in his place as the new Stalingrad Front Commander. When Gordov arrived back at Stalingrad that same day, Chuikov was informed that he was once again in command of the 64th Army.

On the other side of the front line, the German offensive proceeded mostly as planned but was behind schedule largely due to logistical reasons. By mid-July, Operation Blue had reached Phase III of the overall plan. This phase of the summer offensive entailed the occupation of the entire Donbass area, in other words, everything west of the Don River from Voronezh in the north to Rostov in the south. To accomplish this task, German forces were split into two army groups. Army Group B, with the 6th Army Commanded by General Paulus and 4th Panzer Army Commanded by General Hoth, jointly headed toward the Volga River somewhere in the vicinity of Stalingrad. Army Group A, composed of 17th Army Commanded by General Jaenecke and 1st Panzer Army Commanded by General von Kleist, were tasked with converging upon the Rostov area at the mouth of the Don River. Meanwhile, Russian forces generally were withdrawing to the east and south, denying Hitler the huge encirclement battles that formed the basis of the entire summer offensive. Additionally, major difficulties with supplying two army groups far from their supply bases and rail heads were starting to slow and even stop forward movement. Initially, Operation Blue drew supplies from three main supply depots, Kursk, Kharkov, and Stalino. The Kursk and Kharkov depots held about 50 percent of the accumulated supplies and supported Phases I and II of Operation Blue. The Stalino Depot took over logistical support for Phase III, the clearing of the Don River bend, and the drive toward

Stalingrad. Unfortunately, Stalino was some 150 miles from Rostov and about 375 miles from the fighting in the Don Bend. German forces at this point were suffering from the effects of being at the end of a long supply line. (see Map #6 for Kursk, Kharkov, and Stalino locations)

Hardest hit by the supply shortage was Army Group B advancing directly toward Stalingrad and the lower Don. Throughout the month of July, Army Group B was prevented from advancing quickly mostly due to the lack of fuel and not the scattered resistance of withdrawing Russian formations. But forces of the VVS, the Soviet Air Force, were not idle in this matter and were taking action to further slow the German advance. For instance, on 15 July, the VVS attacked and hit a German fuel dump near the village of Morozovsk. How much fuel was lost is unknown, but any loss would have added to the fuel shortage, further reducing the speed of the German advance.[93,94] This was, of course, not the only VVS attack against the advancing Axis forces. More powerful attacks were sure to come as the VVS built up its strength in this new area of operations.

Light Russian resistance, both on land and in the air, eventually convinced Hitler the Russian forces along the southern front were finished as an organized fighting force. Therefore, on 11 July, Hitler interfered with the general plan for the summer offensive by ordering the 4th Panzer Army south to help capture the city of Rostov at the mouth of the Don River. When it turned south, the 4th Panzer Army also took along with it the bulk of supplies from Army Group B. This order would leave the German 6th Army alone to gradually continue its advanced eastward even as a lack of supplies prevented a more vigorous push into the Don Bend and toward Stalingrad.

Later on, another major alteration of the overall plan came on 22 July, when Hitler again changed Operation Blue. Hitler reassigned the 4th Panzer Army back to Army Group B, then ordered the Panzer Army to cross to the east side of the lower Don River, then turn north and advance on Stalingrad from the south. At about the same time, General Paulus, Commander of the 6th Army, finally was approaching the main emplacements of the Russian forces in the bend of the Don. But General Paulus only had enough supplies to start to put pressure on the Russian troops blocking his way. A full-scale attack would take more time and more supplies to organize.

Meanwhile, Army Group A was ordered south and over the Don. Army Group A was to advance toward the Caucasus Mountains and Maikop oilfields supported by the XLII Infantry Corps transferred directly from the Crimea over to the Kuban/Taman Peninsula. Due to Hitler's alteration of Operation Blue, the two German army groups, B in the north and A in the south, were compelled to move in two different directions. Only if the Red Army was as weak and disorganized as Hitler hoped would these two separate forces be able to reach their distant objectives. Furthermore, the Hungarian 2nd Army, the Italian 8th Army, the Romanian 3rd and 4th Armies also were taking part in Operation Blue. Due to the vast scope of Operation Blue, these allied armies were needed to help guard the greatly expanded front line. This sizeable military commitment by Hitler's allies was justified only if the Red Army could be soundly defeated during the summer campaign. Time would tell if relying upon these relative weak allied armies to hold large sections of the front line was a wise choice.

Meanwhile, units of the 64th Army continued to detrain and move from the Stalingrad area across the hot dry steppes to the west. From about 22 July onward, formations of the Army started to reach and to occupy their assigned positions in the bend of the Don River in greater strength. However, these positions were only a line on a map. The troops would still need to dig in and to actually build a defensive line. These units, at first only the 214, 229 RD and 154 NRB, filled in the line on the left flank of the 62nd Army.[49] During this time, the Commander of the 62nd Army, General Kolpakchi, ordered the withdraw of his 196 RD from its position south of Surovikino. The 229 RD then slowly started to relieve units of the 196 RD at Surovikino. This replacement process started on the 24 July and did not finish until 25 July, and then only with partial forces.[95] These 64th Army units were arriving at the front just in time because the forward detachments of the German 6th Army started to encounter the main defense line of the 62nd and 64th Armies on 23 July.

23 July

Fortunately for the 64th Army and General Chuikov, the initial German attack hit the far-right flank of the 62nd Army near the village of Kletskaia on the Don River; this was some 90 miles north of the nearest 64th Army position. Because of this, the 64th Army was temporarily spared an early direct assault. The Germans were somewhat slower in coming up to the center and southern part of the line in force, thus giving the newly arrived units of the 64th somewhat more time to fortify their positions.

At an infantry level, Russian field manuals laid out how long an infantryman would need to entrench. The Russians did not use two-man fighting positions like most other armies. Instead they dug individual fighting positions, just a hole in the ground. These positions might start out as a quickly dug "hasty position" that barely provided any cover at all. About 8-12 minutes were required to finish this position, but it was normally only used when under fire. If time allowed, this hasty position would evolve into a deeper hole, where a solider could kneel, requiring 25-30 minutes more. If a solider could spend another 50-60 minutes, he could dig a full height standing hole. All these times depended upon soil type and other battlefield conditions and could vary greatly.[96] Other, larger fighting positions for machine guns, mortars, AT guns or artillery, would of course take much longer to organize. The individual fighting positions eventually would be joined together into a trench system that would develop over a period of 4-5 days. These linked entrenchments, along with the supporting heavy weapon positions, were vital in forming a stable defensive line; without them, troops and weapons would be exposed to enemy fire and observation and easily defeated.

The 62nd Army had a real advantage in defending because they had occupied their defensive line for over a week. Units of the 62nd should have been fully entrenched and ready for battle. On the other hand, units of the 64th Amy arriving late to their positions found nothing but open ground. Most units had to start entrenching right from the march. The 299 RD was fortunate when it replaced the 196 RD in the front line. The 196 held that position for several days and handed over their already prepared entrenchments to the arriving 229 RD. The rest of the 64th Army troops needed to dig their own fighting positions. However, approaching German forces did not allow them the needed 4-5 days to develop their trench system and to fully prepare for battle. Substantial forces of the 64th Army had finally arrived at the main defense line, but only just before the Germans arrived also. The 64th Army had won the race to protect the Don River crossings and to support the 62nd Army to the north, but they barely arrived in time.

Stalin wasn't the only one interfering with the battle. Acting on Hitler's orders of 22 July, the 4th Panzer Army of General Hoth crossed the Don further south, near the village of Tsimlyansk. This maneuver placed the Panzer Army on the east side of the Don, with a clear path north to Stalingrad. Originally the 4th Panzer Army was to strike directly toward the Volga River with the 6th Army, but Hitler changed the plan and sent the 4th Panzer Army south to assist the 1st Panzer Army of Army Group A. But Hitler again changed the plan for the entire summer campaign. Instead, 4th Panzer Army was ordered to outflank the Russian defenses in the bend of the Don River and to capture Stalingrad from the south.

Following Stalin's orders, units of the 64th Army continued to send out advanced detachments far in advance of the main defense line. They could not have gone too far because German forces were finally approaching the main Russian defensive line in strength. Scattered combat already was taking place within the 64th Army's area of responsibility.

Also of note during this day, General Chuikov was shot down when flying along the front line in a PO-2, a two-seat unarmed biplane. He was examining the Army's forward positions when a German JU-88 bomber found Chuikov and his pilot, which as he said, started "a cat-and-mouse game" with the German aircraft using its cannon and machine guns. After ten or so firing passes, the PO-2 hit the ground and burst into flames. Only then did the German bomber fly away to the west. Chuikov and his pilot were very lucky to have survived with only minor injuries.[97] This was the second time Chuikov cheated death; the first was a car accident soon after arriving at the 1st Reserve Army.

24 July

With Chuikov back in command of the 64th Army, he wasted no time ordering the 112 RD, 66 NRB and 137 TB back over to the west side of the Don to cover the Nizhne-Chirskaya area.[98] These units were needed as a reserve for the right flank of the Army. (see Map #8)

Meanwhile, the German 6th Army continued its attacks on the far-right flank of the 62nd Army, but they were about to strike hard at the 64th Army, too. Later during the day, advancing German units of platoon strength finally made contact with the 229 RD, which started a lively exchange of machine gun and mortar fire. The 64th Army was now in the fight.

25 July

The Germans launched a major attack against the right flank of the 64th Army using the 297 and 71 Infantry Divisions, along with the 24 Panzer Division. This attack struck positions of the 64th Army south of the Chir River. The 229 RD reported that almost half of the units of the Division at the start of battle were not in position; others were forced to join the battle directly from the march. The 229 RD only had five battalions in the line with the remaining four still on the way. The 229 RD lost a few positions, but the line held. Further to the south, the 214 RD also was attacked, but it too held its positions. During the day, the 64th Army HQ reported the Army's combat strength as 65,816 men. With the battle intensifying by the hour, strength totals would surely start to fall more quickly.[99]

On this same day, further north, after breaking through Russian defenses, the German 6th Army surrounded several units of the 62nd Army. Fighting raged all night.

26 July

Germans forces renewed their attacks against the 64th, using both infantry and armor, supported by air and artillery bombardment. Again, the focus of the attack was the 229 RD. Despite receiving mortar and artillery support from the neighboring 214 RD, units of the 229 started to fall back. Meanwhile, Chuikov realized that as the 229 RD retreated to the Chir River, the Germans might break through toward Kalach and the main Don River Bridge crossings located in that general area. This would be a catastrophe for the 62nd and 64th Armies as they would be cut off from supplies and reinforcements. Realizing the danger, Chuikov ordered the 112 RD, located on the right bank of the Don, to move up and plug the opening gap in the front line along the Chir River. The 137 TB also was brought over the Don to reinforce the defensive along the Chir River line. The TB contained 35 operational tanks, (5 KV-1, 10 T-34, 20 T-60) but most of them ran out of fuel before reaching the Chir area. Only a few KV-1s arrived. The 66 NRB also was moved up, and part of it managed to push south of the rail line and move toward the mouth of the Chir.[100] But other reinforcements were arriving to strengthen the 64th Army. On or about this day, the 126 RD, which had been transferred from the Far Eastern Front into Stavka reserve and then on to the Stalingrad Front, had finished regrouping in and around Gumrak Station. Its units had arrived on 18 trains and were at full strength with 12,533 men.[101] The 126 was assigned to the 64th Army and quickly sent off to reinforce the 64th Army's far left flank near the 74 km and Abganerovo Stations. The 126 RD would do good work in this area of the front in the near future.

27 July

During the morning, the Germans continued their attacks and forced the 229 RD to the north and completely over the Chir River. Following this success, the 24 Panzer Division was able to cross the lower Chir and briefly capture Novomaksimovskii cutting the main rail line there. The 24 Panzer also was able to block the bridge over the Don at Verkhne (upper) - Chirskaya. The 24 Panzer must have found a ford or an intact bridge to be able to cross the river so quickly. In this area, the Chir was about 100 feet wide with maybe 20-30 feet of marshy banks on each side. Infantry forces would be able to cross without much effort, but vehicles would need some type of engineering support, unless they found a ford. At this time of the year, the water level would have been rather low anyway, so crossing might not have been much of a problem. A German account notes how some of their infantry followed retreating Russian soldiers and found a low water crossing point. Seizing the opportunity, they immediately attacked and formed a small bridgehead on the north side of the river. The 24 Panzer Division would be a constant danger to the 62nd and 64th Armies for the next several days.

At the same time, further south along the front, the German 71 Infantry Division attacked the 214 RD and 154 NRB. At first the 214 RD and 154 NRB were able to hold their positions. But when the 229 RD withdrew northward, it exposed the right flank of the 214 RD along the front line. Little could be done about this because there were few reserves available to fill the opening gap in the line. At this point, according to General Chuikov, panic broke out in the rear troops "among the medical ambulance battalions, artillery park, and transport units. Someone reported that German tanks were a mile or two away and the rear units rushed for Don crossing in disorder." General Chuikov went on to say that he sent several staff officers that were with him at the time and Major General Braut, the 64th Army Artillery Commander, to the crossing to stop the rush to the river. The Luftwaffe was quick to spot this mass of troops and equipment and started to bomb them. General Braut, Lieutenant Colonel Sidorin, the Operations Commander; Colonel Burilov, Commander of Army Engineering; and several other officers were killed in the bombing.[102] To make matters worse, that evening German aircraft attacked and sank the floating bridge over the Don at Nizhne (lower) - Chirskaya. The 214 RD, 154 NRB, parts of the 66 NRB, as well as parts of the 137 TB were isolated on the west side of the Don without a crossing. The 24 Panzer Division had crossed the Chir and was blocking access to the bridge at Verkhne (upper) - Chirskaya and threatening the main RR bridge over the Don. Later that evening, several officers at 64th Army HQ ordered these same units to cross over to the left side of the Don. General Chuikov did not learn of these orders until he returned to the Army HQ later that night. "I was horror stricken at the thought of what would happen when they reached the river during the night, without a single crossing to use." Acting quickly, the command staff of the Army ordered the withdrawing units to form a defense on the west side of the river, in essence a bridgehead with their backs to the river. By the next day, they were firmly holding positions along the west side of the river, but they still had no bridge to use, only a single ferry. (see Map #9)

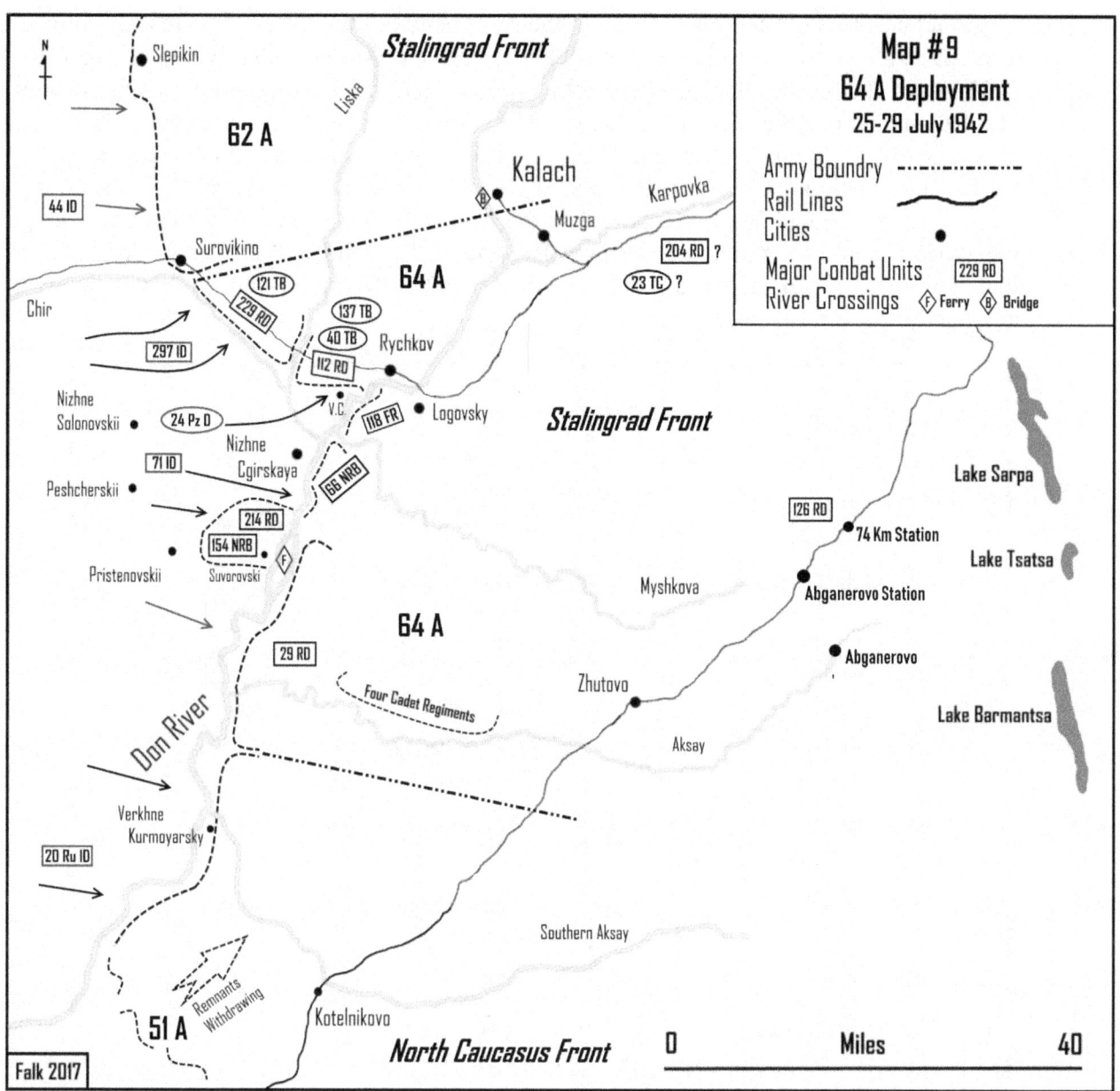

To complicate matters, the topography of the great bend in the Don River is really composed of the Don Plateau with the lower Volga Plateau on the east side of the river. (see Figures 1 & 2) Within the bend of the Don River, the land on the right, west side of the river, is about 300 feet higher than the east side of the river.[103] So any German troops occupying positions along the cliff tops would be able to observe the lower-lying left bank, allowing them to easily view and attack any Russian defenses positions located along the river. In turn, the Volga Plateau is about 200 feet higher than the Caspian Depression, which is located along the Volga and to the south and east. Note: The left and right sides of a river are defined as if you are looking down stream. So, in the bend of the Don River, the right bank is to the west and the left bank is to the east.

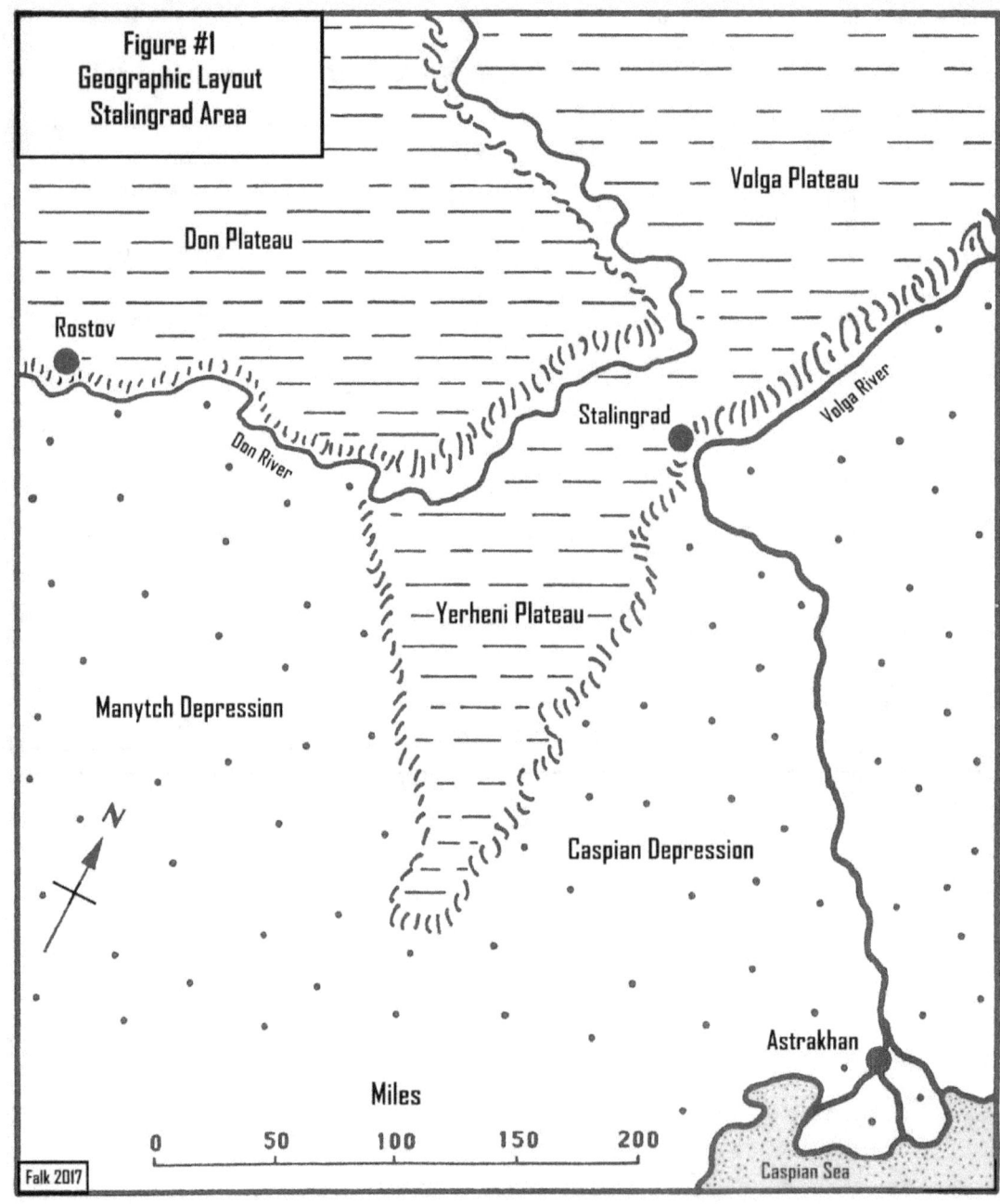

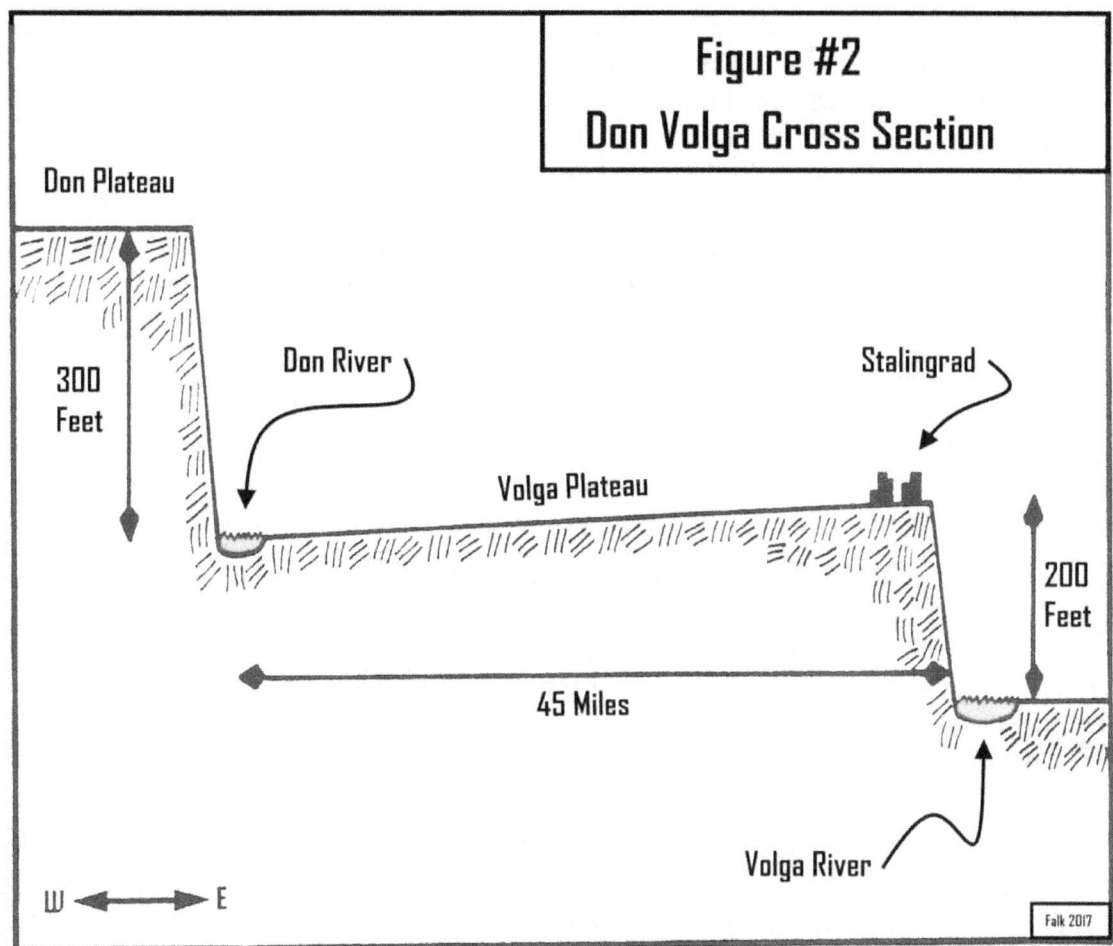

As fighting continued in the north, the surrounded units of the 62nd Army engaged the German 6th Army in fierce combat. Things were not going well at the command level either because on or about this date, General Kolpakchi, commander of the 62nd Army, called Chuikov on the phone. He informed Chuikov that he had been relieved of command and that General A. I. Lopatin was taking over.[104] Replacing army commanders during a major battle was not a good sign.

28 July

At this point of the battle, Major General M.S. Shumilov arrived at 64th Army HQ with orders to take over command of the Army. This was another command change right in the middle of a critical battle. Chuikov had no warning of this change of command, and at the same time, he was ordered to report to General Gordov at his Stalingrad HQ as soon as possible. General Gordov was of course, the new commander of the Stalingrad Front.

Despite continuing attacks against their bridgehead, the 214 RD and the 154 NRB were able to hold out on the west side of the Don. During this period, the 66 NRB apparently had been split up and managed to move both north toward the mouth of the Chir and over to the east side of the Don. The 137 TB was similarly able to reach the north side of the Chir. Over the next four days, the 214 RD, 66 NRB, and 154 NRB crossed over to the east side of the Don using improvised means, joining the 29 RD defending on that side of the river. But this withdrawal over such a wide river, without a bridge or adequate crossings equipment, caused major losses of men and weapons. For example, "On July 25, the 214 RD counted 12,267 people under its command; on July 30 only 7,762 troops remained."[105]

This was a loss of 4,505 men from the 214 RD during just five days of combat. Many of these losses were due to limited medical support which was a problem all along the Army's front. The large numbers of wounded simply overwhelmed the available medical care, which started directly behind the front line and continued on to the rear area hospitals. Many men died because no one was available to provide even simple medical treatment, like stopping the bleeding, covering the wound, and removing the wounded from the combat zone. Day after day these losses mounted, and the strength of the 64th Army melted away at an alarming rate. The German 71 Infantry Division continued to clear Russian troops from the right bank of the Don and reported capturing much equipment at two different locations; however, Russian forces up to battalion strength were still holding out along the river. (see Map #9)

On the right flank of the 64th Army, the 229, 112 RD, 66 NRB, and the 137 TB held their positions along the Chir River and the main rail line. The 112 RD even was able to launch an attack, attempting to regain the region near the mouth of the Chir to drive back the 24 Panzer Division somewhat. The 112 RD recaptured the town of Verkhne-Chirskaya and advanced near to the mouth of the Chir.

The 118 FR (Fortified Region) also came up from Stalingrad Front reserve and was defending along the east side of the Don around Verkhne-Chirskaya. Fortified Regions were small, normally regimental-sized combat units that contained several machine gun-artillery battalions, a mortar platoon, a 76mm field gun platoon and a 45mm AT platoon. A FR had high firepower with many crew-served weapons, but they contained very low manpower, little more than the gun crews themselves. FRs were used to guard quiet, less threatened sectors of the front and could produce impressive amounts of firepower for their small size.

At 14:00 hours, the 64th Army HQ received Directive #170535 from Stavka via Front HQ, stating the 64th Army was to take control of the 204 RD and 23 Tank Corps. The order went on to state that with these forces and along with the 62nd Army, a simultaneous attack was to be launched toward Verkhne Buzinovka. This town was located on the far-right flank of the 62nd Army, more than 43 miles from the 64th Army positions! This attack was scheduled for 02:00 the next day, which was in 12 hours' time. No wonder Chuikov said, "that much of this short order was incomprehensible." No one knew where these units were located, so Chuikov and several other officers spent all night and part of the next day looking for them. Needless to say, the attack never took place because only parts of these formations were located by the stated attack time and the new 62nd Army commander would not be supporting the attack. [106]

This was also the day that Stalin issued his infamous Order #227 or the "No Step Back" order. Much could be said about this order, but it did stiffen Russian resistance all along the front line. Stalin was finished giving up territory to the German summer offensive; the Red Army would now stand and fight to the death!

29 July
The 229 RD was holding the line of the Chir River while the 112 RD continued to attack south of the main rail line toward the lower Chir, with the support of a few tanks. Opposing German forces spent most of the day regrouping and bringing up supplies.

30 July
During the evening, Chuikov finally was able to officially turn over command of the 64th Army to General Shumilov; he then left to report to the Front HQ located at Stalingrad as ordered. [107] Chuikov didn't know it yet, but he was once again to be the deputy commander of the 64th Army. This account would also imply General Shumilov waited two days to take over command of the 64th Army. He

must have wanted to familiarize himself with the situation before officially taking over. Also, during the day, the German 71 Infantry Division, with the help of the 1 Reinforced Romanian Regiment of the 20 Romanian Infantry Division, captured small villages occupied by Russian stragglers along the right bank of the Don. The next day, the final scattered remnants of the 64th Army were cleared from the right side of the Don River south of the Chir. (see Map #10)

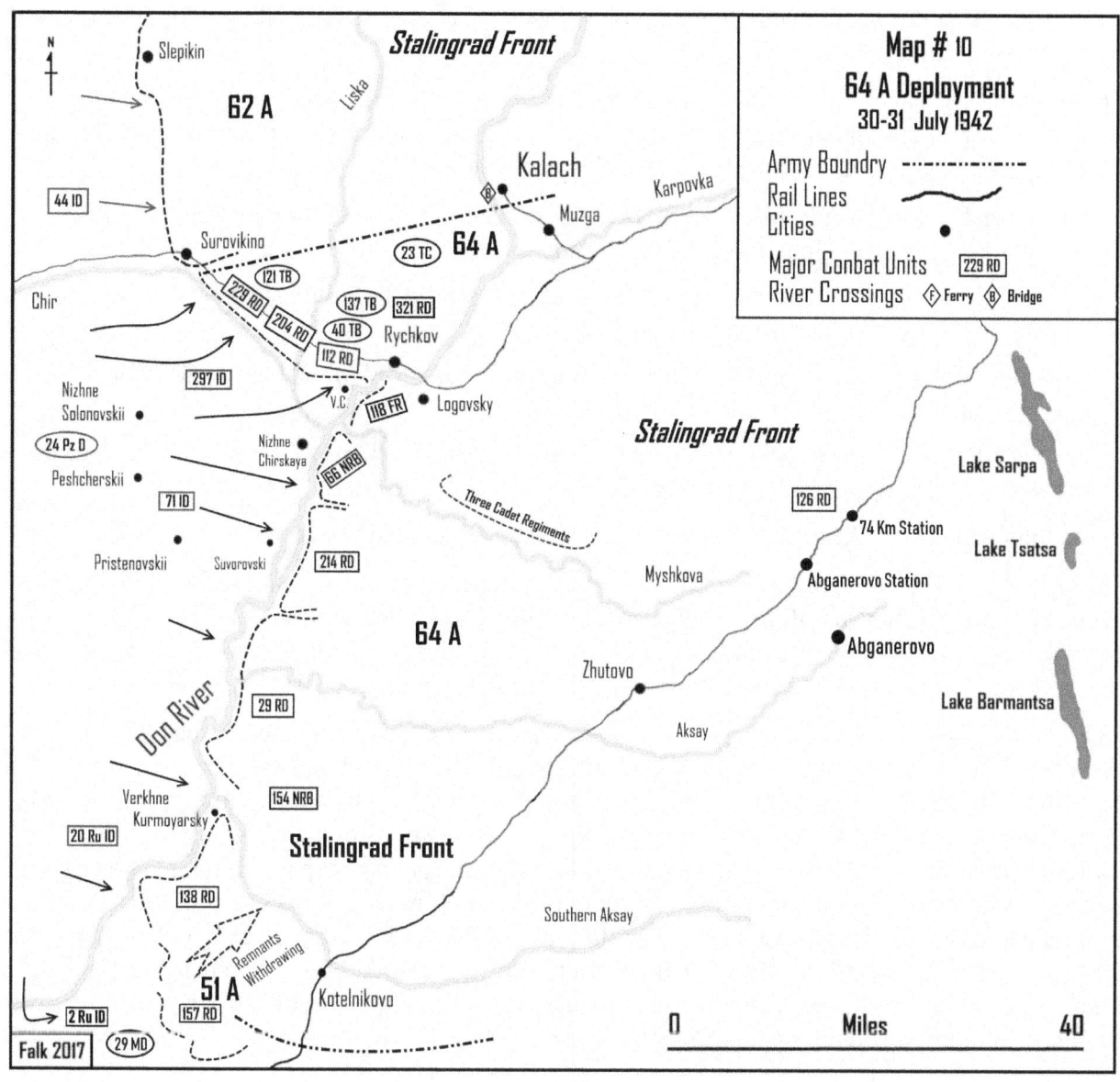

31 July

In spite of repeated German attacks, the 64th continued to hold its right flank positions along the Chir River and the rail line west of the Don. Although German forces were able to clear the west bank of the Don south of the Chir, they were unable to break through the 64th Army positions and push north into the rear of the 62nd Army. In turn, the 62nd Army was barely holding out and desperately needed reinforcements to continue the struggle. Despite conflicting statements, German forces contained the attacks of the 112 RD and retained a small bridgehead on the north side of the Chir near the village of Nixhne-Chirskaia.[108] With the battle raging on the west side of the Don, Stalingrad for the moment was safe from direct attack.

The rifle units of the 64th Army fighting on the right side of the Don had lost substantial amounts of crew-served weapons and artillery when they were forced to withdraw over the Chir and Don Rivers. Between 25-31 July 1942 (6 days) these units lost the following:

- 229 RD lost – 13 AT Rifles, 57 MGs, 77-50mm Mortars, 74-82 mm Mortars, 2-120mm Mortars, 5-45mm Guns, and 13-76mm Guns.
- 214 RD lost – 18 AT Rifles, 12 MGs, 7-50mm Mortars, 8-82 mm Mortars, 2-120mm Mortars, 8-45mm Guns, 12-76mm Guns, and 4-122mm howitzers.
- 154 NRB lost – 4 AT Rifles, 3 MGs, 6-50mm Mortars, 4-82 mm Mortars, 3-45mm Guns, and 10-76mm Guns.
- 66 NRB lost – 6 AT Rifles, 5 MGs, 4-50mm Mortars, 13-82 mm Mortars, 1-120mm Mortar, 8-45mm Guns, and 2-76mm Guns.[109]

Tank units assigned to the Army also suffered heavy losses in equipment.

During the evening, the German 77 Infantry Division attacked at 17:00 hours and finally was able to clear the last few remnants of the 214 RD and the 154 NRB from their bridgehead on right side of the Don near the village of Suvorovski. Because of this final action, the only units of the 64th Army on the right side of the Don were located around the Chir River.

Also, on this date, far to the south of the 64th Army, powerful attacks by the German 4th Panzer Army and the VI Ru Corps (German 29 MD and the 2 Ru ID) forced units of the 51st Army to fall back to the north and south from its positions along the lower Don River. Because of this, the 51st Army was split in two and the northern group was pushed away from its higher headquarters, the North Caucasus Front. To resolve this problem, the retreating northern parts of the 51st Army were soon to be reassigned to the Stalingrad Front. German and Romanian units had broken out of their bridgehead near Tsymlyanskaya and headed due east, but soon they turned north toward Kotelnikovo and the road to Stalingrad. The 64th Army and the entire Stalingrad Front were being outflanked by this bold German attack. Quick action would be required to avert disaster from this new threat. Would the Stalingrad Front and the 64th Army react in time?[110]

So ends the first month of combat for the 64th Army. Overall, this was not the easiest introduction to battle for a new army. The Germans were harsh teachers, but the 64th survived these early battles and quickly learned the art of war.

1 August 1942	
64th Army	
Assigned – Stalingrad Front	
Major General Mikhail Stepanovich Shumilov Commanding (see Appendix 11)	
Rifle, Airborne and Cavalry Units	29, 112, 126, 204, 214, 229 Rifle Divisions 66, 154 Naval Rifle Brigades Three Cadet Rifle Regiments
Artillery Units	1111 Cannon Artillery Regiment 1251, 1252 AT Regiments 140 Mortar Regiment 748 AA Battery 76 Guard Mortar Regiment (rocket)
Armored and Mechanized Units	121, 137 Tank Brigades
Engineer Units	1363 Separate Sapper Battalion
Reported Combat Strength Reported Ration Strength	61,606 men 68,769 men[111]

BSSA part 2

1 August

General Shumilov, having taken command of the 64th Army at the end of July, kept General Chuikov on as deputy commander. The start of the new month would find the Army deployed in the same awkward position, split in half by the Don River. Lined up along the left bank of the Don, from south to north were the 154 NRB, 29 RD, 214 RD, 66 NRB, and 118 FR. Along the Chir River, on the west side of the Don, were the 112 RD, 204 RD, 229 RD along with 121 and 137 TB. In reserve, the three cadet regiments presumedly remained arranged along the Myshkova River in the rear area, but their exact location was unknown. In support the 126 RD also was moving up on the Army's far left flank. The 126 RD was assigned to the Army on 1 August and suffered losses from the Luftwaffe attacks while crossing the Volga. The 118 FR was added to the Army sometime in late July or early August.[112] Events were to quickly alter the 64th Army's situation at the start of the month. (see Map #11) The 62nd Army received the 23 TC and the 40 TB from the 64th Army, and the 321 RD went to the 1st Tank Army. The extreme need for tanks and troops further north was draining away the strength of the 64th Army.

On the right flank of the 64th Army, scattered fighting continued along the Chir River line. Further north, the 62nd Army was fully engaged fending off powerful attacks by the German 6th Army on its own right flank. With the help from the 1st Tank Army, the 62nd Army also was trying to rescue several units that had become surrounded.

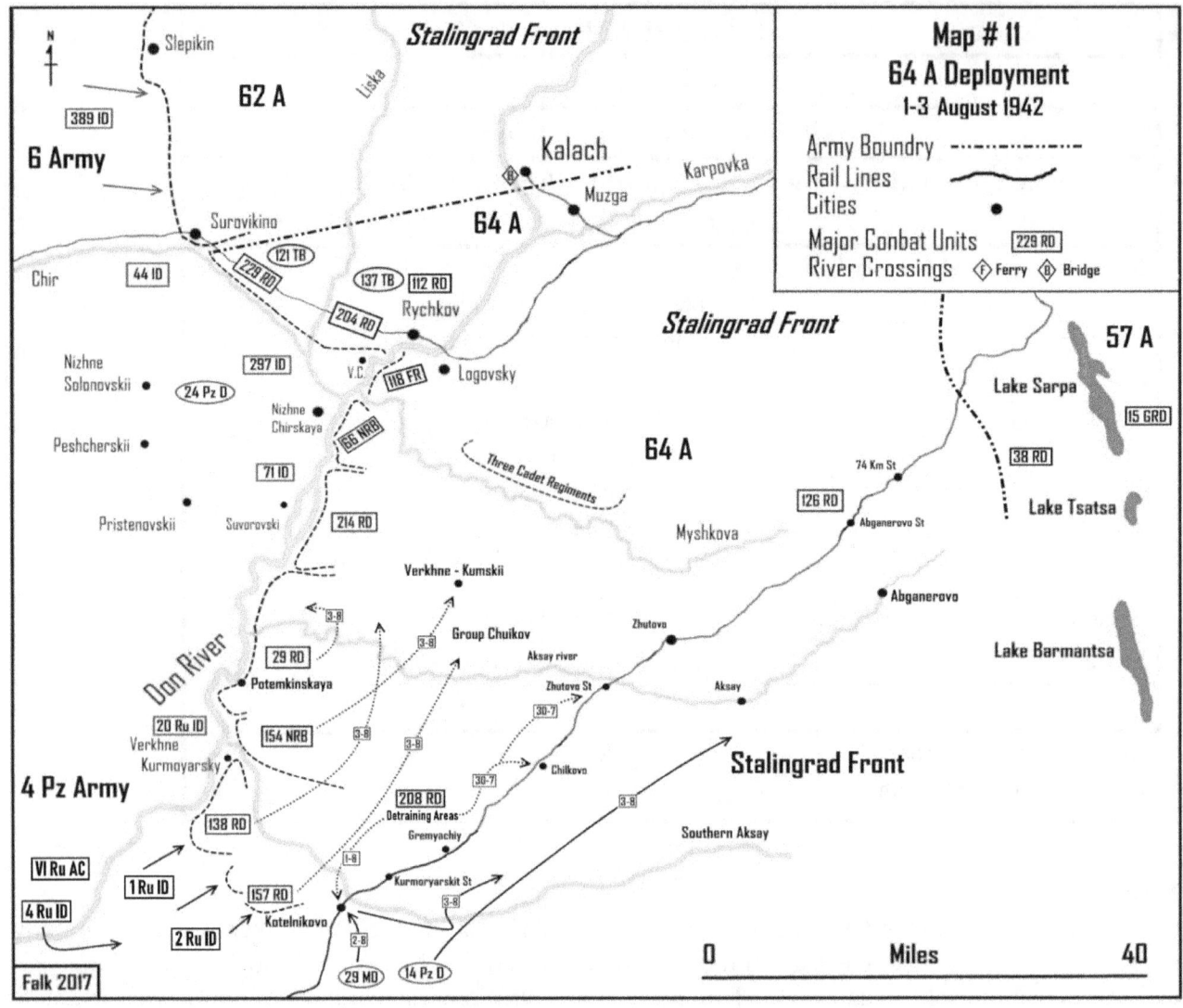

By this date, information started to filter into the Stalingrad Front HQ and to General Shumilov about the situation further south along the Don River. Several days earlier, near the town of Tsimlianskaia, the 4th Panzer Army had crossed the lower Don River in force. The 4th Panzer Army had orders to rapidly push to the north and to cut off Russian forces that were blocking the 6th Army's advance toward Stalingrad. The defending 51st Army was no match against the 4th Panzer Army and had to fall back from its river line position. In response to these movements, General Shumilov sent the 154 NRB south, along the Don, to the far-left flank of the 64th Army. The 154 NRB took up positions south of the Aksay River to block any enemy advances in this area.

Meanwhile, Chuikov arrived back at the 64th HQ after talking with the Stalingrad Front Commander General Gordov. He then started working on a written report that Gordov requested about earlier fighting along the Chir River.

The 8th Air Army (AA) of the VVS, or Red Army Airforce provided air support for the entire Stalingrad Front. The 8th AA only flew 400 sorties this entire day. The focus of VVS efforts were against German 6th Army operating in the Don River bend who were engaged in heavy combat with the 62nd Army and the 4th Tank Army. Few air resources were left to counter the far more numerous German aircraft in action on other areas of the front.

Finally, late on 1 August, General Gordov, Stalingrad Front Commander, ordered the reforming 57th Army out of its reserve position near Stalingrad; it was to move directly south of Stalingrad and support the 64th Army in this area. With its two Rifle Divisions, the 15 Guards and 38 RD, the 57th Army was ordered to "new defensive positions…extending from Raigorod on the Volga toward Krasnyi Don on the eastern bank of the Don River."[113] (see Map #12) This defensive line was the outer-most Stalingrad defensive line, or the "O" line. This assigned section of the "O" line was at least 100 miles long. So with just two divisions, the 57th Army was going to cover this defensive line and back up the 64th Army! On the surface this order sounds absurd. But further research revealed the 57th Army had other units attached to it. Specifically:

- 13 Destroyer Brigade (AT Guns)
- 1104 Gun Art Regiment (AT Guns)
- 1159 Gun Art Regiment (AT Guns)
- 122 Engineer Battalion
- 175 Engineer Battalion
- 525 Sapper Battalion
- 871 Sapper Battalion
- 872 Sapper Battalion
- 1602 Sapper Battalion.[114]

So, with these six Engineering and Sapper units, and three AT Artillery units, it appears the 57th Army was not sent south to defend some 100 miles of the "O" line behind the 64th Army, but to rebuild or upgrade key defensive positions along it. The 57th Army would not be employed in this manner for long; it would soon be put directly into the front line on the far-left flank of the 64th Army. The defensive positions surrounding Stalingrad had been built mostly by civilian labor working on and off for months since the start of the war. Later on, Soviet commanders realized the outermost positions were not very effective but the inner defensive lines would prove to be of more use as the battle moved toward Stalingrad proper.

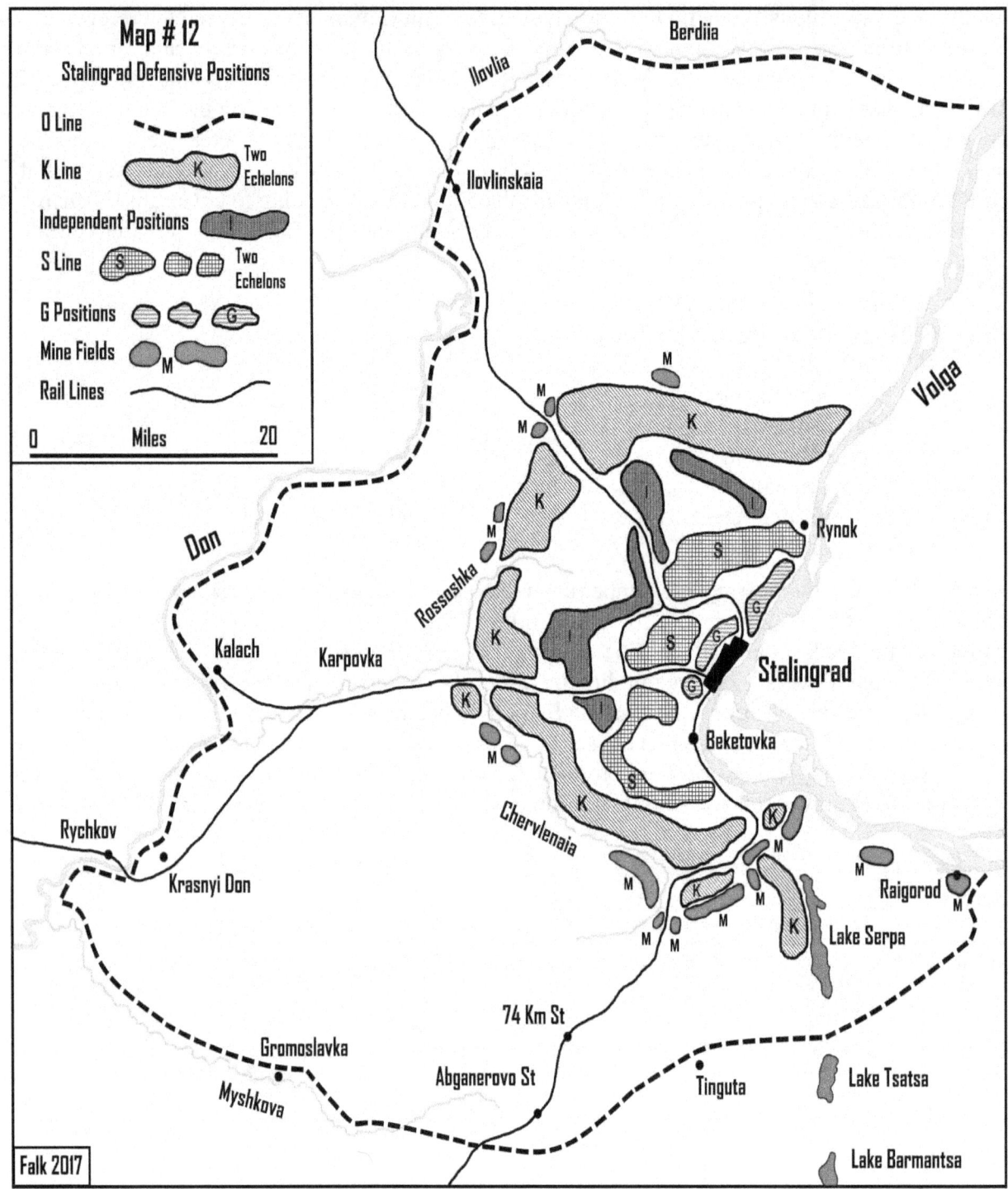

2 August

In the morning, General Shumilov called General Chuikov to Army HQ. Chuikov "found the whole Army Military Council discussing a report by the Chief of Staff on the position on the left flank." Both General Shumilov and the Front Military Council were worried about the far-left flank of the 64th Army and the 51st Army that was covering that area. They wanted Chuikov to "go to the southern sector, clarify the situation, and take such measures on the spot as the situation might demand." Chuikov willingly agreed to take on this new task and was pleased to get out of writing the report for General Gordov.[115] In essence, Chuikov was given an open command opportunity to take control of any and all units found in the south and deal with the local situation. His improvised command would later be called Operational Group Chuikov or The Sothern Group, and it was directly subordinate to the 64th Army and General Shumilov.

Leaving quickly, Chuikov set off in three trucks accompanied by a few guards and several staff officers. One of the trucks was equipped with a radio. The first stop was at the 214 RD HQ, which was defending along the Don with the 66 NRB and the 118 FR. They were along the east side of the Don, starting just south of the Don River railroad bridge. Talking with the 214 RD command staff, Chuikov learned that everything was normal. The German 71 Infantry Division across the Don from them was quiet; they were not even trying to force the Don or to conduct reconnaissance.

Moving on south, Chuikov arrived at the headquarters of the 29 RD by early evening and found a more confused situation. The 29 RD was located astride the Aksay River from where it reached the Don east to the village of Novoaksaiskii. The 29 RD's left flank should have been covered by the 138 RD of the 51st Army, but it might not have been supported by anyone. From the 29 RD HQ staff, Chuikov learned that to the south of the Aksay River, aligned along the Don, was the 255 Separate Cavalry Regiment of the 51st Army. Flanking support for this cavalry regiment was being given by the 154 NRB, further out on the steppes away from the river. These units were the far-left flank of the 64th Army. Somewhere to the south lay the rest of the 51st Army and an unknown situation. Chuikov's job was to find out what was going on in this area and to take control.

Later during the day, the German 71 Infantry Division launched a brief reconnaissance operation over the Don River. This was most likely a typical everyday type of action, but it might have been in support of the developing situation on the left side of the Don, further south in the 4th Panzer Army's area of responsibility.

3 August

On the morning of 3 August, after spending the night at the headquarters of the 29 RD, Chuikov went south to explore in the direction of Kotelnikovo on the main north/south rail line. He had borrowed two extra trucks filled with escort troops from the 29 RD HQ to expand his small group to five trucks. So, Chuikov was currently in command of what amounted to a small mobile headquarters. Visibility in the steppe was good at about five miles as they approached the village of Verkhne (upper) Yablochny, at the boundary between the 64th and 51st Armies. This village was located about 20 miles directly north of Kotelnikovo. His men "noticed approaching from the south two columns of infantry and artillery." They were retreating units of the 51st Army, the 138 RD (4,200 men in total) under the command of Colonel Lyudnikov, and the 157 RD (1,500 men in total) commanded by Colonel Kuropatenko. These two divisions had been attacked by the German 4th Panzer Army when it crossed the Don, thus splitting the 51st Army into two parts. Out of touch with 51st Army HQ and having suffered heavy losses, the two divisions decided to retreat north toward Stalingrad. These fragmented infantry units also brought with them two regiments of Katyusha rocket launchers; they were most likely the 18 and 19 Guard Mortar Regiments, commanded by Deputy Army Artillery Commander Major General Dmitriev.

Quickly assessing the situation, Chuikov took the remains of both divisions and the two regiments of Katyusha rocket launchers under his command. He sent them off to the north with orders to dig in along the north side of the Aksay River and to "pull themselves together." Due to the hot, dry summer conditions, the Aksay was only about 160 feet wide, but it did have steep banks and marshy ground along the river. So, over the next several weeks, the river would provide a useful barrier to enemy movement and anchor the defensive position of Group Chuikov. Unfortunately, the commanders of these two divisions could say little about conditions further south. But they must have informed Chuikov that two Romanian Infantry Divisions (Ru ID, common usage), the 1 and 2 Ru IDs, were pushing north along the east side of the Don. This information was previously unknown to the Russian command staff. The Romanian VI Corps (containing the 1, 2, 4 and 20 Infantry Divisions) had recently started to cross the Don and was providing left flank support to the advancing 4th Panzer Army.

With this new information, Chuikov decided to change the deployment of the units in the southern area. Having sent the 138 RD and 157 RD to the Aksay River, he then ordered the 154 NRB to fall back and to dig in behind these two shaky divisions as a second line of defense. He also decided to set up a Southern Group HQ at the village of Verkhne-Kumskaya, just behind the Aksay River. Chuikov also must have ordered the left flank of the 29 RD to fall back to the Aksay River. This would bring all of the 64th Army/Group Chuikov forces to the north side of the river. This quick command decision was definitely the correct choice and most likely saved all those rifle units. With two nearly full-strength Panzer Divisions and at least two Romanian IDs operating in the area, Group Chuikov needed to defend behind the Aksay River to have a chance to survive.

Having decided to locate his HQ in the village of Verkhne-Kumskii, Chuikov soon arrived at the village and settled in. Chuikov then attempted to contact 64th Army HQ by radio but was unable to get through. So he reported on the day's events, in detail, to the Stalingrad Front which he was able to reach. It was then he learned the new full strength 208 RD was arriving in his area by rail, supposedly at Chilikovo and Kotelnikovo Stations. Chuikov was instructed to add this unit to his Southern Group although the Stalingrad Front HQ could not tell him where the division HQ was located. The 208 RD, along with many other rifle divisions, were coming from Siberia, having received transfer orders from Stavka on 16 July. These early reinforcements to the Stalingrad battle would play a vital role in the coming fight. Several of these new units were to serve with the 64th Army.

Also, on this busy day, the Stalingrad Front Commander General Gordov, acting under orders from Stavka, relieved General Kolpakchi of command of the 62nd Army. Replacing him in this position was General Lopatin, former commander of the shattered 9th Army, which was currently reforming in the rear. Stalingrad Front Commander General Gordov moreover planned to shift the army boundary between the 62nd and 64th Armies southward, effectively giving the 62nd Army control of all 64th Army units on the west side of the Don River, including three Rifle Divisions, 112, 204, and 229 RDs. This was intended "to allow the commander of the 64th Army to concentrate all of his attention in repelling an enemy attack on Kotelnikovo direction." This shift in focus for the 64th Army certainly was needed as the German attack by the 4th Panzer Army and the Romanian VI AC was rapidly developing from the south. General Shumilov needed to focus all his ample skills on this dangerous area.

4 August

In the morning, General Chuikov once again set off to reconnoiter to the southwest toward the village of Verkhne Yablochny. Along the way he continued to come across scattered troops belonging to the 138 RD and 157 RD. Arriving at Verkhne Yablochny, villagers told him that Romanian troops were crossing the Don at Verkhne Kurmoiarkaia, about eight miles away. Chuikov turned east toward Gremyachiy Station which was about 12 miles north of Kotelnikovo, and there he found more Russian infantry units retreating north along the rail line. These troops were the remnants of the leading elements of the 208 RD. After questioning a low-ranking officer, Chuikov was able to piece together the story of what happened. During the days before 1-2 August, four trains carrying four battalions of the 208 RD were detraining at the Kotelnikovo Station when they suddenly were attacked by enemy aircraft and then by 40 tanks supported by an infantry regiment. Chuikov was unable to find any divisional or regimental officers within this retreating group of men to clarify the situation.[116] But it was clear to him these detraining troops, for the most part, had been destroyed by this overwhelming sudden attack.

A veteran of the 208 RD was able to fill in some more details of this battle. Gregory Haritonovich was assigned to the 376 Combat Engineer Battalion of the 208 RD. His unit was detraining at Zhutovo Station on 30 July from one of the first trains to arrive. (This might in fact have been the Chilikovo Station.) His platoon then was ordered to carry out a reconnaissance south, along the rail line toward Kotelnikovo. On or about 1 August, after walking south for most likely two full days, he was surprised to see four trains full of troops from his division drive past him heading to Kotelnikovo. By this date he was close enough to Kotelnikovo Station to observe what happened next. As men from these four trains were detraining, 40-50 German aircraft arrived and bombed the station. After the air raid, German tanks attacked followed by a long-range artillery barrage. After this attack, Haritonovich stated, "Our trains and four battalions were broken." His platoon and the survivors of the attack retreated north later that night along the railroad, most likely the night of 3 August.[117]

A short account from the German 29 Motorized Infantry Division reveals a few more details of this Kotelnikovo Station battle. The Germans stated the trains of the 208 RD unloaded during the night of 1 August at the Kotelnikovo Station, and only after "prolonged fighting" and "tough sustained resistance" was "this well trained and equipped Far East Division thrown out of the city."[118] David Glantz provides us with the last few bits of information stating the 29 Motorized Infantry Division attacked at 11:00 hours with some 40 tanks and were able to clear the 208 RD from the town by 15:00 hours, capturing several trains in the process.[119] After the battle, the 29 Motorized Infantry Division sent out patrols to the northeast.

Understandably these different accounts of the same battle contain slight discrepancies in dates and story. A small amount of speculation on my part might clear up this issue.

Gregory Haritonovich most likely detrained at Chilikovo Station and not Zhutovo Station on 30 July. Then, after marching south for two days, on 1 August, Haritonovich saw four trains from his division drive past his unit and on toward Kotelnikovo. These trains reached Kotelnikovo Station that evening and the troops started to detrain immediately.

The next morning, 2 August, German reconnaissance aircraft noticed these newly arrived Russia battalions in and around the station. Acting quickly, units of the nearby German 29 Motorized Division staged a surprise attack. First the Luftwaffe bombed and strafed the detraining troops; then a tank and infantry assault backed up with an artillery barrage hit the disorganized Russian forces at the train station and within the city itself. After fighting for at least four hours, the few survivors retreated along the rail line to the north that same afternoon and into the evening.

On 3 August, the survivors continued to retreat to the north pursued and harassed by German motorized reconnaissance patrols. The next day, 4 August, General Chuikov found these 208 RD men moving north along the rail line and took them under his command.

After hearing about this "serious news" on 4 August, Chuikov and his small group of men drove north toward Chilikovo station where he hoped to find the 208 RD HQ.[120] Arriving at the station, he found troops and equipment from the 208 RD in and around the trains. Finding an officer in charge of one of the trains, Chuikov informed him about the situation in the south. Then Chuikov ordered him to "post strong covering detachments on the high ground at Nebykovo village" to the southeast. He also instructed the officer to move troops away from the trains and station.

Having made contact with the 208 RD, Chuikov's little command group then moved about a mile away from the station, to Dairy Farm #1, and set up the radio to contact the Stalingrad Front HQ. Just after noon, the Front HQ answered. At this time, Chuikov noticed three different aircraft formations, of nine aircraft each, approaching from the north.[121] In early 1942 a VVS *Palk* or Regiment consisted of two 9 aircraft squadrons called *Eskadrilya*. By mid-July, some *Palk's* were expanded to include three squadrons of 9-aircraft each. So, it would appear that a full strength VVS Regiment of three Squadrons, containing 27 aircraft, were headed toward the train station at Chilekov. Chuikov identified them as aircraft of the VVS, or Red Army Air Force of the Soviet Union.[122] Actually, General Chuikov wrote several different versions of this air attack over the years. The first account was censored; the attacking Soviet aircraft were transformed into German aircraft. Another version was published in his book *Battle of the Century* in 1975. This version corrected the story and firmly blamed the VVS for this mistaken attack upon the 208 RD. Chuikov goes on to say, these Russian aircraft then proceeded to bomb and strafe the trains and troops in and around the nearby station. Sending out an emergency radio call to the Front HQ about this mistaken attack failed to make any difference. To make matters worse, one group of aircraft broke off to bomb and strafe the troops and buildings around Chuikov, in the process knocking out his only radio truck. This was Chuikov's third brush with death. The 208 RD suffered another devastating blow, this time at the hands of their own air force.

Chuikov's little HQ group stayed in the area until nightfall; he was finally able to locate the division commander, Colonel Voskoboinikov. After helping to bring some type of order out of the destruction, Chuikov ordered Voskoboinikov to gather the remnants of his division and withdraw to the north of the Aksay River to form a defensive line astride the rail line between the village of Antonov and Zhutov Farm #1. Chuikov was enraged by these events not only due to the senseless loss of life, but also because he was counting on this fresh division to strengthen the forces under his command.[123] At this point several accounts state the 208 RD only had around 1,000 troops remaining. I must assume they are talking about combat troops ready for action and not the overall ration strength of the Division. Despite the conflicting accounts, the loss of men and equipment by these twin attacks was undoubtedly heavy, rendering the remnants of the division combat ineffective for the next several days.

Chuikov firmly blamed the Stalingrad Front staff for these twin disasters. "The staff of front did not have time to reorient the command of division and railroad service, after leaving for 208 division the previously place of unloading assigned."[124] [125] In other words, the front staff lost control of the situation.

First of all, the command staff failed to use sufficient caution when organizing rail movements in regard to the rapidly changing situation at the front. This error allowed four troop trains transporting units of the 208 RD to proceed to the Kotelnikovo Station for detraining, which was too far forward and had become too dangerous. The Stalingrad Front also was unable to provide air cover to defend these troop trains and stations and/or to provide timely reconnaissance of the surrounding area. This was the recipe for disaster. The second attack at Chilikovo Station resulted from the VVS not being informed about the movements or current positions of friendly ground forces in the area. This omission allowed squadrons of VVS aircraft to bomb and strafe friendly units detraining at a rail station behind the front lines. Lack of coordination at the top and the failure to disseminate critical information had just cost the Red Army most of a rifle division. Later that night, when returning to his

HQ at Verkhne-Kumskaya, Chuikov and his group ran into scouting units of the 255 Cavalry Regiment. These units were ordered to scout toward the south and the Kotelnikovo Station area, and then to report their findings directly to the 29 RD HQ located at the village of Generalovski. Chuikov decided to head to Generalovski with his tired men. Arriving at the headquarters, Chuikov found the 29 RD pulling out of its defensive positions; it had been ordered by the Front HQ to move quickly toward Abganerovo Station some 45 miles to the east.

Also, on or about this date, the 64th Army gained control of the 13 Tank Corps which contained at this time the 6 Guards, 137, 254 Tank Brigades and the 38 Motorized Rifle Brigade.

5 August

After resting the night, Chuikov and his men awoke to explosions and machine-gun fire. The Luftwaffe had spotted the marching columns of the 29 RD and subjected them to continuous aerial attack. Chuikov noted, "The withdraw had not been given air or artillery cover, and the division lost more men on the march than it had during battle."[126] In defense of the VVS, the Soviet 8th Air Army, which continued to provide support to the entire Stalingrad Front, only flew 265 sorties during this entire day.[127] Most of these missions were flown, as ordered, against the newly-arrived 4th Panzer Army which was threatening the far-left flank of the front. Any remaining aircraft attempted to support the 62nd Army on the west side of the Don. Chuikov's Southern Group was left in the middle without any air cover. These losses were not light. An account from the 58 Medical Battalion, which was attached to the 29 RD, stated that during 4-5 August the battalion "treated almost 400 injured and contused!" The next day the medical battalion was ordered to quickly follow the 29 RD to the east and the "task was most difficult: we did not have time to evacuate even half of [the] injured, but among those remaining there were many non-transportable." [128] Once again, during this stage of the battle, the lack of medical support condemned about 50 percent of these wounded to be left on the battlefield to die of their injuries.[129] (see Map Set #13)

Assessing the changed situation, Chuikov ordered the 255 Cavalry Regiment to occupy the abandoned positions of the 29 RD along the Aksay River line. Chuikov knew this was only a screening force, just enough to prevent the advancing Romanian forces from crossing the Aksay unopposed. After contacting Stalingrad Front HQ and reviewing recent reconnaissance reports, Chuikov gained a better understanding about events along his part of the front line. The German 4th Panzer Army was swinging wide around the left flank of the Stalingrad Front and the 64th Army, threatening to attack from the south and to cut off the entire front from their base at Stalingrad. He also learned the Romanian VI Corps was providing left flank support for the German advance and was unlikely to attack in force along the Aksay any time soon.[130] Having been strictly ordered by the Stalingrad Front HQ to hold the Aksay River line with all available forces, Chuikov prepared to do just that.

Other important events were taking place at higher headquarters at the same time. Stavka and Stalin issued on 5 August Directive #170554 ordering the Stalingrad Front to be split in two. This was done in part to improve Front level command and control in the Stalingrad area.

Directly north and west of Stalingrad, the Stalingrad Front, still commanded by General Gordov, contained the 63rd, 21st, 62nd Armies along with the 4th Tank Army, and 28 Tank Corps. To the south and west of Stalingrad, the newly-formed Southeastern Front controlled the 64th, 51st, 57th Armies, 1st Guards Army, and the 13 Tank Corps. Colonel General A.I. Eremenko was appointed the Commander of this new front and the new chief of staff was Major General G.F. Zakharov. Zakharov was given five days to form a new Southeastern Front HQ from the HQ staff, personal and installations of the disbanded 1st Tank Army and former Southern Front. The 57th Army had been the reserve of the Stalingrad Front, but with the division of the Stalingrad Front, the 57th Army with its two rifle divisions, the 15 Guards and 38 RD, moved south to the Southeastern Front.

Map #13 - 64th Army Deployment 4 - 7 August 1942

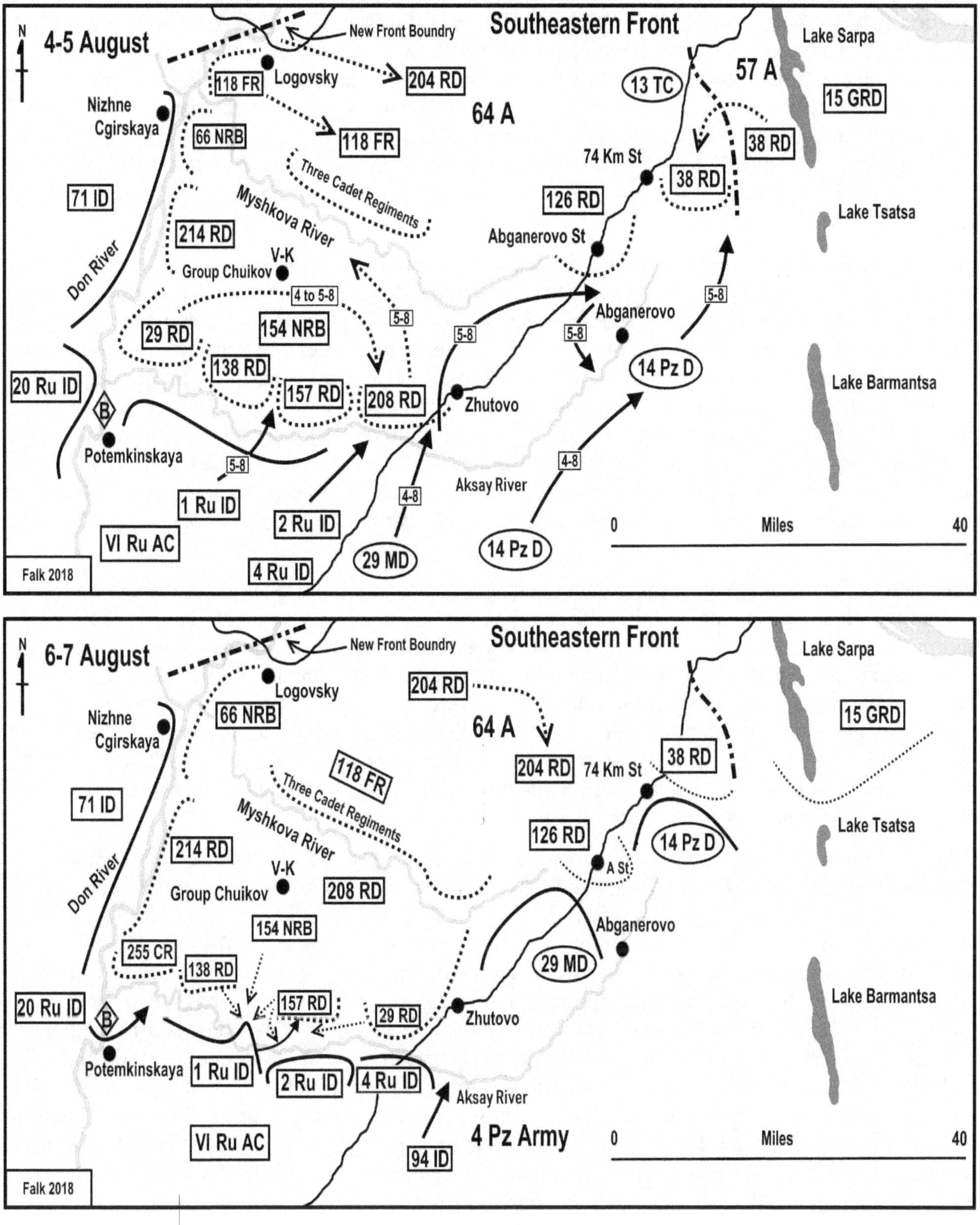

Stavka Directive #170554 also transferred one half of all VVS aircraft from the Stalingrad Front's 8th Air Army to the new Southeastern Front and its yet-to-be-formed supporting 16th Air Army. The creation of this new 16th Air Army was delayed for several weeks, so the 8th Air Army continued as before.[131] An unexpected result of dividing the VVS like this was to weaken the overall air effort. Two different air armies, operating in the same general area, would make it difficult to concentrate the power of the VVS in any one decisive sector. On the other hand, the overall German air commander, General Richthofen, believed in using *Schwerpunkt* tactics or point of maximum effort. This focus of effort allowed the Luftwaffe to quickly concentrate air assets in one sector and overwhelm the VVS. This was one of the reasons why the VVS lost so many aircraft to the Luftwaffe during the Stalingrad battle.

Carrying out Directive #170554 likewise created a new Front Level boundary line right through the middle of the ongoing battle for the approaches to Stalingrad. This splitting of the battlefield into two different commands did not please General Eremenko, but Stalin insisted on this division. This Directive went on to specify that the Stalingrad Front would be in overall command of all forces and be responsible for the defense of Stalingrad itself until 7 August. On that date, the Southeastern Front would control its forces separately. This delay gave General Gordov time to make an importation change at the front. Because the new boundary line went right through the middle of the 64th Army, Gordov simply transferred the entire right flank of the 64th Army, which included all units on the west bank of the Don River, to the 62nd Army.[132] This transfer involved at least the 112, 204, and the 229 RDs and almost certainly the 121 TB and 137 TB that were supporting their operations in that area.

This transfer of forces and the formation of the new Southeastern Front finally allowed General Shumilov to focus the attention of the 64th Army toward the south where the 4th Panzer Army was approaching. This shift in focus already was underway because General Gordov had given Shumilov permission to plan an attack on German mobile troops that were advancing toward the Abganerovo railway Station area. For several days on the Russian side of the line, rifle divisions, artillery, and reinforcing tank units moved toward the east and concentrated near Abganerovo Station. The Germans too turned their long flanking drive to the north, getting ready to cut the rail line in this same general area before they advanced toward Stalingrad from the south. These two forces would soon clash in the first large-scale battle in the south.

Other powerful reinforcements also were being sent to the south. On 5 August, General Gordov transferred the 28 Armored Train Battalion to the 64th Army. This was one of eight Armored Train Battalions sent to the Stalingrad Front from Moscow.[133] The 28 Armored Train Battalion (ATB) consisted of two BP-43 type armored trains. The two assigned trains were #677 (named Uzbekistan,) and #708 (named Komsomol of Uzbekistan.) Each standardized BP-43 train was equipped with 4-76mm cannons each in a T-34/76 tank turret on 4 artillery cars, 4-37mm anti-aircraft guns on 2 flatcars, 1-12.7 mm anti-aircraft machine-gun mounted on the steam engine tender and 12 to 20 heavy machine-guns distributed about the train. Two security flatcars were at the front of the train and two at the back. The steam engine was a PR-43 armored locomotive normally based on the standard OW series steam engine.[134][135][136] Note: The stated configuration of armored trains #677 and #708 appears to be a non-standard type of BP-43 train. They appear to be lacking a 37mm anti-aircraft flatcar and several security cars. This might be due to combat damage or just an omission by the different reference sources. In all, each armored train consisted of at least ten cars and an engine/tender combination. Each train was virtually a self-contained light artillery battery and anti-aircraft battery on rails.

Meanwhile, that same evening, a small probing attack by Romanian forces struck the junction between the 138 and 157 RDs along the Aksay River line. Infantry formations from the Romanian 1 Infantry Division managed to cross the Aksay River and to penetrate into Russian defensive positions forming a small bridgehead on the north side of the river. General Chuikov stated that tanks and German troops were in the area preparing to cross the river the next day, but research does not seem to support this claim.[137] Nevertheless, Chuikov was prepared to hold the river line, so he organized an early morning counterattack for the next day. The objective was to throw the Romanians back across the river. After consulting with his two division commanders, Chuikov planned a modest frontal attack, stating, "To be quite frank, I was afraid of conducting even a simple operation with the troops I had collected during the retreat. I had no idea what they were capable of."[138]

6 August

As dawn broke, Russian artillery, including Katyusha rocket launchers, pounded Romanian troop concentrations within the Aksay bridgehead. This sudden barrage caused troops, transport and even artillery to flee to the south side of the river. Having taken care of the Romanian reinforcements, a much tougher task remained, forcing the dug in infantry out of their positions. Chuikov said this more difficult task of clearing Romanian troops from the north side of the river lasted until almost evening. In the end Chuikov stated, "I had satisfied myself that the retreating troops I had collected had not lost their fighting spirit and fought well."[139] Needless to say, Romanian infantry formations lost a good number of men and much equipment in this first battle against Group Chuikov.

Parts of the 57th Army continued to move south into the zone of the newly formed Southeastern Front, somewhere toward the rail line near the southern front line.

Further east, German troops continued their advance north toward the rail line between the Abganerovo, 74 Km, and Tinguta train Stations.

7 August

At this point, a review of which combat formations were assigned to the 64th Army and Group Chuikov might be helpful. This is after the shifting of the front boundary line.

The 64th Army contained: (Estimated Combat Strength – 52,890 Men) (see Appendix 12)

- 38, 126, 204*, 214 RD
- 66 NRB
- 118 FR
- Three Zhytomyr Cadet Rifle Regiments
- 1st Ordzhonikidzenskih Infantry Training School Regiment
- Krasnodar Machine-Gun and Mortar School Regiment
- 13 TC (6 Guards, 13, 56, 133, 254 TB)
- 28 Armored Train Battalion
- 1111 Cannon Artillery Regiment
- 1251 Tank Destroyer Regiment
- 140 Mortar Regiment
- 76 Guards Mortar Regiment (Katusha rocket launchers)
- 748 AA Battery
- 1363 Sapper Battalion

*Note: At some point General Shumilov was able to retain the 204 RD within the 64th Army and not transfer it to the 62nd Army after the change in Front boundaries.

Group Chuikov contained: (Estimated Combat Strength – 23,800 Men)

- 29, 138, 157 RD and the remains of the 208 RD
- 154 NRB
- 255 Cavalry Regiment
- 18 and 19 Guards Mortar Regiments (Katyusha rocket launchers)

On the right flank of the 64th Army along the Don River all was quiet, but across the river things were different. The German 6th Army launched what was to be the final attempt to encircle and to destroy the remnants of the 62nd Army on the western side of the Don. As the 62nd Army was slowly being surrounded by this new attack, the attention of the front command was focused on this area.

On the far-left flank of the 64th Army, the 126 and 38 RD organized fierce resistance to German attacks along the rail line between the Abganerovo and the 74 Km Station. As the 14 Panzer and 29 Motorized Divisions of the 4th Panzer Army moved north, they seized both stations and cut the rail line. These German attacks also had breached a section of the "O" defensive line. Needless to say, Stalin insisted the lost positions be recaptured. He was worried because this success appeared to allow the Germans to continue their outflanking movement of the Stalingrad Front, but this was not to be. General Shumilov was about to stop them in their tracks.

In the center with Group Chuikov, Romanian infantry formations from the 1 Ru ID once again forced a crossing of the Aksay River, a little further to the east from the last attack. This attack pushed three to four miles deep into the Russian defenses of the 157 RD. At this time, the 2 Ru ID had just come up to the Aksay River, on the right side of the line, so it might have been able to contribute some fire support to the attack, but not much else. Some Soviet military maps display this attack as occurring on 8 August. Chuikov clearly stated it took place on 7 August. Nevertheless, nearly constant attacks took place along the Aksay River line during this time, so the issue was somewhat confused.

Chuikov was forced to move up the 154 NRB, his last intact reserve unit, to fill in the front line. He placed it between the 138 and 157 RD, so uncommitted troops were sparse. Because of this, Chuikov only had the still-reforming 208 RD as a reserve.

In countering this latest Romanian incursion, Chuikov calculated that a late evening counterattack would be best. Attacking at the end of the day would avoid German aircraft interference and surprise the Romanians with a quick response. Feeling more confident with the ability of his troops, this attack was a two-pronged converging flank attack aimed at the base of the Romanian bridgehead. Borrowing a regiment from the nearby 29 RD, Russian forces from all three rifle divisions quickly forced Romanian troops back over the Aksay. [140] [141]

8 August

After their latest setback, the intensity of Romanian attacks along the Aksay decreased over the next several days. This was good news for General Chuikov because he heard reports that a major Russian ammunition dump on the Volga had been blown up. For three consecutive days, starting on 8 August, the Luftwaffe bombed several supply dumps which were located in and around the city of Beketovka, located south of Stalingrad. They also hit the adjacent rail yards and docks along the Volga. A large amount of military cargo was crowded together in this area, loaded on trains and barges; the Luftwaffe was quick to take advantage of this inviting target. Chuikov's worries about ammunition supplies soon were realized as units under his command started to report low levels of munitions. For the first time during the battle, supply trucks returned empty from the rear area dumps. [142]

Back along the Aksay, fighting settled down into a routine of Romanian forces making almost daily probing attacks in the morning, followed by Russian counter attacks in the afternoon. The commander of the Romanian VI Corps, General Dragalina, certainly would have liked to mount a more determined attack, but he was occupied in shifting his infantry divisions further east to support the 4th Panzer Army's left flank. On 8 August, the 4 Ru ID was sent east, on 9 August the 2 Ru ID followed, and then on 10 August the 1 Ru ID moved to the right to cover the gap between the Aksay and Myshkova Rivers. So, by the end of the day on 10 August, only the Romanian 20 Ru ID remained behind holding some 25 miles of front along the Aksay river line. This shuffling of forces took three full days to complete; as a result, the pressure was taken off the meager forces comprising Group Chuikov. With only one Romanian Infantry Division holding the extended river line, any major attack was out of the question. The focus of attention for both armies was shifting further to the east.

On or about this date a commissioner from the 29 RD paid a visit to the 58 Medical Battalion. After an inspection, the commissioner asked what aid was needed, and he was told, "We experienced a constant shortage of the surgical dressings and blood… since many injured from other units entered into the medical battalion." He also learned "that the army hospital does not ensure timely the removal of injured that we are forced to evacuate people into the rear in our transport, but this interferes with the removal of injured with the front." As a result of this visit, "the following day, to us they [the division] delivered a large quantity of bandages, cotton, individual dressing packets, sterile napkins, aseptic bandages, ampules with the blood. Trucks of the army hospital began to come more frequently"[143] to pick up wounded and to move them to army level hospitals. In this case, it paid off to tell the truth to higher authorities.

On 8 August, the only notable action took place on the far-left flank of the 64th Army. The German 14 Panzer and 29 Motorized Divisions, which had earlier gained control of the area around the Abganerovo and 74 Km railway Stations, were preparing for further advances. But they were not allowed to regroup in peace. These two German divisions had to deal with nearly constant Russian low-level attacks and artillery bombardments throughout the day. The 126 and 38 RDs did a good job of keeping the Germans occupied.

Meanwhile, taking advantage of the time gained, General Shumilov put the final touches on the assembly of his counterattack force in the same general area. Shumilov wanted to launch the attack on 8 August, but postponed it for a day because the 204 RD was delayed en route, coming all the way from the Don River. The 64th Army retained this Division after the splitting of the Front. Eventually Shumilov was able to gather together the 38 RD, 157 RD, 204 RD, and the 13 Tank Corps (6 Guards, 13, 133, and 254 Tank Brigades) as the main assault force. Also, in the area were supporting formations from the 208 RD (one regiment in strength), the 422 RD, two student rifle regiments, two artillery regiments, the 76 Katyusha Rocket Regiment, one artillery battalion and one armored train from the 28 Armored Train Battalion.[144] The newly arrived 57th Army was moving into this area as well, taking its place in the front line on the left flank of the 64th Army. General Shumilov also was able to acquire the use of the 38 RD from the 57th Army for the counterattack. Likewise, the remains of the hard hit 208 RD was transferred from Group Chuikov and was available to offer support. (see Map Set #14)

9 August
Desperately trying to support the 62nd Army trapped on the western side of the Don, the 8th Air Army focused most of their attention against the German 6th Army in the Don Bend. During the day it flew a total of 356 combat sorties, comprising 231 Fighter, 74 *Shturmovik* (Storm Bird), and 51

Bomber sorties.[145] Needless to say, few aircraft were allocated to the southern part of the front where the major attack by the 64th Army was about to start.

Attacking at 06:00 hours, German forces in and around the Abganerovo and 74 Km railway Stations were struck almost simultaneously from three sides. General Shumilov had gathered 68 operational tanks and 396 guns and mortars for the counterattack. The Russian attacking force had a 3:1 advantage in men, a 2:1 superiority in artillery, and parity in tanks. Later during the day, the 133 TB joined the attack with its 40 heavy KV-1 tanks, shifting the balance even more in favor of the Russian attack.[146]

On the Russians right flank, the 126 RD and the 254 TB struck positions held by the German 29 Motorized Division dug in around the Abganerovo Station. In the center, the 204, 208 RDs, the Krasnodar Cadet regiment, and the main part of the 13 TC (6 Guards TB and part of the 254 TB) attacked along the rail line from the north and hit the 29 Motorized and 14 Panzer Division from the other side in and around 74 Km Station. On the far-left flank, the 38 RD (borrowed from the 57th Army) and the 13 TB (and later on the 133 TB) hit positions held by the 14 Panzer Division. Providing additional fire support for the 38 RD was armored train #677 from the 28 Armored Train Battalion. Bitter fighting would continue to take place along the rail line for the next two days.

With the VVS fully occupied supporting the 62nd Army, the Luftwaffe was making itself felt in the south. The Luftwaffe claimed to have destroyed an armored train south of Stalingrad on 11 August[147], but several Soviet sources say this attack occurred on 9 August[148]. Whatever the date, the train that was hit was almost certainly #677 as train #708 had withdrawn to the rear for resupply of coal and water. Several rail cars were destroyed with 7 killed and 14 men wounded. Armored Train #677 was out of operation for two days making repairs, but they claimed shooting down a JU-87 and a JU-88 during the attack.

During this same time, Group Chuikov defended the Aksay River line and held off Romanian probing attacks. General Shumilov had taken away the 208 RD for use in the planned counter attack, so Chuikov had few if any reserves.

Late on 9 August, Stavka issued Directive #170562 naming General Eremenko as the overall commander of both the Stalingrad and Southeastern Fronts. This was the change in the command structure Eremenko had been pushing for. General Gordov was to remain in command of the Stalingrad Front while General Golikov was appointed commander of the Southeast Front.[149]

Map #14 - 64th Army Deployment 8-11 August 1942

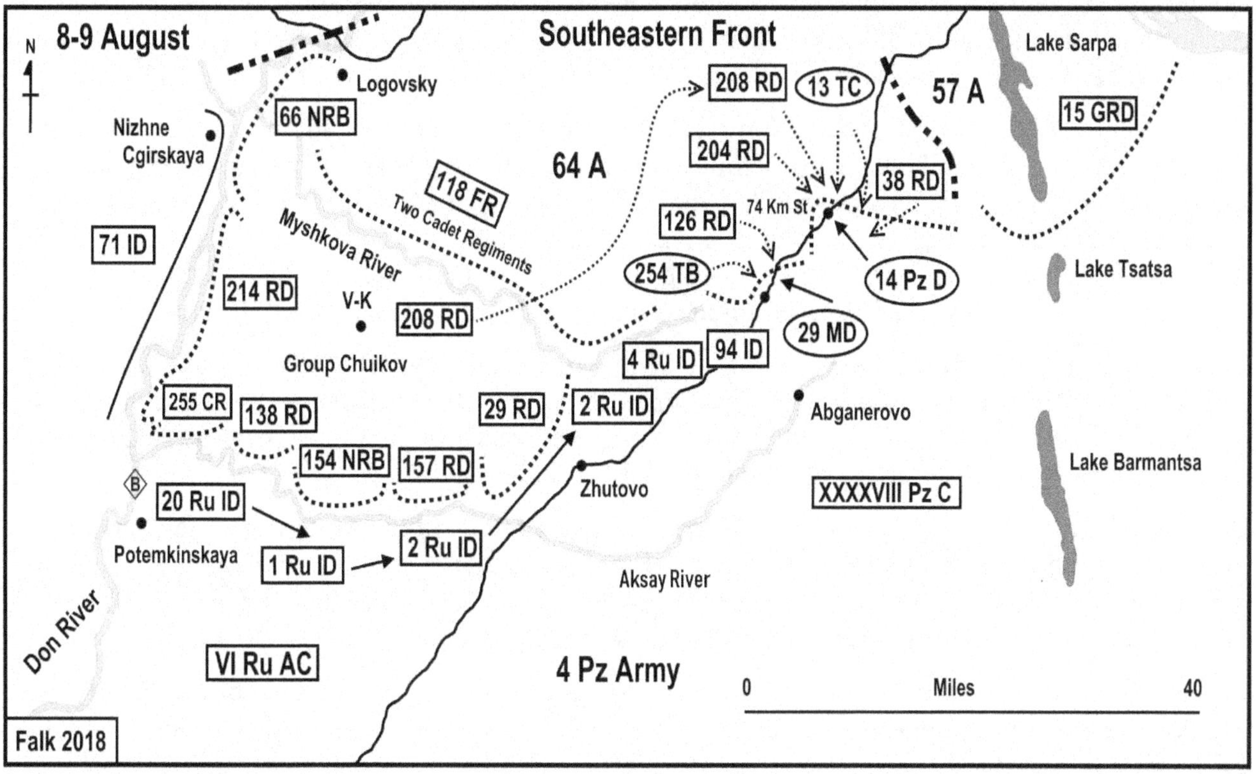

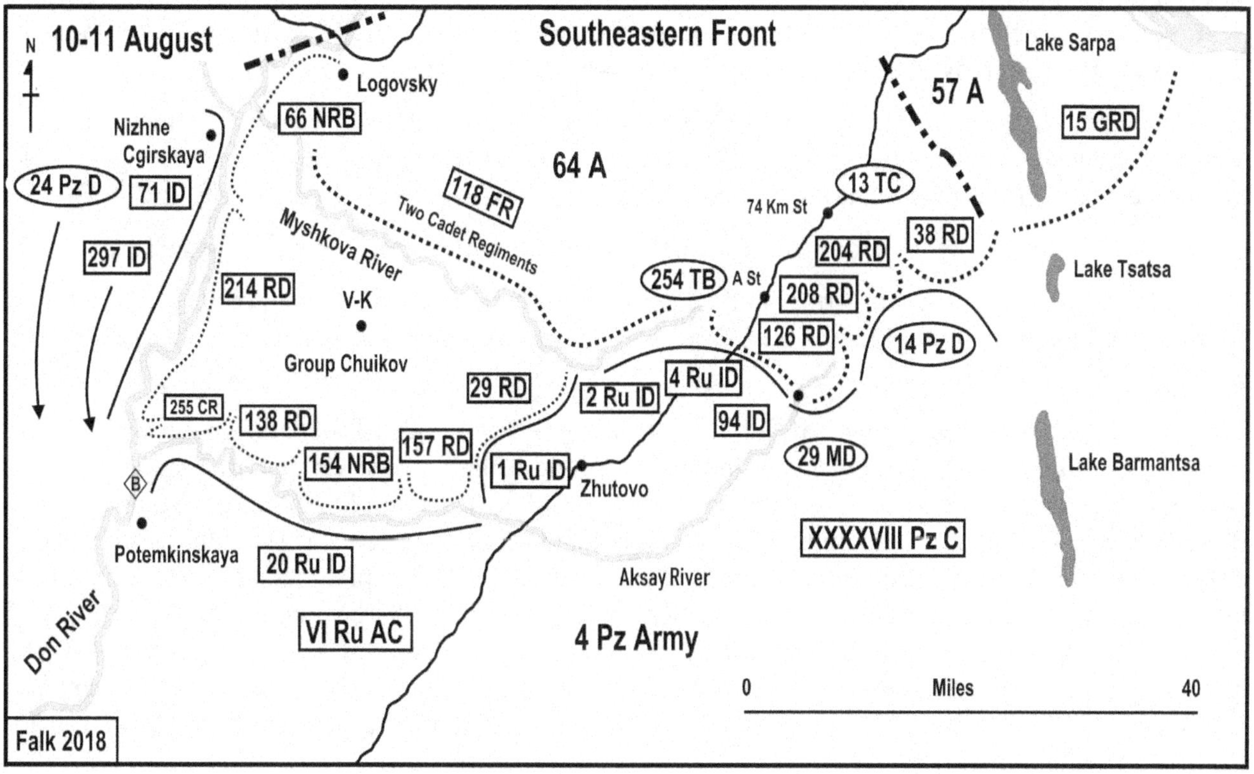

10 August

On the far-left flank of the 64th Army, the counterattack continued in the same general direction, from the north toward the south. Defending German units were forced to give up ground, falling back and looking for a place to make a stand.

By the evening of 10 August, the 29 Motorized Division had been driven out of Abganerovo Station and had been forced to fall back some 8 miles to better defensive positions. During the retreat, the 29 Motorized Division also gave up the village of Abganerovo. The 14 Panzer Division likewise was forced to withdraw some 10 miles from its advanced positions at 74 Km Station, finally crossing to the south side of the upper Aksay River. On the German's left flank, next to the Romanian 4 Infantry Division, the German 94 Infantry Division moved up in support and helped stabilize the front.

Because the 6th Army surrounded most of the 62nd Army west of the Don River, the German 297 Infantry Division was not needed at this location. Therefore, it was released and sent to reinforce the 4th Panzer Army. The 297 Infantry Division took several days to march from the west side of the Don to where it was needed on the German right flank.[150]

Because of continuing attacks and the tenacious defense by units of the 64th Army, the German 4th Panzer Army needed substantial reinforcements before it could resume offensive operations. In two days of attacks, General Shumilov had stopped the 4th Panzer Army and forced its advanced units to fall back to new defensive positions. The defenders of Stalingrad had gained more time.

The same day brought victory of a sort for German Army Group A, when they captured the oil field at Maikop. At least one minor objective of the expanded Operation Blue had been achieved.

11 August

Scattered and sometimes intense fighting continued between units of the 64th Army and 4th Panzer Army around Abganerovo as both sides settled into new defensive positions.

With the 62nd Army on the right bank of the Don being crushed until only remnants remained, the German 24 Panzer Division also was released from the 6th Army. The 24 Pz D was no longer needed on the west side of the river, so it too rushed to reinforce the 4th Panzer Army's right flank.

Group Chuikov continued to occupy positions along the Aksay River line with the same forces.

Due to the collapse of the 62nd Army on the west side of the Don River, Stavka rushed the 1st Guards Army by rail into the Stalingrad area in an attempt to reinforce the left flank of the Stalingrad Front.

12 August

The pocket containing the remains of the 62nd Army on the west side of the Don finally was destroyed.[151] Only scattered remnants of the divisions and brigades that were encircled there were able to escape and to flee over the Don to safety. Those who escaped only amounted to some 3,700 men.[152] While several thousand men were saved in this manner, they left all their heavy equipment behind. German sources claim to have captured approximately 50,000 prisoners, close to a thousand tanks (mostly destroyed) and hundreds of artillery pieces along with large amounts of other equipment. General Lopatin, still in command of a much-weakened 62nd Army, desperately tried to regroup his forces to defend the east side of the Don River. The 62nd Army's hold on this river line was vital because they were all that prevented the Germans on the west side of the river from reaching Stalingrad just 45 miles away.

With the destruction of the 62nd Army pocket, the German 6th Army was now free to move to the north and clear the Russian 4th Tank Army from its remaining positions on the western side of the

Don. Removing this threat would give the Germans total control of the western bank of the Don River and the Don River bend, which would then allow a concentrated direct attack upon Stalingrad.

Meanwhile, the 64th Army was mostly on the defensive, with only scattered fighting on its left flank. German forces of the 4th Panzer Army also were mostly inactive, awaiting further developments and reinforcements. The 24 Panzer and the 297 Infantry Divisions were on the way from the 6th Army and would arrive in several days. They would need to cross the Don at the village of Potemkinskaya, using ferries and a pontoon bridge. Additionally, the 371 Infantry Division was marching up from the south to add further strength to the next series of attacks. Despite recent setbacks, the 4th Panzer Army was not yet finished with the 64th Army.

Group Chuikov was ordered to take control of the 66 NRB that was deployed along the Don River south of Kalach on the far-right flank of the 64th Army. This transfer was part of a plan to redeploy units of Group Chuikov and right flank units of the 64th Army in order to reinforce the far-left flank of the 64th Army. Shumilov anticipated further German attacks in this area and shifted forces to build up his left flank. With the transfer of the 66 NRB to Chuikov's command, it makes sense the 118 FR, defending in the same general area along the Myshkova River, also would have been transferred to his command. If true, this arrangement would have in effect given Chuikov the task of holding the entire right flank of the 64th Army, which covered about 60 miles of front line. Group Chuikov then would be covering the river line starting at the 62nd Army boundary line in the north, then down the eastern side of the Don to the mouth of the Aksay River, then eastward along the Aksay River line to the area around the village of Romashkin. Even with these transfers, the 64th Army apparently still retained control of the 214 RD, which was deployed along the Don River.

13 August
Southeastern Front assigned the 77 Fortified Region, and most likely the 244 RD and 422 RD, to support the 64th Army in the boundary area between the 64th and 57th Armies. At this time these units remained subordinate to the front.

Late on 13 August, Stavka issued Directive #170566 concerning the command structure in the Stalingrad area. This directive formally made General Golikov General Eremenko's deputy, but Golikov still retained command of the Stalingrad Front. This directive also arranged for a group of advisers, sent by Stalin, to oversee and to guide General Eremenko in the defense of Stalingrad. Included within this group was Nikita S. Khrushchev who became a member of Eremenko's Military Council, and [153] General Vasilevsky who would be Stavka's representative. This same Nikita S. Khrushchev would one day become the leader of the Soviet Union.

Group Chuikov continued to hold the Aksay River line, and most likely the Don River line as well, while dealing with minor Romanian attacks from the 20 Infantry Division. Sometime around this date, General Shumilov began thinning out Group Chuikov. With only the Romanian 20 Infantry Division facing it in the south, the 138 and 29 RD were shifted further east. Eventually the 154 NRB also was transferred eastward.

14 August
By this date, the 24 Panzer and 371 Infantry Divisions reached their assembly areas behind the right flank of the 4th Panzer Army in the Aksay area (southwest of Abganerovo). The 297 Infantry Division just crossed to the east side of the Don and needed several more days to reach its assembly area.

Only light action took place along the front for the 64th Army and Group Chuikov. Both groups remained on the defensive and continued redeployment of units to the left flank. The 4th Panzer Army also remained inactive awaiting reinforcements and accumulating supplies of fuel and munitions for the next attack.

15 August

In the north, after regrouping, the German 6th Army launched an offensive against the 4th Tank Army in an attempt to clear the west side of the Don River of Russian forces.

On or about 15 August, the 214 RD was detached from the 64th Army. Most likely the planned withdraw of Group Chuikov to the Myshkova River line was partially intended to release the 214 RD for use elsewhere. The 214 RD briefly was reassigned to front reserve but quickly was reassigned to the 4th TA fighting north of Stalingrad. The 214 RD had been holding a quiet section of the front along the Don, so it would have been considered a fresh, well rested unit.[154] Of special note, it had been originally intended to transfer the 214 RD to the 62nd Army and use it to defend the Don River front in the Vertyachi area, the very area where the German 6th Army would force a crossing of the Don in a few days' time. But on 16 August, the 214 RD orders were changed. It was sent north to reinforce the depleted 4th TA instead.[155]

Another noteworthy event concerned the 126 RD. After taking part in the recent heavy fighting on the left flank of the 64th Army, the 126 RD had been reduced to a reported strength of only 6,495 men.[156] This amounted to a loss of about 6,058 men since it joined the 64th Army two weeks previously. With losses like this, Stavka needed to provide a steady stream of new combat formations to front line armies to keep them up to strength.

16 August

The German 6th Army succeeded in forcing the bulk of the 4th TA to withdraw to the east side of the Don, taking some 13,000 prisoners in the process. With this effort, the entire west side of the Don was under German control.[157] However, a few isolated Russian troops remained on the west side of the river, to be mopped up over the next few days. This was another minor objective of Operation Blue that was successfully completed: clearing the entire western side of the Don River.

Within Group Chuikov, part of the 66 NRB and the entire 157 RD took over sections of the front line currently held by the 138 RD. This allowed the 138 RD to move to the Myshkova River line and eventually to shift to the left flank of the army, to concentrate in the 74 Km Station area, near to where the Germans had been attacking. In turn, at some point, the 118 FR pulled back and covered the Don River area to the right of the 66 NRB. These two units held the far-right flank of the 64th Army and covered the boundary between the 62nd Army to the north.

17 August

Due to the threat of renewed German attacks on the left flank of the 64th Army and the need to consolidate and strengthen its defenses, the Southeastern Front HQ decided to execute the planned withdraw of Group Chuikov. Group Chuikov was holding an increasingly exposed advanced position, so Southeastern Front HQ ordered Chuikov to withdraw his forces north to the Myshkova River, approximately 20 miles away.[158] Front HQ had planned this withdraw for several days, and some German maps appear to show this withdraw taking place in stages starting on 13 August. Acting quickly, General Chuikov successfully accomplished this withdraw during the night of 17 August, thus preventing the Luftwaffe from interfering. (see Map #15)

During the day, German forces launched strong attacks with Luftwaffe and artillery support against the 126 RD between Abganerovo and Abganerovo train Station. Combat continued into the night with parts of two Russian rifle regiments of the 126 RD being surrounded. This battle continued for the next two days. By means of this attack, the Germans tried to clear their left flank, to destroy a strong enemy defensive position, and to widen their base of operations thus gaining freedom of movement.

Also on this date, the 208 RD HQ issued a staff situation report that listed the divisional (combat) strength as 1,678 men. Breaking this down further, the 435 Rifle Regiment was listed as having 498 men, the 760 Regiment with 685 men, and the 578 Regiment with 495 men. This is after the 208 RD took part in earlier combat around Abganerovo Station on 9 August. So, the 208 RD maintained a combat strength of a weak rifle regiment.[159]

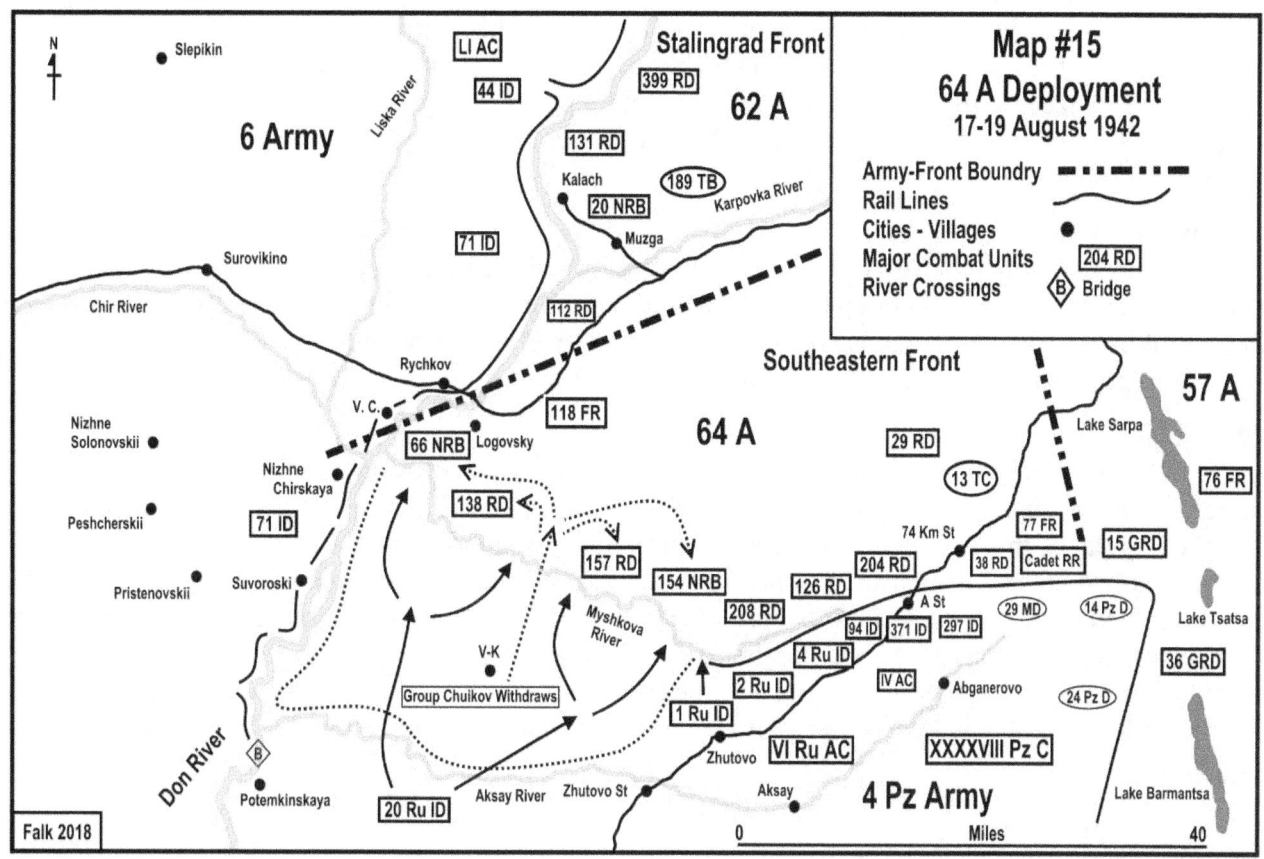

18 August

By dawn, units of Group Chuikov occupied new defensive positions along the north side of the Myshkova River, with stragglers from these formations arriving at the new defensive positions throughout the day. By this date, perhaps only the 66 NRB, 118 FR, 138 RD and the 157 RD remained under Chuikov's direct command, but this would soon change as combat on the left flank of the Army intensified. At first the Romanian 20 Infantry Division did not notice this withdraw and then was slow to push forward to the Myshkova River. A partial reason for this slow response was that Chuikov left behind, in the old Aksay River positions, strong rear-guard formations. These rear guards simulated larger forces by moving about and conducting active firing against the enemy. When threatened, they fell back and harassed any advancing enemy units. In this way, the major formations of Group Chuikov had ample time to retreat and to dig in along the new defensive line.

German infantry units continued to launch attacks against the Abganerovo Station with the 126 RD supporting the defending 204 RD from the right. German Panzer units of the XXXXVIII Panzer Corps also became active by launching attacks against Russian positions held by both the 204 RD and 29 RD around State Farm #2, east of the Abganerovo train Station. In the afternoon the Germans captured Abganerovo Station, but it took until the next day to clear the area of tenacious Russian defenders. The three-day battle in this area was intense. Jason Mark in his book *Panzer Krieg Vol 1* citing German sources, stated the "Falke Division [29 Motorized] captured quite a lot of booty on this day (17 Aug): 450 prisoners, 25 tanks knocked out, one 12.2 cm gun, eighteen 7.62 cm guns, five 4.5 cm or 3.7 cm guns, 10 towing tractors, 5 tractors, 7 lorries and 13 concrete bunkers – and 35 wooden bunkers." These bunkers were, of course, part of Stalingrad's outer "O" defensive line. Mark then goes on to say the next day (18 Aug) another, "[S]even hundred prisoners went into the POW cage." The area between Abganerovo and Abganerovo Station was firmly in German hands. The clearing of this area was critical for future German plans, because General Werner Kempf, commander of the XXXXVIII Panzer Corps, was planning a new assault upon the 64th and 57th Armies. Needless to say, the badly weakened divisions of the 64th Army were in no shape to preemptively interfere with German plans.

19 August
During a meeting at 64th Army HQ, Chuikov was informed the Front Military Council had decided that all the units assigned to Group Chuikov would be officially incorporated into the 64th Army and that Chuikov should once again become the deputy commander of the 64th Army.[160] Chuikov agreed with these proposals, not that he had much say about this issue.

20 August
In the morning, at 07:00 hours, the German 14 Panzer and 24 Panzer Divisions launched attacks directed north against Russian positions of the 57th Army. These attacks were just to the west of a series of salt lakes that extend in a north-south direction, just south of the bend in the Volga River. This attack had entered the Caspian Depression area, which was mostly semi-arid terrain at or slightly below sea level. (see Figure #1) For once, both the Luftwaffe and the VVS provided effective support to their respective ground forces during the day. The IV AC also launched supporting attacks north from the Abganerovo Station area. This was the start of a coordinated pincer attack by both the German 6th Army and the 4th Panzer Army against Russian forces defending the approaches to Stalingrad. (see Map #16)

21 August

At 03:10 hours, without a preliminary barrage, the German 6th Army launched its assault over the Don. Its ultimate goal was to reach the Volga and Stalingrad. By the end of the day, several pontoon bridges were in place. Despite heavy bombing by the VVS and desperate resistance by units of the 62nd Army, several bridgeheads were established and expanded on the east side of the river.[161]

In a coordinated effort, the reinforced 4th Panzer Army continued its offensive north of Abganerovo with the 29 Motorized and 297 Infantry Division pushing forward against the 126, 29 and 204 Rifle Divisions of the 64th Army. The 14 Panzer and 24 Panzer Divisions also continued their attack against the 57th Army. Russian forces fiercely resisted both of these thrusts. To bolster their defense, Southeastern Front reinforced this sector with six Tank Destroyer Regiments. While it's unknown which six TD Regiments were transferred, the 57th Army might have received the 500 and 188 TDRs and the 64th Army might have received the 186, 507, 612, and 665 TDRs.

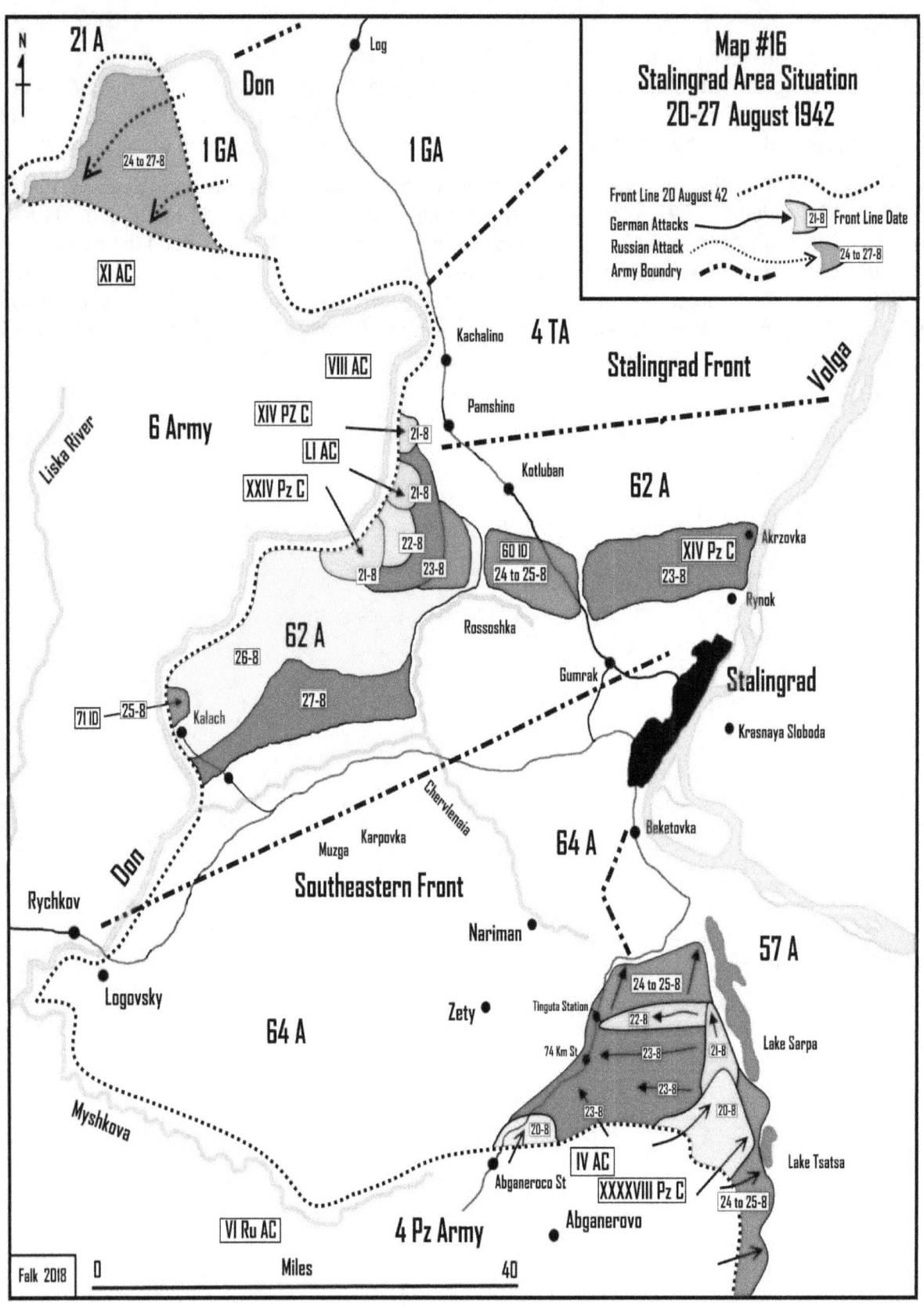

22 August

The German 6th Army continued to develop and expand its bridgeheads with six infantry divisions on the east side of the Don. The 62nd Army and parts of the 4th TA struggled to contain the expanding bridgehead positions, despite launching counterattacks and subjecting the bridgehead to continuous air and artillery attacks. Starting in the evening and continuing all night long, units of the German 16 Panzer and 3 Motorized Divisions moved into assault positions within the bridgehead. Later on, the advanced units of the 60 Motorized Division followed them into position. General Paulus deployed his mobile units into the bridgehead for what he hoped would be the decisive breakthrough attack that would reach Stalingrad!

The 4th Panzer Army continued its offensive toward the north, reaching the area near Tinguta Station.

On the east side of the attack, the 24 Panzer Division broke through between the salt lakes pushing back units of the 57th Army. On the German left flank the 94, 371, and 297 Infantry Divisions pushed north of Abganerovo Train Station and 74 Km Station. To counter these moves, Shumilov launched the 138 RD into a counterattack eastward against the German 297 ID. Further to the north, the 154 NRB and 131 TB made another attack against the 14 Panzer Division in the Tinguta Station area. Fighting continued all day on the left flank of the 64th Army and within the areas defended by the 57th Army. Of note, the 126 RD had been reduced to a reported strength of 3,733 men by this date.[162] This represented a loss of 2,722 men during the previous seven days of combat. Needless to say, losses were heavy on both sides.

Later during the day, German divisions within the 4th Panzer Army adjusted their positions to release forces to support further attacks to the north.

23 August

At 04:15 hours, the German 16 Panzer and 3 Motorized Divisions broke out of the 6th Army bridgehead on the east side of the Don. Their destination was the Volga River and Stalingrad, approximately 45 miles away.

General von Richthofen committed the Luftwaffe's full support to this thrust toward the Volga by "bombing and strafing Soviet pockets of resistance and warding off Soviet aircraft which tried to interfere."[163] The VVS opposed the Luftwaffe, but the 8th AA had far too few operational aircraft to seriously contest the airspace.

To the south, the 4th Panzer Army continued to shift forces around, moving part of the 29 Motorized Division to cover the salt lakes flank, thereby releasing units of the 24 Panzer for further attacks toward the north. Also, the 371 Infantry Division pulled out of the line and moved further north.

Shumilov, in turn, moved into the line the 13 TC, 204 RD, 133 TB, along with several regiments of cadets and artillery units. The commander of the 208 RD, Major Nikitin, informed General Shumilov that his division had only 502 men left in combat formations. He stated he had no artillery, no mortars, and only six 45 mm AT guns left. He asked for fire support and manpower from the 29 RD to his left.[164] The Southeastern Front also positioned the 422 RD, 244 RD and independent artillery units in front of the German attack. Additionally, the 57th Army committed the 15 GRD, 38 RD, 38 MRB, and 77 FR to block the German advance.

In the north, by 18:35 hours, units of the XIV Panzer Corps reached the Volga River just north of Stalingrad. Despite being cut off from the bulk of the 6th Army, and under constant attack, they held their positions throughout the night. One can only imagine the alarm this bold thrust to the Volga caused within the Russian command ranks, all the way from General Eremenko and Golikov at Stalingrad to Stalin and Stavka in Moscow.

To add to the confusion, starting early in the evening, General von Richthofen launched continuous Luftwaffe bombing raids against the entire city of Stalingrad. During the night over 1,000 tons of bombs were dropped. Stalingrad was soon "set ablaze from one end of the city to [the] other."[165] General von Richthofen clearly coordinated this aerial attack with the assault by the 6th Army to the Volga. The bombing also clearly was a terror raid against a mostly defenseless city filled with civilians.

24 August

Against the left flank of the 64th Army, intense fighting continued all along the extended line as German troops pushed north on the east side of the railway line. The 24 Panzer Division was largely occupied with regrouping. On the left flank, in heavy fighting, German infantry divisions slowly widened the base of the attack zone by reaching the rail line from Abganerovo Station to the Tinguta Station area.

In response to these attacks, General Shumilov moved his mobile reserves and created the so-called group of Colonel Sorokin. This group consisted of 126 RD, the remains of 208 RD and the Zhitomir and Ordzhonikidze Cadet schools, as well as a Guards mortar rocket regiment.[166] The 208 RD had been working with and supporting the 126 RD at least since 17 August. This arrangement would continue until the last few remaining troops of the 208 RD were used to reinforce the 126 RD.

In the north, the German 6th Army tried to force a linkup with its units on the Volga. In turn, Russian forces desperately tried to cut off and destroy these same German units. Confused fighting continued all day north of Stalingrad.

The Luftwaffe maintained its support of ground operations, but it also sustained its attack against Stalingrad. All day long and into the night once again bombs rained down on the defenseless city.

Stavka's response to the Germans reaching the Volga north of Stalingrad was to issue Directive #994170 activating the 8th Reserve Army near Saratov, as the new 66th Army. With its eight rifle divisions, the 66th Army under the command of General Malinovsky was ordered to move south along the Volga toward Stalingrad. On or about this same day, Stavka activated the 9th Reserve Army near Gorki (east of Moscow) as the new 24th Army. There appeared to be some confusion with the activation of this army, as if it were not fully formed or ready for combat. In any event, Directive #170588 was issued on 26 August and then Directive #994171 on 27 August confirmed the new 24th Army would be commanded by General Kozlov and would contain five rifle divisions and three tank brigades. Both of these armies were to be used to break through to Stalingrad from the north. Initially these two armies contained the following units: the 66th Army contained the 231, 120, 99, 49, 299, 316, 207, 292 Rifle Divisions and the 24th Army contained the 173, 221, 308, 292, 207 Rifle Divisions, two Tank Brigades (# unknown) and the 217 Tank Brigade transferred from the 66th Army.[167]

For the next several days, further north along the Don River, the Russian 63rd and 21st Armies attacked over the Don and forced back units of the Italian 8th Army holding the river line. A small but significant bridgehead (not shown on a map) was seized on the right bank. The relatively weak Italian 8th Army was unable by itself to push the Russian units back over the river, so the Russians remained on the west side of the Don. A little closer to Stalingrad, over the next several days, the 1st Guards Army launched attacks from its small bridgehead over the Don and forced back the defending

German units. Again, an important Russian bridgehead was enlarged on the west side of the Don. The German 6th Army was too preoccupied with its attack on Stalingrad to eliminate these minor Russian intrusions over the Don.

Also, during this busy day, Stalin sent General Eremenko message #170585 that uncharacteristically suggested he withdraw the 62nd and 64th Armies closer to Stalingrad and into stronger defensive positions. General Eremenko failed to take advantage of this opportunity offered by Stalin to consolidate Stalingrad's defenses.

Also of note on this day, General Shumilov sent General Chuikov to scout the front line along the Myshkova River around the village of Vasilievka which was covered by the 157 RD. Chuikov reported in his book *The Beginning of the Road*, "We suddenly found ourselves [his driver and Chuikov] between our positions and the Germans' – under fire from both sides." Hurriedly driving back to Russian lines from the German positions, Chuikov said, "We were met by our own soldiers with such suspicion that, had I not spoken to them in colloquial Russian, we would probably have been met with hand grenades, and my raincoat would have been riddled with machine-gun bullets." So once again General Chuikov survived yet another close brush with death; this would be his fourth time.

25 August

In the south, a critical point in the battle had been reached when the Germans approached the hills overlooking the Volga and southern Stalingrad. At the time, German General Hans Doerr made note of the significance of the area.

> [T]he Fourth Panzer Army was positioned in the immediate proximity of an important terrain sector that could possibly have decisive significance for the overall Stalingrad operational area – the Volga highlands between Krasnoarmeisk and Beketovka. Here, if one looks downstream, was located the last high ground near the Volga's west bank [this was part of the Volga Plateau]. It dominated the Volga bend at Sarpinskii Island. If in general it was possible to break open the defense of Stalingrad, then the blow had to be struck namely from this place. Krasnoarmeisk was the southern cornerstone of the Stalingrad defense… At no other point would the appearance of German forces be so unfavorable for the Russians as here.

At this time, General Doerr was the Chief of the German Liaison-Staff to the 4th Romanian Army, which commanded Romanian forces south of Stalingrad.[168] Russian commanders from General Eremenko on down also understood this area was of critical importance to the defense of Stalingrad. Stalin had called General Eremenko on 24 August, and among other things, told him to unconditionally hold your positions south of Stalingrad.[169] The Germans had just reached the Volga a few miles north of Stalingrad; therefore, everyone in the south was determined to block German forces from reaching the Volga in this area too.

Unknown to the German commanders, the 4th Panzer Army's attack had reached a section of the "K" defensive line surrounding Stalingrad. (see Map #12) This fortified line had been built in this area for good reason. The "K" line occupied the dominant hilly terrain and contained extensive mine fields, entrenched AT and machine guns, bunker positions, and strong field fortifications. Determined Russian rifle, tank, and AT formations occupied these positions and numerous artillery units backed them up. This is what the next German attack would face.

Moving off in the morning, units of the 4th Panzer Army continued their advance, mainly with the 14 and 24 Panzer Divisions. But soon the German troops were met by "ferocious artillery fire" and "[f]rom the beginning, the enemy overwhelmed the attacking squadrons with the heaviest fire from artillery, mortars, Stalin organs and dug-in tanks."[170] On the left flank, German infantry divisions made little or no headway, and the 14 Panzer Division was under such heavy attack that it was barely able to make any progress. Only the 24 Panzer Division was able to move forward somewhat but suffered heavy losses in the process. After five days of constant combat, the German assault had pushed 15 miles closer to the Volga and Stalingrad but could go no further. By the end of the day, German commanders broke off the attack.

They did not know it at the time, but a significant defensive victory had just been won by forces under the command of General Shumilov and General F.I. Tolbukhin of the 57th Army. They had successfully stopped the major German advance from the south. Their units had proven to be too difficult to dislodge from the defensive positions they had built in and around the "K" line. A victory it was, but nonetheless, their victory was fleeting. General Hoth, commander of the 4th Panzer Army, was master of the *Blitzkrieg* or "lightning war", and lightning was apt to strike at more than one location.

Further north, the German 6th Army was able to force through some vital supplies to the German divisions surrounded on the Volga. However, these units were still in danger of being destroyed by constant Russian attacks.

After three days, the Luftwaffe finished its attack upon Stalingrad. General von Richthofen made note in his diary, "Stalingrad is completely destroyed."[171] This aerial bombardment upon Stalingrad certainly disrupted Russian military command and control operations in the area as well as supply efforts along the Volga, thus making the air attack a valid military action. But the use of indiscriminate bombing also killed thousands of defenseless civilians who were still within the city, thus also making the air attack a war crime. Ultimately, the Germans would soon learn that a destroyed city makes an excellent fortress and that thousands of dead civilians would call out to the Red Army for revenge.

After inspecting front line units during the height of the battle, General Chuikov was welcomed back at 64th Army HQ after being in the field and out of communication for 10 hours. General Shumilov, members of the Military Council and even the chief of staff were glad to learn he had not been killed or captured during the day. Seeing him, General Shumilov shouted, "He's turned up!" The entire command staff was pleased that he was alive even as they scolded him for being lost all day.[172] Apparently, the roving General Chuikov once again had cheated death at the front lines.

26 August
After much discussion between General Hoth, commander of the 4th Panzer Army, and General Warner Kempf, commander of the XXXXVIII Panzer Corps, they decided to shift the focus of attacks to another location. As Hoth put it, "We're bleeding ourselves in front of these damned hills…We must regroup and attack at a completely different spot far from here."[173] Soon orders were sent out to the 14 and 24 Panzer Divisions to hold their current positions and to dig in. Relief forces were on the way. This showed the German command staff at their best. Delay the battle, shift forces to a new location, rest the troops, and gather further supplies. Attack again with new strength and with the element of surprise at a different location.

Meanwhile, troops of the 64th and 57th Armies also were digging in and prepared to resist further German advances, but the next attack never came, at least not along this section of the front.

Once more "Stavka queried Eremenko about the wisdom of withdrawing the 62nd Army's left wing and the entire 64th Army" to a defensive line based on the Rossoshka and Chervlenaia Rivers, in other words the Stalingrad "K" defensive line. "Once again Eremenko demurred."[174] This inaction by Eremenko would have deadly consequences later on for the 62nd and 64th Armies. These armies were the primary defenders of Stalingrad and General Eremenko was gambling he could use them to hold the Germans away from the city. In the end, Eremenko failed in this task, and the fate of Stalingrad and two heroic armies hung in the balance. At this time, General Shumilov and his staff also were debating the need to withdraw the Army to the Chervlenaia River line. Taking it upon himself, Shumilov called the Southeastern Front HQ and stated the opinion of his Military Council, but because General Eremenko was not at his command post, the Southeastern Front HQ made no decision.

27 August

In the south, low level combat continued all along the front lines, which included sudden artillery barrages and raids by aircraft at all hours. Unknown to General Shumilov and General Tolbukhin, German assault troops quietly began to slip away. For the next several nights, Infantry Divisions of the German IV Army Corps replaced the Panzer Divisions of the XXXXVIII Panzer Corps. General Hoth was shifting the focus of his next attack some 35 miles to the west, to a less well-defended location.[175] This redeployment was missed by Russian commanders largely because the Luftwaffe controlled the air. Few if any VVS aircraft were available to fly reconnaissance missions over the southern part of the front.

In the north, the German 6th Army continued to enlarge their bridgehead on the east side of the Don and kept supplies flowing to troops in the newly captured corridor to the Volga. If only the 6th Army and the 4th Panzer Army could quickly link up somewhere outside Stalingrad, then they could surround the bulk of the 64th Army and most of the 62nd Army. Stalingrad would then certainly fall.

28 August

As the divisions of the XXXXVIII Panzer Corps continued to move to the west and concentrate in their new assembly area, the Romanian 4 Infantry Division moved to the east to cover the exposed right flank of the 4th Panzer Army. In response to the current situation, General Shumilov asked Eremenko for help from VVS aircraft, as well as to allow the withdrawal of the 64th Army to the boundary of the river Chervlenaia. "Eremenko said that aircraft cannot be sent, since the 62nd Army's situation is even worse, and regarding the withdrawal of the army, Shumilov was ordered to wait for his instructions."[176] This wait order would cost the 64th Army a great deal.

Units of the 64th and 57th Armies remained focused on defending the area directly south of the Volga. Despite claims to the contrary, it appears the entire Russian command staff, from the Southeastern Front on down to the 64th Army, were only vaguely aware German mobile divisions had been shifting to a new section of the front for the past two days. Apparently, the concentration of German assault units in this new area was only spotted shortly before the impending attack. But, at that point, it was too late. General Shumilov had spent weeks building up his left flank to resist the powerful attacks from the 4th Panzer Army, and now he was caught off balanced and unable to react in time to counter this new threat.

By the end of the day, the 14 and 24 Panzer Divisions and most of the 29 Motorized Division had gathered behind the Romanian 2 Infantry Division and 20 Romanian Infantry Division. To their front were formations of the 126 RD, containing some 3,700 men in combat formations[177] and the 29 RD, a bit stronger with approximately 4,500 men in combat formations. This weak section of the line, just to the west of the Abganerovo train Station, had been quiet for several days, but the next day everything would change. (see Map #17)

29 August

At 06:30, the 4th Panzer Army renewed its attacks against the center section the of 64th Army's front. Clearing the way, the Luftwaffe shifted its ground support operations to the south in full support of this new 4th Panzer Army attack.

It appears that just before, or maybe even during the early stages of the German attack, Shumilov ordered the 29 RD to act as a rear guard. Its job was to delay the German advance as much as possible to allow the rest of the army to withdraw unhindered. The 29 RD would attempt to fulfill this task courageously, but in doing so it would pay a terrible price. Overwhelmed by the sheer power of the surprise attack, the 126, 29, and 138 RDs quickly gave way and were forced to fall back.

By the end of the day, the 14 and 24 Panzer Divisions and the 29 Motorized Division had pushed forward some 12 miles, reaching the village of Zety. The advancing Romanian 2 Infantry Division covered the left flank of the offensive while the Romanian 20 Infantry Division pushed hard on the right flank.

The 64th Army and the Southeastern Front faced a major crisis. Lacking adequate reserves and time, Shumilov was unable to build a new defensive line in front of the German attack. This first day saw the 126, 29 and 138 Rifle Divisions forced out of their defensive positions as German and Romanian units overran their rear areas. Entire formations of the 64th Army, including support units, were overrun. These units suffered heavy losses in men and equipment as the hurried withdraw turned into a rout. As other units withdrew, the 29 RD fought hard to cover their retreat. In the process the divisional command post was overrun with many staff officers being killed and their divisional commander, Colonel V. E. Sorokin, captured.[178] By the end of the day, the 29 RD had been reduced to only 1,054 combat troops.[179]

In their book *To the Gates of Stalingrad Vol 1*, p 374, Glantz and House quote from General Shumilov's report to the General Staff about the day's events: "The units of the 64th Army fought stubborn defensive battles with enemy forces of up to two infantry divisions and 78 tanks." Shumilov went on to say how several different units were fighting with enemy tanks and being attacked by aircraft, but with no changes to the positions of the Army's units. But Glantz and House have a slightly different view: "The collapse of the 126th Rifle Division's defenses also unhinged the 64th Army's defenses along the railroad line, forcing 29th and 138th Rifle Divisions to withdraw northward and posing a threat to both 64th and 62nd Armies rear areas." Their assessment has merit. This brilliant attack by the 4th Panzer Army was the last real chance to defeat the 64th and 62nd Armies outside of Stalingrad proper. If these two Russian armies were allowed to escape into the city, the battle of machines would be all but over, thus depriving the Germans their most powerful weapon, freedom of maneuver. Combat then would degrade into a man vs. man *mêlée* in the ruins of a dead city.

Only at the end of the day did Shumilov report to the front headquarters that "the army's units began to retreat in conditions of encirclement and semi-encirclement."

Map #17 - 64 Army Withdraw 26-30 August 1942

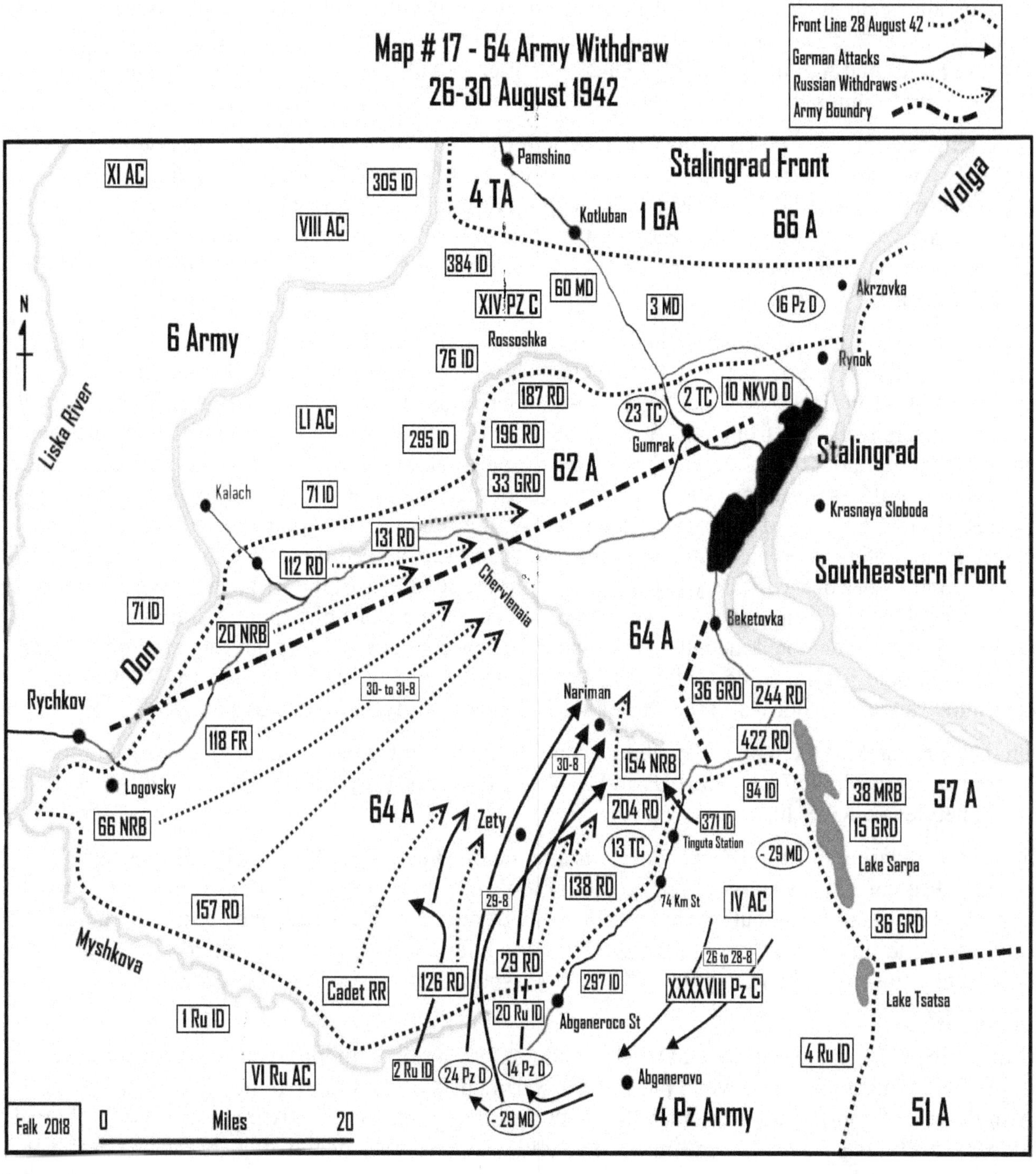

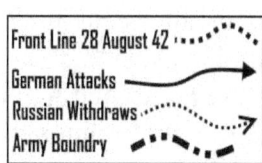

30 August

The next day the German advance continued north to the Chervlenaia River and the village of Nariman. By the end of the day, German troops threatened to cut off all units of the 62nd and 64th Armies west of Stalingrad. At this point, Eremenko was forced to order a general withdraw of his two vulnerable armies. These two armies would attempt to establish a new defensive position along the inner Stalingrad "K" defensive line based on the Rossoshka and Chervlenaia Rivers, just to the west of Stalingrad. The race was on to see whom would occupy these new positions; would the attacker or defender arrive first?

Strategically speaking, General Eremenko should have pulled the 62nd and 64th Armies back to the inner defensive line much sooner. The 4th Panzer Army gave him a few days' grace with the regrouping of their mobile forces, but General Eremenko failed to act, and that error almost cost him the battle for Stalingrad.

During this confusing withdraw, the Commander of the 208 RD Colonel K.M Voskoboinikov and his entire divisional staff were overrun and captured by the advancing Germans. This effectively destroyed what was left of this hard luck division. What few troops remained were used as reinforcements for the 126 RD. Eventually the 208 Division was officially disbanded on 28 November. This disbandment was covered by GKO Order #00248 stating, "The elimination of the Red Army military formations, units and institutions, as not to be restored." [180]

As German and Romanian units pushed further north, more and more retreating formations of the 64th Army became intermixed with the attackers. Needless to say, confusion reigned in this area. One unit, the 154 NRB, never received the order to fall back to new defensive positions. Three different liaison officers were sent from 64th Army headquarters with the order to retreat and all failed to deliver it. So, without orders to withdrawal, the brigade commander, Colonel A.M. Smirnov, ordered 154 NRB to hold its positions. The Brigade was deployed from the village of Zety to the Tinguta rail Station, right in the path of the German advance. As the fighting raged, parts of the Brigade were overrun while others were split into smaller pockets and surrounded. Late that night, Captain G.D. Fokin, commander of the artillery battalion, brought the remnants of the Brigade through German lines and back under 64th Army control. After receiving a report from Captain Fokin, an angry General Shumilov demoted the injured Colonel Smirnov to lieutenant and put Captain Fokin in his place as temporary commander of the Brigade. However brave the actions of the 154 NRB were, its determined stand was not worth the loss of some 80 percent of its strength in a single day.[181]

On the right flank of the 64th Army, things were different; units there withdrew in a much more orderly fashion. The 157 RD, 66 NRB and the 118 FR successfully fell back in several stages to the new Chervlenaia River line.

In the north, during this time, General Paulus and the 6th Army were fully occupied by nearly constant Russian attacks against their northern flank. The thin corridor of German troops that reached to the Volga was holding, but just barely. No units from the 6th Army could be spared to attack southward and link up with the 4th Panzer Army pushing north. Thus, the Germans missed a great opportunity to surround and to destroy the two Russian armies defending Stalingrad.

31 August
Units of the German 29 Motorized Division and the 297 and 271 Infantry Divisions forced a crossing of the Chervlenaia River in several places near Nariman. This act ruptured the "K" line in the 64th Army's area of control and accelerated the withdrawal of 62nd and 64th Army units into Stalingrad proper. Several days later, General Shumilov would say, "August 29 and 30 were the heaviest days for the army for the entire defense period."

In the air, the strength of the 8th AA "had rapidly dwindled to just 193 serviceable aircraft at the onset of September - 57 fighters, 38 day bombers, 51 night bombers, 32 IL-2 [*Shturmovik*], 13 liaison aircraft and 2 reconnaissance." With numbers like these, it's understandable why the formation of the new 16th AA had been delayed.[182] Luftwaffe aircraft formations freely roamed over retreating Russian columns, continuously pounding them. The depleted VVS units could do little to prevent this rain of destruction from the air.

Throughout this month of heavy fighting, the 64th Army suffered substantial losses during its many battles. For example, the valiant 126 RD reported to the 64th Army on or about 26 July and contained a reported 12,553 men. During the fighting in August, this rifle division suffered 10,336 men killed and missing,[183] and these losses did not include the many wounded it suffered. Other units too lost most of their men and equipment during the fighting, while other formations, like the 208 RD, were so badly shattered they simply disappeared or were later removed from the Red Army order of battle. Despite its exposed position at the end of the month, the 64th Army would live to fight on into September. Yet, while many men were wounded or injured, some did return to duty. For instance; "For two months [August - September 1942] the combat situation in the 64th Army were evacuated 39,000 injured, of which about 2 thousand returned to duty…. according to the Surgeon General of the Stalingrad Front, M.G. Gurevicha."[184] This works out to a wounded return rate of just under 5 percent. With permanent losses like these, the Army would need even greater replacements and reinforcements to keep fighting.

So ends the month of August 1942. One major defensive battle had been won and another lost, but the fighting would go on regardless. Finally, all the combatants were gathered together and focused directly upon Stalingrad, the prize, the hope, the burden.

Joseph Stalin

Adolf Hitler

(on right) Colonel General Andrei Eremenko, Commander Southeastern Front then Stalingrad Front. (on left) Nikita Khrushchev, member of Eremenko's Military Council and future leader of the Soviet Union.

(on right) Colonel General von Weichs, Commander of Army Group B.

Field Marshal von Manstein in 1942 – future Commander of Army Group Don.

(L-R) Colonel General Wolfram von Richthofen, Commander of Luftflotte 4 (Air Fleet 4); with Field Marshal Erich von Manstein.

Major General Timofey T. Khryukin, Commander of the 8th Air Army.

(on left) General of Armored Troops Friedrich Paulus, Commander of 6th Army.

(second from left) Lieutenant General Vasilii Chuikov, Commander of the 64th Army and then the 62nd Army. This photo shows Chuikov at his 62nd Army HQ in Stalingrad. Notice the frozen fingers on his right hand. (on right standing) Major-General Alexander Rodimtsev, Commander 13th Guards Rifle Division.

(L-R) General von Weichs, General Paulus, and General of Artillery Walter von Seydlitz, Commander of 51 Corps

(L-R) Colonel General von Richthofen, Commander of Luftflotte 4; Lieutenant General Warner Kempf, Commander of 48 Panzer Corps; Colonel General Hermann Hoth, Commander of 4th Panzer Army.

Lieutenant General Warner Kempf, Commander of 48 Panzer Corps.

(L-R) Field Marshal Paulus, General Schmidt and Colonel Adam as POWs arriving at 64th Army HQ in Beketovka.

Field Marshal Paulus in Captivity - February 1943.

(second from left) Newly promoted Colonel General Konstantin Rokossovsky, Commander of the Don Front, with newly promoted Field Marshal Paulus (on right). This photo was taken at the village of Zavarykin where the Don Front had its HQ. Field Marshal Paulus arrived here, by car, directly after leaving 64th Army HQ at Beketovka.

(second from left) General Chuikov, (third from left) General Shumilov, (on right) Nikita Khrushchev, (in center front, next to Khrushchev) Major-General Alexander Rodimtsev, Commander 13th Guards Rifle Division. This photo is from 4 February 1943 when a mass rally was held within the ruins of Stalingrad. Thousands of soldiers and officers came together to celebrate their great victory despite armed groups of Germans still at large within the city.

Editorial Staff for the 64th Army Paper "For the Motherland."

(on left) 64th Army Senior Sergeant Elsekov is inspecting documents at a checkpoint behind the front. (on the right) A traffic regulator is assisting him. The sign in the background says ПРОВЕРКА ДОКУМЕНТОВ = VERIFICATION OF DOCUMENTS.

A 64th Army solider joins the Komsomol (Young Communist League) during training. This photo most likely was taken in July 1942.

In contrast to the previous photo, this picture shows a group of tough veteran troops of the 64th Army after the battle for Stalingrad, February 1943.

General Shumilov and his driver in their Lend Lease jeep sometime during early 1943.

1 September 1942	
64th Army	
Assigned – Southeastern Front	
Major General M.S. Shumilov Commanding	
Rifle, Airborne and Cavalry Units	36 Guards Rifle Division 29, 38, 126, 138, 157, 204 RDs (208 RD Remnants) 66, 154 Naval Rifle Brigades Remnants of five Cadet Rifle Regiments 118 Fortified Region 77 Fortified Region
Artillery Units	85 Guard Howitzer Artillery Regiment 266, 1104, 1111, 1159 Gun Artillery Regiments 186, 504, 507, 612, 665 Antitank Artillery Regiments 140 Mortar Regiment 18, 19, 51, 76 Guard Mortar Regiments (rocket) 1261 AA Regiment
Armored and Mechanized Units	13 TC (6 Guard, 13, 56, 133, 254 Tank Brigades) 28 Separate Armored Train Battalion
Engineer Units	1363, 1581, 1615 Separate Sapper Battalions
Reported Combat Strength Reported Ration Strength	33,684 men 42,611 men[185]

(BSSA - 42 part 2 and BSA - 42 part 4)

1 September

The start of September saw withdrawing units of the 64th Army begin to reach the defensive positions along the Chervlenaia River line. The 62nd Army was aligned with the 64th Army on the right flank, covering Stalingrad's central and northern districts. The 57th Army was aligned on the left flank, and as before, covered the southern approach to the Volga. Working together, both General Shumilov and Chuikov had an extremely difficult job ahead of them. They had to withdraw their army quickly to these new positions to protect the southern part of Stalingrad. They then had to occupy the fortifications of the existing "K" line, in and near the Chervlenaia River, all the while facing an aggressive ground pursuit by German mobile units compounded by near constant air attacks from the Luftwaffe. (see Map #18)

Rush to Stalingrad - Map #18 - 31 August - 3 September 1942

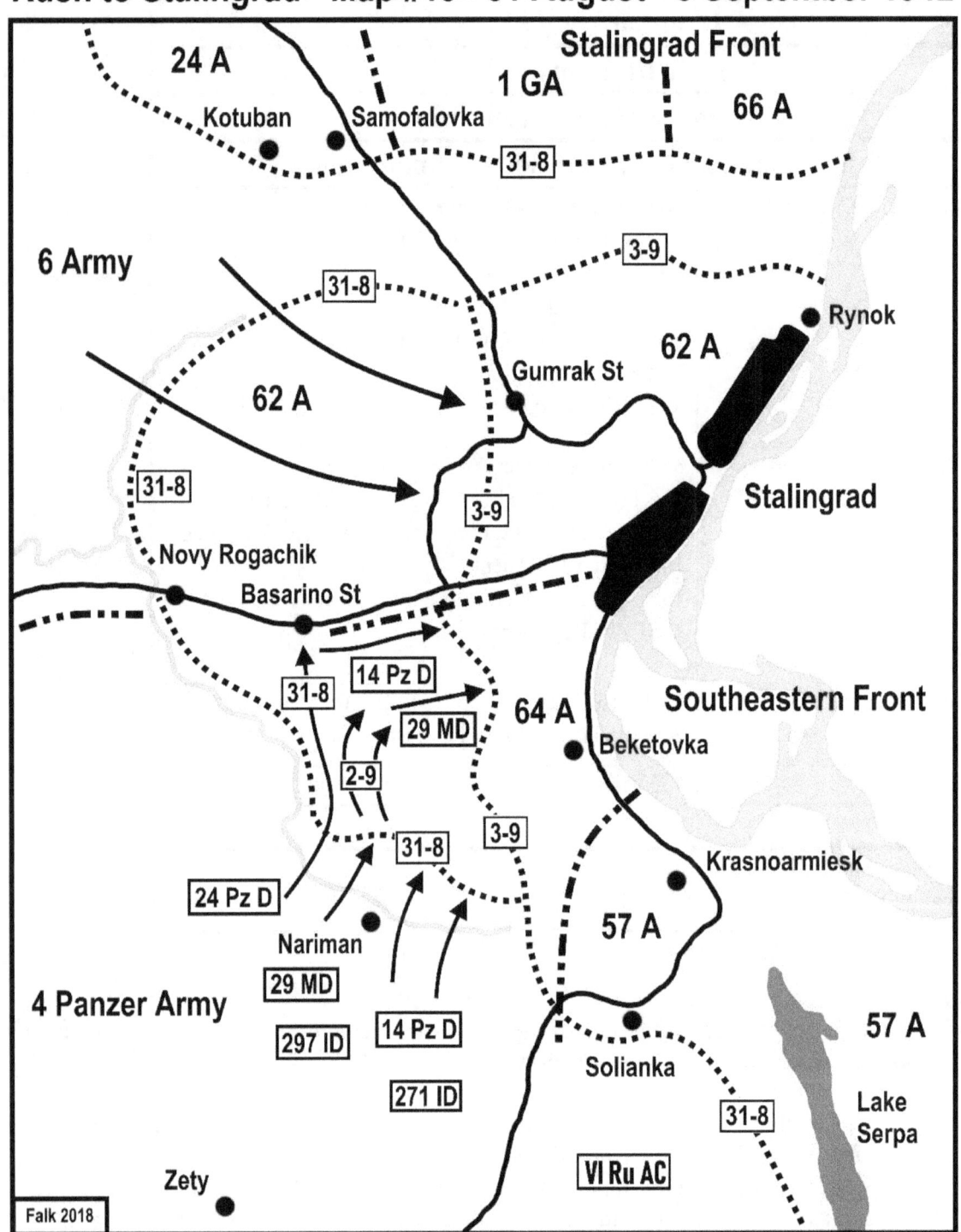

General Chuikov, like normal, moved around the front lines "waiting for the appearance of the units withdrawing to the new defensive positions." He stated "[I]t was clear that our units had not been able to disengage themselves unnoticed from the enemy; the rumble of bombs and shell explosions confirmed our fears."[186] Meanwhile, Shumilov deployed other units of the 64th Army to defend Stalingrad's inner defensive line, using 2nd echelon positions of the "K" line as well as positions of the "S" line, on the outskirts of the city. (see Map #12) General Eremenko supported these efforts through the use of Southeastern Front reserves by distributing tank and AT artillery units among these very same positions. Also during this time, the Southeastern Front Military Council published Order #4 "which required the defenders of Stalingrad to strengthen resistance, not regret this fight with the hateful enemy, not to allow him to the Volga and to defend [the] city."[187] Needless to say, Generals Shumilov and Chuikov had the same goal: prevent the enemy from reaching and capturing Stalingrad, at all costs.

Hampering these efforts were the German units, both Infantry and most notably the 24 Panzer Division spearhead, that had already crossed the Chervlenaia River. By crossing the river, German forces had not only breached the river front section of the "K" line, but their advance also left only a small gap along the rail line, which was still open for Russian units to escape to the east. Scouting toward the north, the 24 Panzer Division reached the rail line near the Basargino Station and destroyed a partially equipped Armored Train.[188] This armored train only had a locomotive with three wagons, so it was not fully equipped. The one picture in the reference book (not displayed) shows a single antiaircraft wagon; it's unknown what the other two wagons were. This armored train almost certainly was assigned to the 62nd Army's 30 Armored Train Battalion.

At this point the 24 Panzer Division was ordered to turn toward Stalingrad. Instead of pushing a short distance further north and linking up with the German 6th Army, thereby cutting off all the Russian forces trapped to the west, the 24 Panzer turned to the east. Higher headquarters decided the gap between the two armies would be closed by units of the 6th Army, which at the time were fully occupied along its northern flank. This failure to linkup would prove to be a mistake. For at least the next 24 hours, by day and night, units of the 62nd and 64th Armies streamed through the tiny remaining gap, escaping encirclement and reaching the outskirts of Stalingrad, but it was a near thing.

After continuing its eastward attack, the 24 Panzer soon was stopped by the antitank guns of the Russian 20 Destroyer Brigade. This unit was positioned on the outskirts of the city by General Eremenko for just such a possibility.[189] The few tank and antitank units placed in the suburbs soon were supported by rifle formations of the 62nd and 64th Armies. The defenders of Stalingrad had gained time to reform.

The combat units that did make it through to the city were fighting formations in name only. They had suffered substantial losses during the retreat. For instance, just before the 29 August attack, the 29 RD had a ration strength of about 8,000 men; at the end of the day on 1 September the division was down to about 1,000 people, with its 77 Artillery Regiment fielding just five guns. Divisional commander Colonel Kolobutin ordered any remaining soldiers from the 128 Rifle Regiment, along with any men from the separate training rifle battalions, to be consolidated into the other two regiments, the 106 and 299.[190] This consolidation would allow these formations to keep fighting in some manner. Other units suffered similar loss rates, and only time and substantial reinforcements, both of men and equipment, would heal the badly wounded 64th Army.

The situation in the air was similar. Despite heavy losses, the VVS deployed in the Stalingrad area 266 aircraft. Of these only 40-50 percent were serviceable at any one time. The Luftwaffe also suffered heavy losses during nonstop operations over the previous few months. Still, they fielded some 950 aircraft, with about 550 of them operational.[191] Clearly the Luftwaffe was still powerful enough to seize control of the air over sections of the front as needed, but the VVS would not go away. The 8th AA and the soon-to-be-activated 16th AA were always able to strike back in the air and against ground targets. When the Luftwaffe was absent, the fighters and bombers of the VVS were very effective. In the coming months, Stavka insured a reinforced VVS continued to pose a growing threat to Axis forces.

2 September

Most of the day saw mixed fighting with moderate German advances. German and Romanian units pushed in toward Stalingrad from all sides and were opposed by harassing artillery fire along with local infantry attacks supported by tanks. Aircraft of the VVS also made an appearance. These air attacks were mostly small strikes aimed at the leading edge of the enemy advance. Losses mounted on both sides. The 24 Panzer Divisions reported at the end of the day 45 men killed with 185 wounded and 1-man missing.[192] Russian losses were much heavier. In just the area of the 64th Army, the Germans reported taking over 2,500 prisoners. The 62nd Army was in a similar position as it too fell back toward Stalingrad.

A great deal of regrouping went on behind the front lines on both sides. On the Russian side, combat capable units were thrust to the front, to stem the German advance, while other disorganized units regrouped to the rear. On the German side, the breakthrough to and over the Chervlenaia River forced German and Romanian combat units and their supporting services to hurry toward the new front line in order to keep the momentum of the attack going.

Meanwhile, further to the south, the Romanian 1, 2 and 4 Infantry Divisions moved into position to protect the 4th Panzer Army's extended right flank. Soon these units of the Romanian 4th Army, commanded by General Constantin Constantinescu, and its subordinated VI Corps were reinforced by additional units of the Romanian Army and Air Force. When they completed this realignment, the Romanian 4th Army held and screened an impossibly long front line stretching 170-190 miles south of Stalingrad.

All around, this was a day of consolidation and reviewing future options. But two critical pieces of information came out of the 24 Panzer Divisions area. "The result of the day was that the connection to 6th Armee [sic] was not established due to the non-appearance of troops from the north ... and it seemed the enemy had not yet completely occupied the forward positions west of Stalingrad."[193] This information compelled German formations to advance toward Stalingrad regardless of losses, fatigue and supply considerations. The prize was too close not to seize the moment.

3 September

During the early morning, units from the 6th Army and the 4th Panzer Army finally linked up just outside Stalingrad; from the north came the 71 Infantry Division and from the south the 20 Romanian Infantry Division. Later in the day, the 24 Panzer Division also moved up and established contact with the 71 Infantry Division. At last the gap between the two attacking German Armies was closed; the battle for the city proper would begin soon.

During the day, the Germans launched a general attack against the entire front held by the 62nd and 64th Armies. Tanks, heavy artillery bombardment, and supported by substantial Luftwaffe activity, backed up German infantry assaults. In a somewhat strange turn of events, General Eremenko ordered General Shumilov to personally lead a counterattack against German forces that had crossed the Chervlenaia River and were advancing on the 64th Army's left flank. This unsuccessful front-line effort by Shumilov wasted several valuable hours of the Army's commander's time, and it also put his life at risk. In turn, General Chuikov was stuck at an advanced command post under heavy Luftwaffe bombardment, barely escaping the destruction of the command post and subsequent personnel attack by a German bomber. This was his fifth close encounter with death. At the end of the day he related a grim picture. "The Germans had broken through the 64th Army defenses ...Things were going no better on the 62nd Army sector"[194] where the Germans had also broken through. Chuikov stayed at this advanced command post until nightfall when General Shumilov summoned the command group to a new location three miles west of the city of Beketovka, directly south of Stalingrad on the Volga. To sum up the situation, Chuikov stated, "The 62nd and 64th Army units made a bitterly-fought retreat to the last positions toward Stalingrad."[195]

Making slow but steady progress, the Germans had a strategy for the next day that could best be understood by the operational order sent to the 24 Panzer Division: "Push into the city." With this order, the crisis facing the defenders of Stalingrad quickly intensified. Impending danger, it would appear, approached Stalingrad from all sides.

Also on this day, strong evidence suggested the last five surviving cadet regiments, out of the nine that had been assigned to the 64th Army, were consolidated into a single composite cadet regiment. These regiments started with between 2,000 and 2,500 men each, but due to extremely heavy losses over the previous two months, they finally were reorganized. It appears this newly combined regiment was named the Krasnodar Infantry School Regiment from this date onward.[196] The five consolidated cadet regiments were Grozny Infantry School, Zhytomyr Military Infantry School, Ordzhonikidze (1st) Red Banner Infantry School, Ordzhonikidze (3rd) Infantry School, and the Krasnodar Machine-Gun and Mortar School. The names of the other four cadet regiments are unknown as well as their movements and actions. This newly formed composite regiment's strength was slightly over 1,000 men, which confirmed an extremely high casualty rate in the five contributing regiments. General Shumilov stated, "Of course, it was difficult for the cadets to fight. They were not particularly well armed, in fact we threw them into the most dangerous axis and they withstood it well. The people were fearless!"[197]

4 September

The morning attack toward Stalingrad did not at first go as planned. In the sector of the 24 Panzer Division, supporting JU-87 Stuka dive bombers struck positions held by German troops by accident. Russian infantry deliberately were firing the same color of signal flares that were used to mark the front line. The confused pilots bombed anyway, causing losses among the German forward troops. Throughout the day only moderate progress was made mostly due to heavy Russian flanking fire and unfavorable terrain. Most of the attacking German divisions held back, with only the 24 Panzer reaching the edge of Stalingrad, the day's objective. The 24 Panzer Division was the first unit to reach the city. Later on, even they were forced to withdraw to better positions for the night.[198] In the evening, most of the other German formations tried to move up to the same line in preparation for the next day's action.

On this date, the 64th Army most likely had in the front line, (from north to south) 204 RD, 157 RD, 29 RD, 38 RD, 133 TB, and 36 GRD. In the second line, were the 66 NRB, Krasnodar Cadet Regiment, 126 RD, 154 NRB, 138 RD, and the remainder of the 13 TC. The 77 and 118 FR and two more unnamed cadet regiments were most likely deployed by the Southeastern Front, somewhere in the rear area. Remaining artillery, tank, antitank and engineer units were in support as needed. The 64th Army formed a weak but functional defensive line, and it held. But the Russians badly needed reinforcements to keep the Germans from entering the city. The estimated combat strength of the 64th Army at this time was around 20,000 men. (see Map #19)

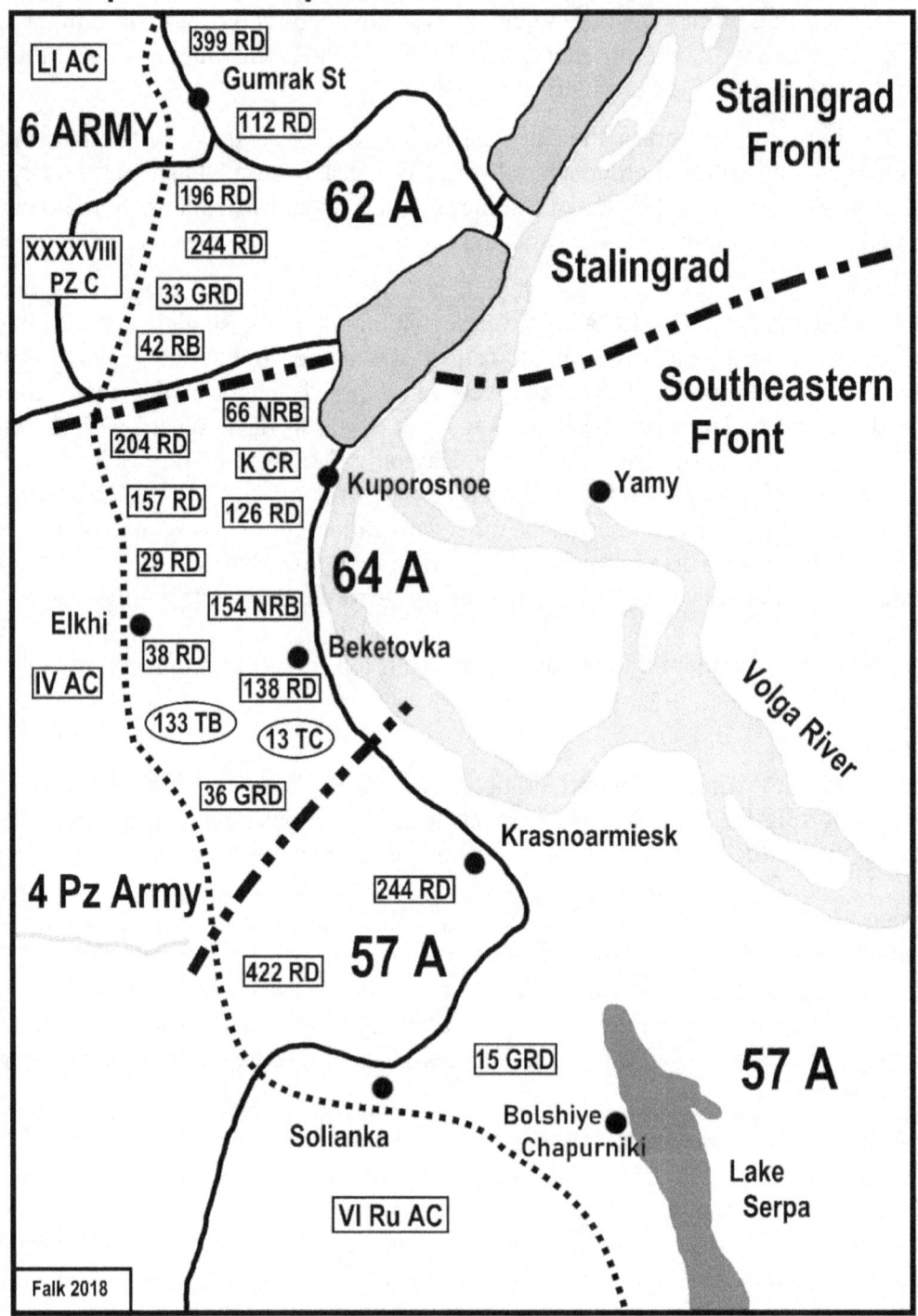

Map #19 4 September 1942 - Overall Situation

Adding to the confusion, Southeastern Front chief of staff General Zakharov issued a strange and desperate order to General Shumilov to rush a tank company with submachine gunners to the north of Stalingrad to assist in the relief of some surrounded troops.[199] This company of tanks and troops were most likely never dispatched, but if they were, they would have needed to drive right through the entire 62nd Army and the full length of Stalingrad, some 30 miles, to reach their objective!

Also, on this day, the 16th AA finally was activated. So the Stalingrad Front at last had its own air army to call upon for support, along with the 8th AA for the Southeastern Front. At this time the 16th AA only had 152 serviceable aircraft, but if we include the entire Stalingrad area, VVS numbers had reached 738 aircraft.[200] With the new command structure in place, and with increased reinforcements, VVS airpower was on the rise.

5 September

In and near Stalingrad, events continued to unfold as German troops tried to push closer to the city, and Russian units attempted to slow them down.

In the northern section of the front, units of the 64th Army delayed the German advance by the use of flanking fire and strong barrages of artillery, Katyusha rockets, mortars, and even an occasional air raid. In turn, supporting German formations took most of the day to move up to the same line as the 24 Panzer Division. Meanwhile, Generals Shumilov and Chuikov strengthened the front line as best they could.

In the center, the 29 RD and the 157 NRB were forced back from the village of Elkhi by an attack of the German 94 Infantry Division. This was right in the center of the 64th Army's front, so the command staff arranged a counterattack to restore the situation. The 10 RB, which had recently been added to the rolls of the 64th Army, now got some action. The 10 RB, along with 66 NRB, part of the 157 RD, Krasnodar Cadet Regiment and the 56, 13 and 133 TB launched a counterattack against the German 94 Infantry Division in the area of Elkhi. Not much was achieved, but the front was stabilized again. The Germans held on to Elkhi and continued to do so for the next three months. In the end, the Germans turned Elkhi into a virtual fortress, repelling all further Russian attacks upon it.

Having recently occupied new positions on the right flank of the 4th Panzer Army, the Romanian 1 and 2 Infantry Divisions were strong enough to launch an attack of their own against the 15 GRD of the 57th Army. This two-pronged attack was directed toward the village of Bolshiye Chapurniki. Most likely the objective of this limited attack was to straighten the front line to improve defensive positions. The Romanians pushed back the 15 GRD and gained some ground but was stopped just short of Bolshiye Chapurniki.[201]

6 September

On the north flank of the 64th Army, the 204 RD held out against attacks by the Romanian 20 Infantry Division, and in turn the 204 RD launched a few attacks of their own. The German mobile units in this area moved away further to the north.

In the center, the German 94 Infantry Division continued to attack near Elkhi. This attack was resisted by most of the same units from 5 September, 29 RD, 157 NRB, 10 RB, Krasnodar CR and parts of the 13 TC. Together they blocked all attempts to force a breakthrough in this area. The German attack was aimed at the city of Beketovka but could not break through the firm Russian defense. Beketovka would soon become very important to the 64th Army and a constant threat to the German advance and capture of Stalingrad. Further south, the 38 RD and the 36 GRD held their positions against German attacks that were supported by tanks.

To give some idea of the level of action that took place over the previous week, the 24 Panzer Division alone claimed the following prisoners and booty taken between 31 August and 6 September. This was almost entirely against units of the 64th Army. The Germans seized 4,910 prisoners, 10 airplanes, 87 guns (artillery), 27 destroyed guns, 46 tanks, 54 anti-tank guns, 21 antiaircraft guns, 36 mortars, 86 anti-tank rifles, 2 Stalin organs, 503 railway wagons loaded with war material, and 25 locomotives (most of them were rendered useless).[202]

However the casualties suffered by the 24 Panzer over the same time period, 31 August to 6 September, were reported as 161 men killed, 938 wounded, and 6 missing, for a total of 1,105 men.[203] These totals illustrate the substantial losses the 24 Panzer Division suffered at the hands of the defending 64th Army. By this stage of the battle, few fresh units were available in the front line for either side.

In addition to manpower losses, the fighting also consumed huge amounts of munitions. For example, during the month of September alone, German forces in and around Stalingrad used 25 million rounds of rifle and machine-gun ammunition, 500,000 anti-tank rounds, and 750,000 artillery shells.[204] With that amount of firepower being deployed, no army could survive long without constant reinforcements.

7 September

Along the 64th Army's front, combat calmed down somewhat as German mobile forces continued to regroup, resupply and attack northward against units of the 62nd Army. These forces maneuvered to deliver a decisive thrust into Stalingrad at the boundary between the 62nd and 64th Armies. Units of the 64th Army used this break in the action at the front to fortify their positions, rest, resupply, and to keep enemy forces under observation. They also sent out reconnaissance missions and generally maintained a steady barrage of fire against enemy positions. The war went on even during a lull in the fighting.

In turn, the 24 Panzer Division also was active and claimed to have captured 50 deserters and prisoners during the day. It was not uncommon to have defectors from the Red Army cross the line and give themselves up to German units. This was also an issue that Russian commanders took seriously and addressed in their usual severe manner. To prevent (or at least slow down) the flow of Red Army troops to the other side, the High Command directed army commanders to create so called blocking detachments. These blocking detachments were placed "behind unstable, unreliable troops in order to increase their resistance, impeding the emergence of panic among them, to prevent escape from the battlefield fighting with desertion." [205] In essence, blocking detachments had the authority to stop and question anyone found/caught behind the front lines. Once intercepted, these individuals had few options. They could be allowed to proceed with their legitimate activities, or they could be sent back to their units at the front. Less desirable options were detention, questioning, being found guilty of something minor and being sent to a penal unit, or the final option of simply being shot on the spot for various reasons. These army blocking detachments were allegedly staffed with the most experienced troops and officers taken from individual combat units and the army in general. In this way, army commanders had the most reliable combat troops overseeing, backing up, and policing less dependable units. There is some debate if the "most experienced troops" would have been used in this manner.[206] Anyway, each blocking detachment could contain up to 200 heavily armed men who were supported by NKVD (secret police) personnel, which guaranteed ruthless enforcement of Stavka orders behind the front lines.

Blocking detachments came about when Stalin issued his famous Directive #227 or the "Not One Step Back" order on 28 July 1942. Directive #227 in turn triggered Stavka to issuing Directive #170542 on 31 July 1942, laying out the formation of blocking detachments, specifically behind the divisions of the 62nd and 64th Armies.[207] In conjunction with the creation of blocking detachments came the formation of penal battalions and penal companies. Penal units were special combat units that contained troops who were detained by blocking detachments and/or were found to be engaged in some type of criminal activity. Penal battalions were used to hold middle and senior officers. Penal companies contained offending junior officers and ordinary soldiers. Evidence shows that at this time the 64th Army had formed at least the 76 and 83 blocking detachments. The 64th Army also maintained the 1st, 2nd, 3rd, 4th, and 5th separate penal companies, each containing anywhere from 150 to 200 criminals. An unknown number of separate penal battalions containing some 800 men each also were at work within the Army.

For an example of blocking detachment activities, let's look at the area of the 62nd Army during 19-20 September. On these two days, blocking detachments detained 184 men. Of these, 21 were shot while 40 were arrested, with the remainder being sent back to their units.[208] The men who were arrested most likely ended up in penal units, where they would be given an opportunity to redeem themselves by atoning for their crimes against the Motherland with their blood. Penal units were deliberately used for the most dangerous tasks and missions, resulting in enormous losses. But, if these criminals lived long enough, or especially if they were wounded, they might be returned to their units at the front, cleansed of the crimes by the spilling of their blood. Working together, army level blocking detachments and the NKVD would strive to maintain control of the Red Army and to guarantee that penal battalions and companies were at full strength.

These severe measures were needed because at this time tens of thousands ex-Russian soldiers willingly served with Axis forces in and around Stalingrad. These so called *Hiwis* (helpers) were mostly working behind the lines, keeping the roads open, moving supplies, and so on. These men worked for food: they also worked to stay out of POW camps and to stay alive. The *Hiwis* knew if they ever returned to Soviet control, the NKVD would ensure their fate would be grim.

One final notable event occurred late in the evening of 7 September. The Southeastern Front Commander, General Eremenko, asked Stalin to relieve General Lopatin, commander of the 62nd Army, of his post. Lopatin had lately become so despondent about the situation of his army that he began to withdraw units from Stalingrad. There was a question of his ability and resolve to hold the city to the last man.[209] Stalin acted quickly by issuing Directive #170603 early the next morning; this directive came right to the point:

DIRECTIVE Stavka number 170603

Commander of the Stalingrad Front to dismiss from his post as commander 62nd Army

September 8, 1942

04 h 40 min

The decision of the Military Council of the Stalingrad Front to dismiss Lopatin from his post as commander of the 62nd Army

Supreme Command approved.

But Stalin did not act quickly in appointing another commander in place of Lopatin. This decision would take a bit more time and consideration. Meanwhile, General N.I. Krylov, the 62nd Army's current second-in command, took command of the Army in the interim. Also, another small staff change, Colonel General I. Laskin had been transferred from the 62nd Army on 3 September and he became chief of staff of the 64th Army on 7 September.

8 September

Increased pressure on the left flank of the 62nd Army forced the transfer of new 10 RB, and other small units, from the 64th to the 62nd Army. Meanwhile, the 29 Motorized and 14 Panzer Divisions struck hard at the right flank units of the 64th Army. The 204, 126, 138 RD and the supporting 133 TB were pushed back. This attack almost destroyed the 343 Regiment of the 138 RD, and later on in the day forced General Shumilov to withdraw the badly weakened 126 RD to reserve positions.[210] The rest of the Army's front remained fairly quiet. This powerful German attack along the boundary line between the 62nd and 64th Armies signaled another emerging crisis; the Germans nearly reached the Volga on this day. Two weeks previously, German forces had reached the Volga north of Stalingrad, severing outside supply and communication links to the 62nd Army in that area. Failure here in the south would allow the Germans to totally isolate the 62nd Army within Stalingrad by cutting its last land link to outside logistical support. This was something the 64th Army had to prevent at all cost.

9 September

The daily report on combat activities summed up the action nicely. "The right flank of the 64th Army conducted bitter defensive fighting against infantry and tanks of the enemy. The 204 and 138 RD were forced back as enemy tanks were wedged in our defenses."[211] The infantry and tanks came from the 29 Motorized and 14 Panzer Divisions which continued their advance from the previous day. Their goal was to split the front in two and to reach the Volga. Along the way, the 157 RD also was forced to withdraw as Shumilov's forces tried to prevent a breakthrough to the river. Only strong artillery and mortar fire placed upon the enemy slowed their advance. Some of this supporting artillery fire came from the east side of the Volga. But, soon the deteriorating strength of the Army, especially in infantry, would decide the issue.

Late in the evening, Stalin finally acted upon a proposal from General Eremenko about a new commander for the 62nd Army. Directive #994200 issued at 23:10 hours, named General Chuikov as commander of the 62nd Army with General Krylov as his Chief of Staff.[212][213] The account of this change of command is an interesting one, and relevant to the 64th Army because with this transfer, General Shumilov lost his very experienced second in command.

On 11 September, Chuikov stated that he was summoned to the Front HQ at Yamy, on the left bank of the Volga. After saying goodbye to staff officers at 64th Army HQ, he set out on his journey. After crossing the Volga on a ferry that night, he finally reached the remains of the village of Yamy at midnight. Searching the destroyed village, he finally located the Southeastern Front HQ, but he was told to come back the following day, as it was very late. The next day, 12 September, Chuikov reported to the Front HQ at 10:00 hours where he quickly met with General Eremenko and Nikita Khrushchev. After a brief conversation, and with little elaboration, they told him that he had been appointed commander of the 62nd Army. Khrushchev then asked Chuikov, "How do you interpret your task?" Chuikov replied, "We cannot surrender the city to the enemy…. We will defend the city or die in the attempt." Both Eremenko and Khrushchev then declared that Chuikov understood his task correctly.[214]

General Chuikov would command the 62nd Army for the rest of the war, and it was eventually renamed as the 8th Guards Army in honor of its heroic defense of Stalingrad. Stalin and General Eremenko had finally found a worthy commander for the 62nd Army. Chuikov would ruthlessly feed into battle every man, every weapon, and every round of ammunition that he could lay his hands on. In essence, he had been given the task to sacrifice the entire 62nd Army if need be, to hold onto Stalingrad. In this, Chuikov was not going to disappoint his superiors. Note: It appears that Colonel

Ivan Andreevich Laskin, the Chief of Staff of the 64th Army, took over Chuikov's place as second in command of the 64th Army.

10 September

In the morning, the German 29 Motorized Division broke through to the Volga near the southern Stalingrad suburb of Kuporosnoe.[215] But with only one battalion of infantry actually reaching the river, the question was, could they hold it? With German troops on the river bank in the southern part of Stalingrad, the 62nd Army was isolated within the city itself. Land contact with the 64th Army had been severed, despite all attempts to prevent this from happening.[216] Because of this, the 62nd Army only had the Volga to provide a means of supply, reinforcement, or escape. The current layout of the 64th Army prevented any swift reinforcement of this area, and Shumilov quickly organized only a minor counterattack against these German units on the Volga.

Later on that night, acting against this incursion, the 131 RD and 35 GRD of the 62nd Army launched furious attacks from the north against the German battalion on the river. From the south came assaults by the 126 RD from the 64th Army.[217] Eventually, the Germans were overrun and the battalion was forced to withdraw. For a moment it looked like a path had been opened to Stalingrad from the south, but that was not to be. The next day, the German XXXXVIII Panzer Corps was ordered to start into the southern quarter of Stalingrad and take it "piece by piece"[218] and that is what they intended to do! (see Map #20)

The weakness of his army prevented Shumilov from interfering effectively with these German plans. By 10 September, the composition and ration strength of the 64th Army had improved somewhat from the first of the month, but the 64th Army lacked any real offensive power. The minor increases in strength came from the return of any stragglers to the Army, any sick and wounded discharged from the hospital, and the few replacements received since the first of the month. Also included were any new units assigned to the Army. In all, the ration strength totals were not that impressive.

Rifle Unit	Men
36 GRD	7,149
29 RD	1,856
38 RD	3,435
126 RD	2,036
138 RD	2,123
157 RD	1,996
204 RD	2,500*
208 RD	200 (remnants)*
66 NRB	1,334
154 NRB	876
77 FR	1,400*
118 FR	1,430
Krasnodar Composite Cadet Regiment	1,043
Remnants from four other Cadet Regiments	1,500 (estimate)

(*see Appendix 13)

Total for rifle units = 28,878 men[219, 220]

<u>Supporting Combat Arms</u> (estimates)
Artillery units   5,000 men
Armored units   2,500 – with about 44 tanks
                (some not operational)
Engineer units   400

Total for supporting combat units = 7,900 men.

Army HQ staff, Army level Supply, Medical and other support units would add approximately 13,000 people. This would bring the combined Army ration strength, on this date, to approximately 49,778 men and women. The Army's actual combat strength, specifically men within combat positions, was much lower than this value.

With his battered army, the best Shumilov could do was to prevent the Germans from expanding their positions along the Volga to the south. If there were to be any hope in relieving the 62nd Army surrounded within Stalingrad, the 64th Army had to hold onto every inch of the front line along the river, staying as close to Stalingrad as possible.

Map #20 4-12 September 1942

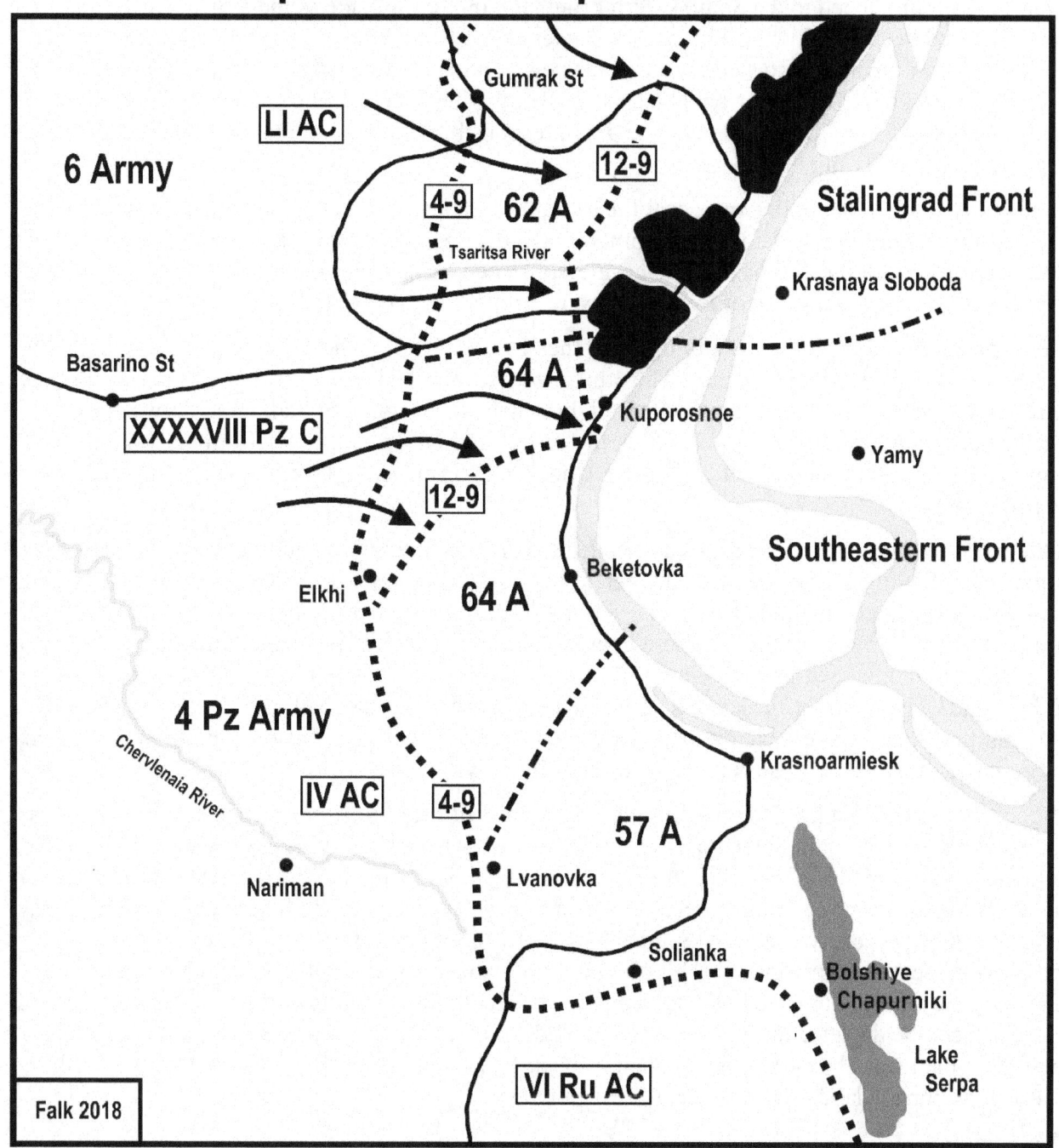

11 September

While that struggle unfolded, the personal story of General Chuikov continued. Lieutenant-General Chuikov was summoned to the Southeastern Front HQ, on the east side of the Volga, where he learned he had been appointed the new Commander of the 62nd Army,[221] but it would take another two days or so for Chuikov to actually reach his new command. Meanwhile, German forces from both the 6th Army and the 4th Panzer Army attacked all along the front into the outskirts of Stalingrad proper. Chuikov had willingly taken on the grave responsibility to defend Stalingrad to the last, and he was going to carry out his orders.

On the right flank of the 64th Army, near the river, the 126 and 138 RD along with support from the fresh 56 TB and the Krasnodar Cadet Regiment, held the line against the 14 Panzer Division's push to the south; this attack did not make much progress.[222] The Germans discovered the growing Russian defenses in and around the city of Beketovka were stronger than expected.

The Germans called the Beketovka bridgehead the "Beketovka Bell" because of its general half bell-shaped outline when drawn on a map. The 4th Panzer Army was directed to prepare plans, under the code name Autumn Journey, to crush this Bell, but attacks into southern Stalingrad drew off the required forces. Later on, the German 6th Army transferred major units away from the 4th Panzer Army to assist with the capture of Stalingrad. Once again, insufficient forces remained to carry out Autumn Journey.

In the end, the German High Command intended to carry out the following sequence of events. First, fully occupy Stalingrad. Second, launch operation Autumn Journey to clear the Beketovka Bell of the enemy. Third, launch operation Heron, the capture of the city of Astrakhan some 300 miles away at the mouth of the Volga. Ultimately, these operations were repeatedly postponed because the required forces were tied down within Stalingrad. The Beketovka Bell and the 64th Army would remain a threat to the Germans right flank south of Stalingrad for the rest of the battle.[223] The original code name for the attack on the Beketovka Bell was Autumn Leaf, which was also the code name for active gas defense, so the name was changed to Autumn Journey.

Aligned along the front line, starting from the right flank of the 64th Army, which was anchored on the Volga river, then south to the boundary with the 57th Army, at the village of Ivanovka, were the following units: 126, 138, 204, 157, 29 RD, 66 NRB, 154 NRB and the 36 GRD. All other units were in supporting positions or refitting behind the front lines. At this time, the frontage of the 64th Army defensive zone was approximately 18 miles long and 9 miles deep at the most. The Germans had compressed the 64th Army into a relatively small bridgehead position with not much room to maneuver. So troops that were in the rear area were actually just occupying the next fortified defense line. Even so, the depleted units of the Army would find it difficult at times to hold onto even this small but vital area. The 64th Army too had its back to the Volga, but the 57th Army to the south guarded the narrow land route to the rear.

12 September

The Germans continued their enveloping attacks against Stalingrad's suburbs. By the evening, they had captured most of the high ground outside the main city center and industrial areas but had not yet reached the Volga in the south. In support, Luftwaffe kept up constant attacks mostly against the Stalingrad area, flying some 938 sorties per day. The VVS opposed these raids as best they could but were only able to mount about 354 sorties per day. The air war was still weighted in favor of the Luftwaffe, but VVS reinforcements were on the way.

The position of the 64th Army was mostly unchanged. German and Romanian units conducted probing attacks all day long. Formations of the 14 Panzer and the 20 Romanian ID failed to make much headway against the 126 RD and the 138 RD, gaining only small amounts of ground near Kuporosnoe, just short of the river. [224] The 64th Army held fast to its positions with very weak forces and dug in. Shumilov was not going to offer the Germans any weak spots to exploit, if he could help it. In his atlas *The Struggle for Stalingrad City*, Vol 1, on page 30, David Glantz stated that on this date the 13 TC contained 23 tanks with only 11 of them operable, 4 T-34, 6 T-70, and 1 T-60. With so few tanks at his disposal, General Shumilov could not undertake any major offensive actions in the coming days.

On a related topic, while meeting with Stalin in Moscow between 12-13 September, Stavka representative and Deputy Supreme Commander General Zhukov and the Red Army Chief of the General Staff, General Vasilevsky, concluded that direct attacks against German positions near Stalingrad were not going to decisively reverse the situation there. Together, with Stalin's consent, they started to plan for a larger solution to the Stalingrad problem. This would become the blueprint to encircle the entire German grouping in and near Stalingrad. Eventually this plan was called Operation Uranus. But it would take months to implement this idea, and the troops defending Stalingrad needed to hold out in the meantime.

Before the fighting actually reached the city proper, and to understand the battle more fully, one must understand the geographic layout of Stalingrad, its surroundings, and a bit of history. The city's original name was Tsaritsyn (yellow water), and it was founded in 1589 as a military outpost on the lower Volga by Tsar Theodore Feodor I, the last Rurikid Tsar of Russia. After the Bolshevik revolution and the following civil war between the Reds and Whites during 1917-1920, the Soviet Union was formed. Eventually Joseph Stalin seized control of the Communist Party and he also took credit for the successful fighting that occurred in and around Tsaritsyn during the civil war. Later on in 1925, Stalin renamed Tsaritsyn to Stalingrad, to honor himself for his supposed great victory there in 1918. With Stalin's support, Stalingrad was further developed as a major industrial city and a center of armaments production. By this time, Stalingrad had assumed its modern form, a long narrow city stretching north and south for some 25 miles along the 200-foot-high western riverbank overlooking the Volga. In contrast, the built-up part of the city was only about 3 miles wide, with further suburbs of wooden buildings reaching out onto the steppes toward the Don River some 45 miles away.

Stalingrad itself was roughly divided into north and south sections by the deep Tsaritsa River gorge. To the north of this gorge lay the great industrial area with the Tractor, Barricades, and the Red October factories. Also located in this area was the Lazur Chemical plant and a dominating hill called Mamayev Kurgan, which was actually an ancient burial mound. South of the Tsaritsa gorge laid the city center with its political and commercial buildings, grain elevators, and several railway stations and river landing platforms. The city was further cut into even smaller areas by a number of deep, dry gullies called *Balkas* reaching down to the riverbank. In all, Stalingrad contained many different housing areas, city services, and industrial sectors, all of which became one huge battleground. Also of note, when the Germans finally reached the city, the Tsaritsa River gorge was chosen by the German command staff as the dividing line between the 6th Army in the north and the 4th Panzer Army in the south.

At Stalingrad the Volga River was nearly at sea level and made an abrupt 90° turn to the southeast flowing to the Caspian Sea some 310 miles away. The Caspian Sea itself is about 90 feet below sea level. Directly south of Stalingrad, a series of marshes and dry lakes mark what once was the channel of the Volga. The east bank of the Volga consisted of flat marshy lowlands forming a vast flood plain some 10-20 miles wide with many islands and smaller channels. In places the main river channel was over a mile wide. Needless to say, during the summer this lowland area was not easily crossed, but the winter freeze allowed for easier travel. [225]

13 September

At 06:30 hours, the Germans launched a major attack toward the city center and the southern part of Stalingrad. Stubborn resistance by Russian troops limited German gains. Heavy bombing and artillery fire from both sides pounded the city. Having reached his new command, Chuikov was able to speak with the front commander only once during the day due to the many broken telephone lines.[226] That night, under heavy fire, General Chuikov and his command group were forced to move the headquarters from the area around Mamayev Kurgan hill south to a larger bunker in the Tsaritsyn River gully.

In the Beketovka bridgehead, the 64th Army was subjected to continuous probing attacks. But late in the day, General Eremenko ordered both Chuikov and Shumilov to mount counterattacks against German penetrations in the city.[227] By the time Chuikov made preparations for these counterattacks, it was early the next morning, and they failed to achieve anything. It's unknown if the 64th Army was even able to launch supporting attacks from the south. If Shumilov was able to attack at all, it would have been somewhere near the river using the 126 and 138 RDs with supporting units. By nightfall German troops finally had reached the Volga just south of Kuporosnoe. This German foothold on the river was very small, but it steadily grew over the next few days forcing the 62nd and 64th Armies apart, and Chuikov or Shumilov could do little about it. The 62nd Army was effectively isolated within Stalingrad.

14 September

Recognizing the need to reinforce front line armies, on or about 14 September, the Front commander reassigned the 422 RD to the 64th Army from the 57th Army.[228] The 57th Army at this time was under little pressure, so this transfer made sense. In the middle of the front line, the 29 RD and 66 NRB defended against a regimental-sized attack by the 20 Romanian ID. This was another probing attack against the 64th Army, as German and Romanian forces continued to search for a weak spot in the defenses guarding Beketovka.

Within the Stalingrad area, concentrated German attacks pressed forward, despite intense resistance from the dwindling number of defenders. Later that night, Chuikov and his 62nd Army too received reinforcements in the form of the 13 GRD, some 10,000 men strong. This rifle division had to be laboriously ferried across the Volga over the period of two days and only at night, because the German artillery and Luftwaffe attacks made any daylight crossing suicidal.

15 September

The German directed minor scouting activity against the front of the 64th Army. After resting and refitting, the 118 FR and the 38 RD took up positions directly adjacent to Beketovka, covering the north and western approaches to the city.

Within Stalingrad, an early morning assault struck the southern part of the city. Breaking through tough Russian defenses, with the help of major Luftwaffe support, German forces made significant headway, reaching Railroad Station #2.

16 September

On the left flank of the 64th Army the 422 RD finally took its place alongside the 36 GRD. These two units covered the vulnerable junction between the 64th and 57th Armies.

The 64th Army continued to face substantial enemy forces arrayed along its front. From north to south, the 29 Motorized Division, 14 Panzer, 20 Romanian, 297 Infantry and parts of the 377 Infantry Division manned the line. These formations certainly took advantage of the low level of activity along this section of the front to rest and to refit. But combat continued with daily probing attacks and scouting missions launched by both sides. Artillery and mortar fire was a constant reminder to troops to keep their heads down and to dig in even more.

Much harsher conditions existed for the surrounded 62nd Army as German forces attempted to crush the defenders while making further gains in the built-up sections of the city. Relentless pounding by artillery of all types, from both sides, along with nonstop bombing raids made for appalling losses. Most of the defending artillery supporting the 62nd Army was located on the far side of the Volga. There was simply too little space for artillery to operate within Stalingrad itself, and it was much easier to supply them with ammunition on the far side of the river. Further reinforcements provided to Chuikov during the night were minimal, one naval rifle brigade and a tank brigade. These units were ferried over the river that night and quickly were put into the front lines.

17 September

Little activity was reported along the 64th Army's front.

Further north, fighting raged in the city as positions changed hands multiple times during the day. With his headquarters under constant fire, Chuikov again was forced to change its location. This time he moved the HQ north to the factory district near the riverbank. Because German troops blocked the land route to this location, the HQ staff used boats to make the move.[229]

18 September

All was quiet along the 64th, 57th and 51st Armies' front. The Germans were occupied with Stalingrad and focused their attacks against the northern end of the city.

Meanwhile, the commander of the Stalingrad and Southeastern Fronts, General Eremenko, along with Stavka representative General Zhukov, planned a coordinated attack against German forces around Stalingrad. From the north, the Stalingrad Front would attack with three armies, the 66th, 24th, and the 1st Guard Army. The Southeastern Front would launch supporting attacks with the 62nd Army from within Stalingrad and the 64th Army from the south. The plan was ambitious and far-reaching.[230] The 62nd and 64th Armies were much too weak to have a large effect. The Stalingrad Front forces were far stronger, but they also had attacked in the same general area several times before over the previous three weeks. The defenses on the Germans northern flank were very resilient and built in depth. Still, the attack from the north began in the morning and suffered heavy losses.

19 September

The northern assaults continued, and the 66th, 24th, and the 1st Guard Armies gained little ground. By day's end, the Russians called off the offensive. During the period of 1-20 September, these same Stalingrad Front armies had launched several different offensives against the German northern flank. Over this time they lost the following number of troops:

The 1st Guard Army suffered about 36,000 casualties, the 24th Army about 32,000, and the 66th Army about 20,000 for a total of some 88,725 men killed, wounded or missing.[231] This was a high price to pay for so little gain. This recent attack only temporarily diverted German ground and air strength away from Stalingrad.

The 62nd and 64th Armies were able to contribute little or nothing to support these efforts. As ordered, starting at 12:00 hours, the 62nd and 64th Armies went over to the offensive. Despite making an effort with limited forces, the 62nd Army's attack accomplished little. General Shumilov in turn used the fairly fresh 422 RD (minus one regiment), and the 36 GRD. Both shifted from the far-left flank of the Army's front to make the assault. The attack headed due north toward Stalingrad, in the area just to the west of Kuporosnoe, on the Volga. Again, little or nothing was gained.

20 September
Early in the morning, the 422 RD and the 36 GRD continued their supporting attack to the north. Meeting strong resistance from the German 14 Panzer Division, they made little progress.

Within Stalingrad, German assaults continued into the city undisturbed by the minor efforts of the 64th Army. Units of the 62nd Army, bitterly fighting in the south, were forced back to the northeast and closer to the Volga. The Germans were gaining more ground along the river-front, which aided them in their efforts to bring effective fire against Chuikov's lifeline over the Volga. Within the city, over the next several days and weeks, one attack followed another, with counter attacks and bombardments blending into one huge mass. The exact ebb and flow of this fighting is beyond the scope of this account and will only be related if details are relevant to the 64th Army's position.

21 September
The German 14 Panzer Division launched a counterattack against the 64th Army to regain what little ground it had lost over the previous two days. Shumilov's troops were forced back to their original positions in the Kuporosnoe region. The 14 Panzer was content just to reclaim its original positions.

22 September
Protracted fighting continued over minor defensive positions on the right flank of the 64th Army. Scattered artillery and mortar barrages prevented much real fighting. Moving about above ground level during the day was extremely dangerous.

23 September
The Stalingrad Front resumed attacks against German positions in the north with the 1st GA and the 24th Army. This attack was directed along a new axis further to the west from previous attacks. Once again, this effort achieved little success with heavy losses both in men and in equipment.

Fighting continued within Stalingrad unabated, while the 64th Army's front settled down again with little action.

24 September
The 1st Guards and 24th Armies resumed their attacks from the previous day, with the 1 Guards making a small breakthrough against the German 113 Infantry Division. On the left flank, the 66th Army launched its own attack and made minor progress. These attacks continued over the next few days, finally ending in early October after the Germans launched heavy counterattacks backed up with ample air support.[232] For a short time, some of the pressure was taken off the defending 62nd Army within Stalingrad.

Units of the 64th Army defended against small reconnaissance efforts of the enemy.

Relentless fighting continued within Stalingrad.

25 September

On orders of the front commander, the remaining four original cadet regiments assigned to the 64th Army were consolidated to form yet another composite regiment. The strength of this new regiment was 1,016 men and was commanded by Colonel S. P. Zatul.[233] The four consolidated cadet regiments were Vinnitsa, Zhitomir, Grozny and Ordzhonikidze. This was the second consolidation, and it appears to have included the previously consolidated Zhytomyr regiment (with a slightly different spelling). So, by this date, out of the nine cadet regiments that had been assigned to the 64th Army over the summer months, only two composite regiments remained, each approximately 1,000 men strong. Note: The names of these two composite regiments are unknown.

In the center of the 64th Army's front line, two rifle companies from the 38 RD attempted to reach hill 128.2 but were forced back by heavy German fire.

26 September

Units on the right flank of the 64th Army launched a small probing attack toward hill 145.2. They were met by heavy fire and a German counterattack.

27 September

General Chuikov launched a spoiling attack in the north against German forces in the factory district. In the south, General Shumilov was ordered to launch a supporting attack with his 36 GRD to capture the village of Kuporosnoe. This attack drew off some German strength but was unable to capture Kuporosnoe which remained in German hands.[234]

28 September

Shumilov continued his unsuccessful assault with the 36 GRD.

To improve the command and control situation around Stalingrad, the two front commands were reorganized. General Rokossovsky, the commander of the Stalingrad Front, had his front renamed to the Don Front. General Eremenko retained control of the Southeastern Front, which was renamed the Stalingrad Front.[235] This change of front names was ordered by Stavka Directive #994209. So, the new Stalingrad Front actually now covered the entire city of Stalingrad. Changes like these appear minor, but it helps to clarify the situation for everyone at all command levels. So, the 62nd and 64th Armies were finally both fighting under the same front command. Despite this change, fighting continued in and around Stalingrad more or less as before. (see Map #21)

For several weeks, General Shumilov constantly shifted his forces in and out of the front line, thus making it extremely difficult to determine the exact location of the various units and formations as they moved about. Exhausted formations were shifted to the rear area (a few miles back) to allow some rest, and to receive replacements and resupply of equipment. Hindering these refitting efforts were constant orders by the Front HQ to attack German positions in an effort to support the 62nd Army in Stalingrad. Shumilov, responding to these attack orders, sent these somewhat restored units back to the front line once again and committed them to make desperate attacks against prepared enemy positions. Not surprisingly, under these conditions, the Army's combat strength wasted away.

The only reason the 64th Army and other armies could keep fighting like this was the growing logistical support provided by field supply depots on the east side of the Volga. As can be seen on Map #21, an entire series of bridges and ferries were put into place to bring supplies to fighting Red Army troops on the west side of the river. For the most part, these bridges and ferries allowed sufficient supplies to reach the troops, despite constant attack by the Luftwaffe and German artillery. Bridges were damaged and rebuilt, ferries were sunk and replaced, but the supplies continued to be delivered to where they were needed. The one major exception was the very long 3,917-foot bridge, just south of Kuporosnoe. This bridge had been in operation for only four days before being destroyed. The Red Army logistical organization adjusted to this loss, and the struggle to supply the front lines continued unabated.

29 September

The 57th and 51st Armies launched long-range diversionary attacks against Romanian forces of the VI Corps on the far-left flank of the front (not shown on a map). Some progress was made against the Romanian 1 and 4 Infantry Divisions, but the Russian troops achieved no breakthrough.[236] Most likely in response to these attacks, the 14 Panzer Division was ordered into a reserve position behind the line. The 371 Infantry took over its positions opposite the right flank of the 64th Army. The 29 Motorized Division also withdrew from the front lines into reserve. The German command recognized these mobile units were not the best ones to be fighting from a static position or within an urban area like Stalingrad. Anyway, they both needed a rest and refit after prolonged fighting.

The 64th Army continued to hold its current positions and prepared to launch further attacks to the north. Understandably, during this time, both the 62nd or 64th Armies would endure nearly constant combat operations with little pause for rest or for the opportunity to recover their strength.

30 September

General Eremenko ordered Shumilov to launch yet another diversionary attack northward into prepared German defenses.[237] This attack was intended to be much larger than normal, so Shumilov began careful preparations. He shifted his meager forces around to give maximum support to this new effort. These diversionary attacks, both from the north and south, drew off German forces to their flanks and caused all sorts of problems for the defenders, but there was little hope of a major Russian breakthrough.

The 14 Panzer Division continued to move into a reserve position to the southeast behind the Romanian VI Corps. This movement was undoubtedly in response to attacks by the 57th and 51st Armies against Romanian forces in this area. Army Group B was starting to take notice of its exposed position at Stalingrad.

At the end of the month, the 62nd Army found itself closely surrounded by German units within Stalingrad. Only by crossing the wide Volga River could the 62nd Army gain logistical access to the east for resupply. To the north and south of Stalingrad, the existing front lines became more fixed in place.

Both sides were firmly dug in and both sides lacked the strength to change the situation. Only within Stalingrad itself was German pressure increasing in a desperate attempt to crush the 62nd Army and to capture the city before the coming winter called an end to major combat operations. To the south of the city, a badly weakened 64th Army continued to do all that it could to prevent the fall of Stalin's city and the destruction of its sister army trapped within the ruins.

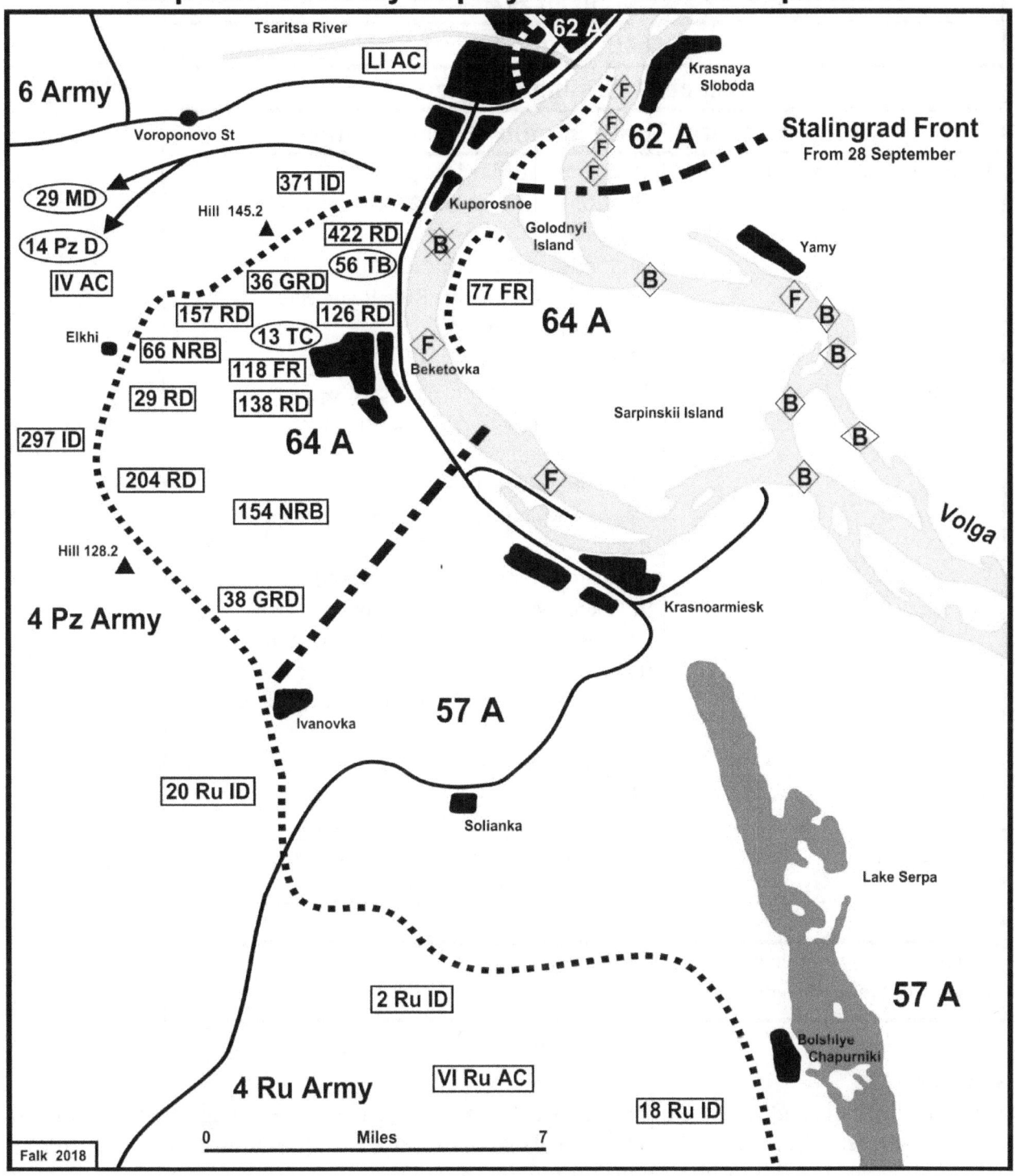
Map #21 64 Army Deployment - End of September

1 October 1942	
64th Army	
Assigned – Stalingrad Front	
Major General M.S. Shumilov Commanding	
Rifle, Airborne and Cavalry Units	36 Guards Rifle Division 29, 126, 138, 157, 204, 422 RDs (minus 1334 Regiment) 66, 154 Naval Rifle Brigades Remnants of two Cadet Rifle Regiments 118 Fortified Region 77 Fortified Region
Artillery Units	1104, 1111, 1159 Gun Artillery Regiments 186, 500, 504, 507, 612, 665 Antitank Artillery Regiments 140 Mortar Regiment 4 Guards Mortar Regiment (rocket) (minus the 111 Battalion) 19 Guards Mortar Regiment (rocket) 91 Guards Mortar Regiment (rocket) 3 Battalion of the 2 Guards Mortar Regiment (rocket) 346 Battalion of the 76 Guards Mortar Regiment (rocket) 1261 AA Regiment
Armored and Mechanized Units	13 Tank Corps (13, 56 Tank Brigades) 28 Separate Armored Train Battalion
Engineer Units	43, 329 Separate Engineer Battalions 175, 1363 Separate Sapper Battalions
Estimated Combat Strength Estimated Ration Strength	39,216 men 51,916 men

BSA part 4 and BSSA part 2

1 October

All around the Stalingrad area, the front lines became stationary, and any resulting combats started to assume a more repetitive nature of frontal assaults with little ground gained or lost. Only the battles within Stalingrad appeared to have significance for each side.

For all practical purposes, Stalingrad already had been destroyed as a center of transportation, armament production, and as a functional supply center for the Soviet war effort. Yet Hitler had become obsessed with the capture of the entire city, despite no real military necessity to capture every building or patch of ground. Hitler must have Stalin's city!

Stalin too was fixated upon his city, but in a different way. Stalin and the Red Army High Command used Stalingrad as bait. Plans for the great counteroffensive were progressing; all that was needed was enough time to accumulate and to deploy the required forces and supplies.

To the south of the city, General Shumilov continued to gather and to organize troops of the 64th Army for the next attack. At the same time, the German 6th Army launched heavy attacks into the factory district of northern Stalingrad.

In the southern section of the front directly facing the 64th Army, the German 20 Motorized and 14 Panzer Division remained in reserve. These two mobile units were pulled from the front line for much needed rest and refitting. So, in the sector covered by the 4th Panzer Army were only three divisions in the front line: the German 371, 297 and the Romanian 20 Infantry Divisions. The removal of these two mobile divisions from combat operations in the southern part of Stalingrad hardly went unnoticed by Russian military intelligence. This reduction in front line strength was viewed as an opportunity to attack, and General Shumilov was ready to test the new German defenses.

2 October

During the early morning, at 04:30, General Shumilov used four rifle divisions to begin his supporting attack northward. The 422 RD and 36 Guards RD attempted to break through the opposing German 371 Infantry Division. Further west, the 157 RD and 138 RD attacked the German 297 Infantry Division. These efforts were an attempt to link up with the closest units of the 62nd Army which were only 4.5 miles away. Any disruption of German operations within the city would have been considered an additional benefit.[238] The 66 NRB, the 29 and the 204 RDs also became active along the front. After fighting the entire day, Shumilov's forces failed to make the expected breakthrough. This was the first serious attack launched by the 64th Army from the Beketovka bridgehead position, and it should have been a warning to the Germans, as more attacks were sure to follow.

Of special note, during this attack, units of the 64th Army undoubtedly continued to receive supporting fire from the two armored trains of the 28 Armored Train Battalion that were still assigned to the Army. But new fire support units also had been added to the Army; they came from the Volga Military Flotilla. The 2nd Brigade of river ships was assigned directly to the 64th Army to provide fire support. The 2nd Brigade consisted of three gunboats: *Kirov*, *Fedoseyenko* and *Tchors*. Each boat was armed with artillery of caliber 100-152 mm. The Brigade also contained four separate smaller armored cutters. "These armored cutters were armed with one or two tank turrets each with a 76-mm gun, also by heavy machine guns, and sometimes mortars."[239] In all 14 of these armored cutters fought at Stalingrad, but only two of them mounted two tank turrets; the remainder only carried one 76-mm gun and assorted machine guns. Despite the added firepower from these gunboats, the attack failed.

Within Stalingrad, an unsettling event occurred to disrupt the defense of the city. During the day German aircraft bombed some empty oil storage tanks directly up slope from Chuikov's HQ near the river. The problem was these storage tanks were full of oil. Chuikov described a burning, flowing mass of oil that gushed across the dug-outs that held his HQ. This fiery mass continued on its way down to the Volga, scorching everything in its path.[240] Despite some initial panic and considerable danger, the HQ group remained among the flames and continued to direct operations of the Army. This was Chuikov's sixth close brush with death; he must have been living a charmed life.

3 October

After the previous day's efforts, the 36 Guards Rifle Division continued its attacks against the German 371 Infantry Division. Other units of the 64th Army held their positions with little action on either side.

4 October

The 64th Army continued to hold its previous positions but undertook a partial regrouping of forces along its left flank. This was almost certainly in response to the recent combat operations. The 66 NRB launched probing attacks against the 297 Infantry Division. On or about this date, the 77 and 118 FRs moved to Sarpinskii and Golodnyi islands in the Volga. This relocation was to ensure the Germans would not be able to cross the Volga. These two units also brought heavy flanking fire with their heavy machineguns and mortars against any enemy forces along the river.

5 October

The 64th Army continued regrouping its forces. Shumilov clearly was getting ready for further action along his front. The 138 RD withdrew from the 64th Army into the Stalingrad Front reserve for rest and refitting. The entire division only had 2,646 men remaining, so a long rest behind the front line was the right choice.

But Stalin was not satisfied with the performance of his armies at Stalingrad or with General Eremenko. Stalin sent Eremenko a message claiming that he didn't "see the dangers which threatens the forces of the Stalingrad Front." Stalin went on to say that, "the enemy intends to take your crossings [over the Volga], encircle the 62nd Army and take it captive, and, thereafter, encircle the southern group of your forces, 64th Army and other armies, and also take them captive."[241] One can only wonder at Eremenko's response to this and other calls to action from Stavka and Stalin. Eremenko knew full well what the fighting within Stalingrad was like, and he continued to send new forces to the 62nd Army to strengthen the defense of the city.

6 October

General Shumilov started to focus his attention against the German 371 Infantry Division which was holding the front line directly south of Stalingrad. In other words, this single enemy infantry division was standing in the way of the 64th Army moving north and linking up with the trapped 62nd Army within Stalingrad itself. But not every day is a big day, so only a simple artillery bombardment was directed against these German positions, with little or no ground action taking place. Yet, even a seemingly minor action like this artillery bombardment still caused losses to the enemy. The 371 Infantry Division reported to higher headquarters that it had suffered 40 combat casualties this day, 4 killed and 36 wounded.[242] Note: In a military sense, a casualty was a combat loss. This was normally anyone who was killed, wounded, or missing due to enemy action. Soldiers who were not fit for duty due to sickness or noncombat injury were usually tallied separately.

7 October

The 64th Army held its previous positions with only minor harassing action taken against positions held by the German 371 Division. The 371 Division reported 11 casualties for the day, 2 killed and 9 wounded.

The 8th AA only had 188 operational aircraft to support the Stalingrad Front's efforts, so most of these limited resources were focused upon supporting the 62nd Army within Stalingrad.[243]

8 October

The 64th Army held its previous positions with normal levels of activity along its front. The opposing German 371 Infantry Division reported 35 casualties, 4 killed and 31 wounded.

General Eremenko and Khrushchev submitted report #2889 to Stavka. This was a more detailed plan of the Stalingrad Front's part of Operation Uranus, the winter attack against the German and Axis forces in and around Stalingrad.[244]

9 October

People's Commissar of Defense (NKO) issued Order #307, "In accordance with the Decree of the Presidium of the Supreme Soviet on October 9, 1942, on establishing the full unity of command and the abolition of the institution of military commissars in the Red Army." Removing military commissars from the chain of command of the army meant the leadership of the army was finally in the hands of its commanders and not party overseers.[245] [246] This order, in essence, restored a professional military command structure to the Red Army. During battle, commanders could make quicker decisions, based upon sound military reasoning, and not on excessive political considerations. This decision also signified Stalin's growing confidence in the leadership of the Red Army.

A lull in the fighting within Stalingrad took place, as both sides were exhausted.

The 64th Army held its previous positions and continued minor actives against the opposing 371 Infantry Division, which reported 21 casualties, 5 killed and 16 wounded.

10 October

The newly formed 7 Rifle Corps (RC) arrived at the Stalingrad Front as a reinforcement.[247]

One regiment of the German 29 Motorized Division was assigned a position at the front against the 64th Army, between the 297 Infantry Division and the 20 Romanian Infantry Division; this was directly opposite the 66 NRB. The remaining units of the 29 Motorized Division were engaged with replenishment and training activities.

The 422 Rifle Division launched several probing attacks against the 371 Infantry Division, which reported suffering 19 casualties, 4 killed and 15 wounded during the day. Of course, formations of the 64th Army also suffered similar losses nearly every day. But the limited amount of publicly available Soviet records from World War II do not reveal the scope of these losses.

In a static situation like this where the front line hadn't moved in weeks or months, troops on both sides suffered from a monotonous and deadly routine. Mandatory scouting parties went out almost every night to learn what the enemy was doing. Troops dug trenches and bunkers for protection, and placed land mines and wire obstacles between the lines to prevent the enemy from approaching unnoticed. Others brought food and ammunition to the most forward positions and then carried the wounded and the dead back to the rear area. In between these various activities, one must always be on guard against random artillery barrages, harassing mortar fire, and sudden bursts of fire from hidden machinegun nests spraying bullets at suspected targets. Keeping one's head down also saved soldiers from that single aimed bullet, fired by a sniper, that often found its target.

Within Stalingrad, Russian sniper Vassili Zaitsev described what occurred after his sniper team killed three Nazi officers and an enemy sniper in a trench behind German lines.

> Ten or fifteen minutes passed, but no one approached the slain Fritzes. We were growing bored, when out of the blue, a massive artillery bombardment rained down on us. We crawled underground, waiting anxiously as shells exploded closer. When a shell

approaches the earth, its scream is so intense that it feels like your head is going to explode. You lie there, feeling like your intestines are being slowly yanked out by a winch. Then the Luftwaffe arrived…You know you must have taken out some serious Nazi brass for them to call in the air power. One of their bombs hit our trench and we were knocked around by the shock wave. For over two hours, bombers, artillery, and mortar fire relentlessly pounded our positions.[248]

Zaitsev reported his sniper team escaped this bombardment with only minor injuries, but for so many men on both sides, such an everyday experience of bombs or bullets led only to death or serious injury.

11 October

The German 29 Motorized Division gave up its small section of the front line to another unit and was once again fully in reserve, where it continued to receive replacements of men and equipment.

The 64th Army held its previous positions and ended its harassing attacks against the 371 Infantry Division. Combat operations tapered off along the entire front held by the 64th Army.

12 October

The 64th Army held its previous positions and conducted reconnaissance near the village of Elkhi.

Despite the minimal combat in the 64th Army area, Stavka was very active moving new units and reinforcements behind the lines to build up the assault forces. One of these units, the 169 RD, was assigned on 16 September to the Stalingrad Front as part of the newly formed 7 Rifle Corps. Then on 12 October, this division arrived separately in the Stalingrad area and remained in reserve; its movements and actions are somewhat unclear during this time. However, later in October the 169 RD was assigned to the 64th Army, but evidently never committed to the front line, always remaining in reserve positions. Then on or about 10 November, the 169 RD was transferred from the 64th Army to the 57th Army so that it could take part in the Russian winter offensive. Later on, in December, this Division would once again be serving with the 64th Army. The somewhat strange movements of this Division illustrate the shifting of forces from place to place that was needed to fulfill the changing requirements of the Stalingrad Front during the planning and combat phases of the upcoming offensive.

13 October

The 64th, 57th, 51st and 28th Armies all held their previous positions. The 64th Army also had control of the 90 TB on the east side of the Volga near Svertlyi Iar. This tank unit was part of the overall defenses on the eastern bank of the river and not of much help to the 64th Army in that location, some 18 miles from Beketovka.

Even as Joseph Stalin reduced political oversight of the Red Army and allowed his commanders more flexibility, Adolf Hitler was doing the opposite by changing the command structure in regards to the Eastern Front and General Staff. Newly appointed on 24 September, General Zeitzler, as the head of the General Staff, acting in accordance with Hitler's wishes, removed command authority for the Eastern Front from the OKW. So only the OKH was able to draft strategic directives concerning the Eastern Front. OKW was the German Armed Forces High Command, which had command authority over all land, sea and air forces; the OKH in turn was exclusively the German Army High Command. So, in this instance, the German Army High Command took over responsibility for the entire Eastern Front.[249]

14 October

The 64th Army held its previous positions with little or no activity at the front.

Adolf Hitler, now acting through the OKH, issued Operations Order #1 which covered all forces along the Eastern Front. This order also was known as the "Winter Standby Directive" and stated in part, "This year's summer and fall campaign, excepting the operations currently underway and several local offensives still contemplated, has been concluded."[250] The 6th Army at Stalingrad was exempted from what amounts to a winter stand down order. The attack to capture Stalingrad would continue even while the rest of the German Army began to prepare for the coming winter.

General Eremenko, in turn, planned another series of local supporting attacks from the north and south of Stalingrad to relieve the pressure on the 62nd Army, but General Paulus struck first. Paulus launched a major attack against the Dzerhezinsky Tractor Factory and a brick factory in the northern part of the city. The Germans reached the Volga in this area by the end of the day.

15 October

Once again, the 64th Army held its previous positions and conducted reconnaissance.

Stalingrad Front Commander, General Eremenko, issued an order assigning the 7 RC and the 90 TB to the 64th Army as reinforcements. These new units were to be used in an attack to link up with the 62nd Army in Stalingrad. The 57th Army also was ordered to conduct supporting operations in the south. This attack order was in response to Stavka Directive #170669 calling for an attack from the Beketovka area and further south along the front. Preparation for this attack took longer to organize and to launch than specified by this order. Eremenko wanted Shumilov to attack by 19 October, but this attack actually did not take place until 25 October.[251]

The transfer of the fresh 7 RC with its three Rifle Brigades, 93, 96, 97 RB, was a major reinforcement for the Army. The 7 RC was not listed as being within the 64th Army on the 10 November listing of BSA data but shows up on the 20 November list. So, the 15 October order issued by General Eremenko was the first indication of the 7 RC being assigned to the 64th Army. These three rifle brigades should have been at full strength with some 5,000 men each. If we include the Corps HQ and sub units, this represented an addition of about 15,500 men to the 64th Army's ration strength.

The introduction of the 7 RC into combat also was a significant development for the Red Army. During the battles of 1941, the Rifle Corps as a command formation was found to be too unwieldy. So, they were mostly disbanded by Stavka Directive #1 issued on 15 July 1941,[252] and by the end of 1941, only six remained. However, during operations in 1942, experience showed that field armies, in general, were having problems controlling the large number of formations and sub units that were directly subordinate to army HQs. Therefore, by late 1942, the Rifle Corps gradually were reintroduced into the army command structure, with the 7 RC being one of the first to enter combat. This change also was a direct indication of the growing proficiency of senior commanders within the Red Army, as they were able to effectively use these new corps in battle. Placing the 90 TB under direct control of the 64th Army, on the west side of the river, greatly boosted its offensive power. This was a fresh tank brigade and should have contained 32 T-34s, and 21 T-70s.

The German attack in the north continued to pound the defenders within Stalingrad. The 62nd Army, losing ground and large numbers of troops, fell back in the process. General Chuikov reported that two of his divisions lost 75 percent of their remaining strength on this day alone.[253] By the end of the day the tractor factory was lost as well as sections of the city center and the brick factory on the bank of the Volga. Eremenko acted to support the 62nd Army by assigning the 138 RD to Chuikov. At this time the 138 RD contained 2,646 men (combat strength) and a large amount of artillery. Its artillery remained on the left side of the Volga, but it still took at least 24 hours for the first regiments of the division to ferry across the river into the city.[254]

16 October

The German 29 Motorized Division moved south a few miles to a new assembly area. It was once again assigned a small section of the front facing the 64th Army, between the 297 Infantry Division and the 20 Romanian Infantry Division.

The 64th Army held its previous positions while the 422 RD conducted some local fighting, but this small amount of activity was unable to help the 62nd Army, which was desperately fighting within Stalingrad.

During the evening, Chuikov was surprised by two visitors to his command post on the banks of the Volga. The Front Commander General Eremenko and his deputy General Popov arrived to see for themselves the situation within Stalingrad. They only stayed until dawn but upon leaving promised to deliver more ammunition and men in order to keep the 62nd Army fighting.[255] Chuikov was of course pleased for the visit and encouraged by the promise of reinforcements. Unknown to Chuikov at this time, Stalin had ordered General Eremenko to go into Stalingrad and to see for himself conditions within the city. (see Appendix 14 for more info about this visit)

The weather conditions started to change as forecasters mentioned frost at night for the first time. Rain, mud and further cooling continued for the rest of the month.

With the changing of the weather, leaders on both sides needed to address the logistical setup in and around Stalingrad. By this stage of the battle, both the Germans and the Russians had maneuvered themselves into a logistical nightmare. Each army needed to overcome major logistical problems in order to supply their forces with the required provisions and resources needed to fight and to win a protracted campaign.

At this time, the Germans found themselves at the end of a very long and fragile supply line. The fighting directly within Stalingrad was being supplied largely by a single rail line starting far to the west, at a rail junction located on the western side of the Donets River near the town of Lichaja. This single rail line, capable of supporting a maximum of 12 pairs of trains a day, first had to cross the newly repaired, 800-foot-long Donets River Bridge. Then the rail line traveled east another 130 miles to the Chir River area. Because the Russians had completely destroyed the 2,500-foot rail bridge over the Don during the fall, the rail head was located at the Tschir Station about 10 miles short of the Don. At this point all supplies were unloaded and transferred to trucks or horse drawn wagons. The supplies then were driven northeast about 24 miles to a pontoon bridge over the Don and on to the village of Kalach. At the Kalach Station, supplies were loaded onto captured Russian trains running on a section of restored Russian Broad-Gauge track. These trains covered the last 40 miles to the 6th Army in the Stalingrad area where they were unloaded. This entire transportation route was time consuming, inefficient and lacked sufficient carrying capacity. Additionally, a second lesser rail route supplied the area south of Stalingrad. This route originated at the city of Rostov located at the mouth of the Don River. This rail line traveled south of the Don via Kotelnikovo and then on to the front south of Stalingrad. This line only handled 4 to 5 pairs of trains per day and delivered supplies directly to the 4th Panzer Army and the 4th Romanian Army.

Because of these railroad limitations and other issues, the Germans and their Allies were barely supplied at Stalingrad. Bombs and bullets were reaching the front in adequate amounts, but food, clothing, POL (Petroleum, Oil, Lubricants), spare parts and a host of other supplies were lacking.

During the summer months, minor deficiencies were not too difficult to overlook, but with the oncoming winter, and the unrelenting combat, the ability of Axis forces to conduct operations would only get worse. The lack of proper provisioning and logistical planning had an increasingly negative influence upon the conduct of the battle. Of special note, the German 6th Army was forced to send the bulk of its horses away from the front for the winter due to inadequate supplies of fodder. Without these horses, most of the German infantry formations largely became immobile. This was not much of a problem as long as the front line remained static, but if mobile operations were required, these same infantry formations would be forced to leave most of their heavy weapons behind and to fight with greatly reduced capability and firepower.[256]

At first, Soviet troops were in a similar position; they were forced to depend upon an inadequate supply line. But from the very start of the battle, not a single person, horse, train or truck was spared in the effort to supply the Red Army. Stalin himself said supplying armies and fronts required an "iron discipline" and that the new deputy commanders for rear services "must be dictators in the rear zone" of their fronts.[257] The battle at Stalingrad quickly became an all-out struggle against both the Germans and nature. Critical improvements to roads, river facilities and rail lines were started, which made all the difference in increasing the flow of vital supplies. In the end, the Russian logistical effort was successful while the German effort failed.

Red Army forces in and around Stalingrad were divided into roughly three separate supply areas, depending upon the logistical effort needed to supply each area.

First, the armies north of Stalingrad located within the Southwestern and Don Fronts were the easiest to supply because of their geographic location. They benefited from a direct land/road access to the interior, a nearby fully developed rail system, and no major rivers to cross, except for several bridgeheads on the south side of the Don. While the logistical effort was difficult in its own right because of the large amount of tonnage that had to be moved, the effort to supply these forces was achievable without extraordinary measures. Routine logistical improvements behind the lines were adequate to meet the needs of these forces.

The second area included the Stalingrad Front armies that were located south of Stalingrad but west of the Volga; these were the 64th, 57th and 51st Armies. These armies had few direct rail lines reaching their section of the front, and they were fighting with their backs to the Volga; therefore, it fell to road and river to carry a greater part of the burden. The resulting logistical effort was substantial and required upgrading the entire supply system, covering rail, road and river access. With great determination and much hard work, Russian troops and civilian workers solved the essential logistical problems in this area. The overall effort insured these armies had not only the means to survive but also the strength to actively participate in the battle for Stalingrad.

The third supply area was within Stalingrad itself, making the 62nd Army by far the most difficult force to supply. Stalingrad posed a challenging geographic setting of fighting within a destroyed city, perched along the top of a cliff next to the wide Volga. Surrounding enemy forces needed to be taken into account when moving supplies, and even the weather conspired to dislocate the best logistical planning. Only through extraordinary effort and sacrifice by the supply services was the 62nd Army able to hold out within the ruins of Stalingrad. Fighting against seemingly hopeless odds, and suffering tremendous losses, the 62nd Army lived or died subject to the flow of men and supplies that reached them over the river. This was truly a heroic effort not only by those within Stalingrad but also by those outside it as well.

In general, the most important means of transporting military provisions within Russia during the war were the rail lines. During the course of the war approximately 68.5 percent of all war-making materials flowed along the rail lines, verses 20 percent by road and 9.3 percent via water.[258] Each of these methods tried to bring needed cargos as close to the front as feasible. For example, the centralized storage of supplies in the interior meant that railroads were used to the greatest extent possible to move needed goods to front level depots. From there, rail transport was also the preferred method to transfer these supplies closer to the front and to deposit them at army-level depots, but in some cases, this was not possible. From army-level depots, supplies normally traveled via military roads in trucks or wagons to divisional-level storage areas close to the front lines and then on to the troops. Because the Volga split the battlefield in two, river transport also played a critically important role. Ultimately it was road transport, both by truck and horse drawn wagon, which finally delivered goods to the army and to divisional supply centers.

In the Red Army, supplies flowed from the army-level depots to the divisions and then to subordinated units. The army and not the divisions contained the major supply and support services that made this arrangement possible. "As in all other armies, the field army was a flexible organization, but because the Soviet field army service units delivered supplies rather than the division picking them up in army depots as was done in the German army, the tie between a Soviet division and its field army was much closer."[259] Basically, a rifle division was severely limited, in a logistical sense, if it were not directly attached to a controlling headquarters from which supplies flowed.

To more easily understand the vast scope of the logistical exertion needed to meet requirements at the front, let us examine the situation of the 64th Army in more detail. At this time the 64th Army contained seven rifle divisions, five rifle brigades, and two cadet regiments, along with numerous supporting artillery and tank units. Statistics taken from fighting in December 1943 illustrates a typical daily requirement for an average rifle division. This information can be directly applied to the current 64th Army situation to estimate the Army's logistical needs. In December 1943, an active (fighting) rifle division during an offensive was allocated daily "311 tons of munitions, 19 tons of rations, 15 tons of fodder for the horses, and 13 tons of fuel, for a total of 358 tons." A division holding a reserve position, or one that was inactive, would naturally require fewer munitions, thus reducing its daily requirements.[260] With most of the divisions within the 64th Army being understrength, and with the Army generally on the defensive, using a reduced allocation of some 160 tons per day, as an average requirement, was more representative of the actual needs of the Army at this time. This 160-ton amount per division was based upon averages for units in similar positions. This quantity would include 110 tons of munitions and 50 tons of food, fuel and fodder, per day, per rifle division.[261] So the 64th Army's total daily estimated requirements would look like this:

> Seven rifle divisions, so 7 x 160 tons = 1,120 tons.
>
> The five rifle brigades were likely equal to four reduced strength RDs, so 4 x 160 = 640 tons.
>
> HQ staffs, other rifle formations, and the supporting combat arms would equal two reduced strength RDs equivalents, so 2 x 160 = 320 tons.
>
> Logistical and support units would need few munitions, but large amounts of fodder and fuel to do their job, so add in another one reduced strength RDs equivalent, 1 x 160 = 160 tons.

So, the 64th Army's daily tonnage requirements would be 1,120 tons plus 640, 320, and 160 for a total of about 2,240 tons per day just for maintenance of troops while mostly on the defensive. Conducting even modest offensive operations quickly would increase ammunition consumption by hundreds of tons per day. If the army were trying to stockpile munitions, food, clothing, and other provisions in order to support some future offensive operation, then an even greater daily tonnage was required in order to build up this reserve amount.

To put this level of day-to-day consumption into perspective, let us look at the required transportation demands. A fully loaded Russian supply train, consisting of some 40-50 cars each, pulled by a single steam engine, could normally transport between 600-900 tons of cargo with an average load of 13-18 tons per railroad car.[262] So, 3-4 such trains would be needed to arrive every day to just keep the 64th Army operational. Using fewer cars per train and/or loading them with less cargo would increase the total number of trains required to meet these supply requirements. Of course, overloading cars and trains would increase the tonnage delivered, but that also might damage the rails, cars and engines. If we add in the transportation of replacements, reinforcements, miscellaneous equipment, spare parts, and stockpiling for future operations, then the Army easily could require at least another complete supply train to deliver the required tonnage. So, to support the 64th Army in the field and to build up a reserve for future offensive operations, the Army's total logistical requirement would rise to about 5 supply trains per day, or their equivalent via other means. The other armies stationed in and around Stalingrad needed their daily tonnage to stay combat ready too.

A huge number of variables influenced supply trains and the overall logistical effort. For instance, a pure troop train required more rail cars than an ammunition train verses a mixed train carrying both men and equipment. It all came down to how many cars were needed to carry the maximum load and how heavy or bulky a particular cargo was. A bulky cargo like food and fodder needed more cars to carry the required tonnage than a heaver, denser cargo like fuel, which would require fewer cars to reach the max weight limit. A single steam engine was able to pull only a certain amount of weight, so logistical planners tried to optimize each train to carry about the same overall tonnage. (see Appendix 15 for steam engine info) Variations in loading, the type and amount of cargo loaded and the availability of different car types made each train unique. Also, train speeds were dependent upon rail and roadbed conditions, grade of the tracks, steam engine performance, weather or signaling issues and, of course, enemy actions that were designed to prevent any logistical movements at all. All of these variables and more certainly affected calculations about requirements and tons delivered per day, so any values presented here should be considered approximations.

But this level of logistical effort only delivered the supplies to the nearest railhead. To reduce the burden placed upon the railroads, the Volga was used as much as possible to transport supplies, so not everything went by rail. A river port or dock acted in a similar manner as a train station or railhead.

For the 64th Army, the nearest railhead was 70 miles away at Akhtuba, on the east side of the Volga River. A new short section of rail line was under construction in this area during September. This new rail line started at the Akhtuba railhead and headed west toward Stalingrad, stopping a few miles from the Volga. This section of line eventually was able to carry eight pairs of trains per day and greatly eased the logistical problems for this section of the front.

River transport arriving from the south also unloaded somewhere between Akhtuba and Stalingrad. No matter where the railhead or river dock was located, once the supplies were unloaded everything was transported by truck or cart to army-level supply bases, located mostly on the east side of the river. But even that was not the end to the difficulties. The 64th Army was fighting on the west side of the Volga, so supplies had to be transported over the river either by boat, barge, ferry, or by road across one of the few bridges. In the area of the 64th, 57th and 51st Armies, several sets of bridges crossed the Volga. These bridges took advantage of two islands, Golodnyi and Sarpinskii, to provide for a reduced river crossing length. Even with using these islands as staging points across the river, the length of these bridges was considerable. As mentioned in an earlier chapter, one such bridge was over 3,900 feet long, but was used only for four days before being put out of action by the enemy.[263] Only a massive combined logistical effort, via rail, road, and river, was able to support multiple armies in various areas fighting along the entire Stalingrad Front.

In the case of the 62nd Army, the logistical effort had an added hazard, the intense combat action in and near the city of Stalingrad. Once supplies reached the supply depots on the east side of the Volga, everything crossed the river to the city by boat; no bridges crossed the entire Volga here. In fact, a few short bridges reached the islands of Spornyi and Zaitsevskii in the area of the 62nd Army. These bridges and islands helped the supply effort, but the final crossing used some type of boat to carry supplies into the city. With German troops overlooking the river from the land, and the Luftwaffe dominating of the air, most crossings were undertaken at night. Adding to all this struggle was the changing weather. Wind, rain, mud and eventually drifting ice on the Volga hampered the work of the logistical organization. Ultimately, the combined labors of tens of thousands of people, both military and civilian, ensured that proper amounts of provisions reached the front-line armies.

But for the troops actually doing the fighting, things were a bit different. NKVD special sections, in action behind the lines, were reading the soldier's mail in order to judge troop morale, and things were not looking good. Common soldiers were complaining in their letters home about the small amounts of food issued, or improper food, or just no food at all reaching the front lines. Adding to these grievances were statements about not having proper clothing, never being able to clean themselves, not being able to bathe in weeks, having a lack of soap, being full of lice (a real danger due to typhus), and having a shortage of arms and ammunition. Taking these grievances seriously, a NKVD Senior Major, N.N. Selivanovsky, took action. He relieved the chief of the 64th Army's Food Department, then the chief quartermaster in the 13 GRD, and even arrested the chief of the 62nd Army's Administrative-Economic section.[264] One would like to think these actions helped the common solider at the front, but the reality was, probably not. The fighting went go on as before, shortage or no shortage.

All this activity by road, river and rail would of course attract the attention of the Luftwaffe, but there were too few aircraft and too many targets. Logistical efforts occasionally were disrupted, men and shipments were lost, ships were sunk and trains destroyed, but the overall flow of supplies never stopped. The buildup for the great offensive continued almost unhindered and almost unnoticed by the German High Command.

17 October

The 64th Army held its previous positions with the 422 RD involved in local fighting against the 371 Infantry Division. One or both armored trains from the 28 Separate Armored Train Battalion supported the attacks by the 422 RD. The armored trains suppressed two enemy mortar batteries and several machine gun positions.[265] Despite this effort, the attack gained no ground.

Within Stalingrad, General Chuikov was forced once again to move his Army HQ, this time south to a position near the dominating Mamayev Kurgan hill. The chosen site was located right along the river bank, near the center of the city. One can imagine the danger and stress of constantly moving the HQ staff, records and communications setup from place to place as the Germans compressed the Russian forces within Stalingrad into an even smaller sliver of land along the Volga. Chuikov didn't know it at the time, but this location was to serve as his final HQ through to the end of the fighting within

Stalingrad.[266] Time was running out for the Germans to capture the city as Russian forces continued to stand firm within the rubble.

18 October

The 64th Army held its previous positions, conducted reconnaissance, and exchanged artillery fire with the enemy.

The German 6th Army was prevented from continuing its heavy attacks for several days, due to substantial rain and light snow showers that interfered with the delivery of critical supplies from the railhead west of the Don. Within Stalingrad, Chuikov made use of this pause in operations to further prepare his defenses. The battle, of course, continued but at a lower level of intensity.

Of special note, on or about this date, General Shumilov and member of the Military Council of the Army K.K. Abramov made a special request of the engineers of the Stalingrad chemical plant Khimprom #91 located in Beketovka. General Shumilov asked if they would be able to produce underwear and soap for the men. After several attempts, they successfully produced many tons of soap. The workers at this chemical plant also produced 150 thousand bottles of flammable mixture, and large amounts of smoke grenades. Additionally, they repaired artillery, machine guns, and motor vehicles, and produced field kitchens, mugs, pots, toilet and laundry soap, and even cologne as a gift to veterans.[267]

19 October

Once again, the 64th Army held its previous positions.

Continuing bad weather limited the effectiveness of German attacks in the city. One of the attacking German Corps reported losing a total of 186 men this day, 34 killed, 137 wounded and 15 missing. The two supporting panzer divisions also reported the loss of 29 tanks.[268]

Acting on orders directly from Stalin, General Eremenko once again launched a major attack against entrenched German forces north of Stalingrad. For the next seven days, units of the Don Front, the 24th and 66th Armies, threw themselves against these extensive German defenses and gained nothing. A German communiqué from the OKW, High Command of the Armed Forces, about these attacks reported, "In the sector of the front between the Volga and the Don, enemy infantry and tanks units transferred to this region launched diversionary attacks from the north which were repulsed by our forces. The enemy suffered significant losses."[269]

20 October

The 64th Army held its previous positions and conducted reconnaissance. Note: Reconnaissance missions both large and small were a regular activity of front-line units. Around this time Russian forces also were dispatching infiltrators, men, women and children, on long range missions behind German and Romanian lines in an effort to gather intelligence and to identify targets of interest in their rear areas. These infiltrators (spies) knew that if they were identified and caught, only torture and death awaited them at the hands of the Germans.

21 October

The 64th Army held its previous positions while preparing for further attacks.

22 October

The 64th Army launched small, low level attacks against German forces. The 422 RD hit the 371 Infantry Division while the 36 GRD hit the 297 Infantry Division. The rest of the Army's front remained quiet. These attacks were probing attacks looking for weakness or changes in the enemy defenses.

23 October

The 64th Army continued to hold its previous positions, as fighting from the previous day tapered off. Nothing much was gained or lost.

Paulus and the 6th Army launched another major attack against the 62nd Army, using four divisions, two infantry and two panzer. Each side suffered heavy losses, but units of the 62nd Army were hit very hard and forced to give up ground.

24 October

The 64th Army held its current positions and was almost ready, once again, to make a significant supporting attack to the north. Despite little chance of actually breaking through German defenses, General Shumilov stressed, during a commanders' meeting, the effort had to be made to divert some enemy strength away from the desperate 62nd Army.

The German 6th Army continued its assaults starting at 03:00 hours. By day's end, exhausted German infantry had taken a few more buildings and inflicted more losses on the weary defenders of the 62nd Army.

25 October

Finally, the 64th Army was able to launch a heavy attack toward the north against the German 371 Infantry Division. General Shumilov carefully planned this attack and made an effort to amass substantial firepower to back it up. Leading the attack was the newly arrived and fresh 7 Rifle Corps with two of its three rifle brigades, the 93 and 97 Rifle Brigades. Secondary attacks were made by the 422 and 126 RDs. Supporting these forces were 50+ tanks from the 90 TB, massed artillery and rocket launcher regiments, and even a few aircraft from the 8th AA. Once again, the 28 Separate Armored Train Battalion was on hand with its two armored trains. Even ships from the 2nd Brigade of the Volga River Flotilla joined in to provide additional fire support. After an hour-long barrage, combat troops moved off at 09:00 hours against German positions. Heavy fighting in and around the village of Kuporosnoe lasted all day and Russian troops started to gain ground. Within Kuporosnoe itself only a few hundred yards were taken, but on the left flank of the attack, near the village of Novo Peschanka, a mile-deep wedge was driven into the German defenses. (see #Map 22)

Within Stalingrad, the Germans continued offensive operations with the same tired divisions. The 6th Army had no fresh units in reserve to turn the tide of the battle.

Also, on this day, to celebrate the 25th anniversary of the Great October Socialist Revolution, the factory workers from the chemical plant Khimprom prepared gifts for the defenders of Stalingrad. "Soap-boiler" engineers V.N. Antonov and E.N. Hoffman released a large batch of soap and cologne to the troops.[270]

26 October

In order to tie down German forces, inflict losses, and in general force them to respond to the actions of the 64th Army, the Army renewed and even intensified their attacks from the previous day. Resuming combat operations at 11:30 hours, the 7 Rifle Corps and its 93 and 97 Rifle Brigades continued to lead the way with the 422 and 126 RD supporting. But General Shumilov added the 36 GRD to strengthen the attack, thus increasing the pressure on units of the German 371 Infantry Division. These efforts forced back German troops a bit more inside the village of Kuporosnoe and broadened the deep wedge near Novo Peschanka. This of course was the first noteworthy advance by the 64th Army against the 371 Infantry Division since the front stabilized weeks before.

Meanwhile, along the Don Front north of Stalingrad, General Eremenko finally gave permission to the 24th and 66th Armies to suspend attacks and to go over to the defense in this area. Needless to say, the Russian forces involved in this area suffered heavy losses, but they also forced the Germans to react to these attacks and thus not gather forces elsewhere.

Of note, this attack by the 64th Army was under assessment by several high-level commanders. "On October 26, 1942, the head of the General Staff of the Soviet Army, Colonel-General Vasilevsky, the deputy commander of the front, Lieutenant-General Popov, arrived at our division (422 RD) at the observation post on the second floor of the house." The Commander of the 64th Army General Shumilov was also in attendance during the start of this major attack.[271]

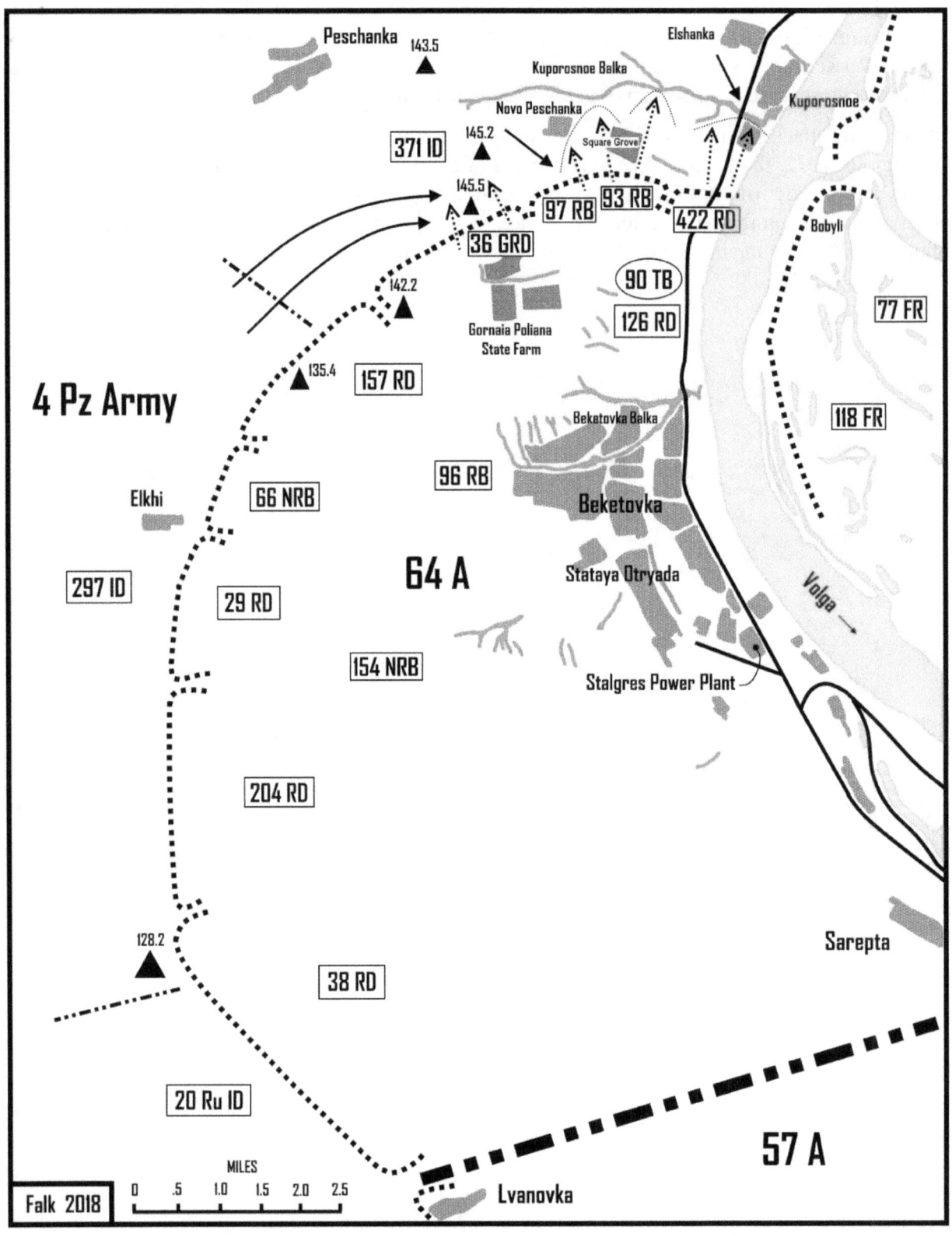

27 October

Heavy fighting continued against positions held by the German 371 Infantry Division, especially in the area of Hill 145.5, otherwise known as "Bald Mountain". Units of the 7 RC and the 36 GRD repelled four enemy counterattacks on their hill positions alone. At the end of the day, units of the 64th Army held onto their gains. For the 4th Panzer Army, this situation was far from ideal. If the 371 Division gave up too much ground in the face of constant attacks by the 64th Army, it might compromise the operation to capture Stalingrad. So, any Russian success in this area was treated as a serious threat. Therefore the 29 Motorized Division, the only reserve formation south of Stalingrad, was compelled to stand ready to intervene anywhere in the Beketovka Bell area; thus, it could not be employed elsewhere.

For the German 6th Army, the reserve situation was even worse. In the immediate area around Stalingrad, the entire Army's reserves consisted of one regiment from the 79 Infantry Division. All other formations were committed to holding the front line or attacking within Stalingrad itself.

28 October

The German 297 Infantry Division joined the fight in the south and drove back attacks launched by the 36 GRD toward Hill 145.5. The 371 Infantry Division also launched powerful counterattacks against Russian units in and around Kuporosnoe, seizing the Kuporosnoe *Balka*, Topor Grove, and Kvadratnaia Grove just to the west of Kuporosnoe. The 7 Rifle Corps, 422 RD and the 126 RD lost all the ground they had captured over the previous few days.

In their book, *Armageddon in Stalingrad Vol 2*, p 538, Glantz and House relate how the German OKW daily report stated, "As before, the diversionary attacks by large enemy forces supported by tanks against the German positions south of Stalingrad were repelled. The enemy suffered heavy losses." Undeniably, Russian forces lost heavily in these attacks, but German forces suffered losses too. These unrelenting diversionary attacks were depleting German strength all along the front and preventing them from concentrating exclusively on the capture of Stalingrad.

29 October

The German OKH's Foreign Armies East intelligence report stated, "There is nowhere any sign of preparation for a major attack, but the entire area needs continued observation."[272] This report covered the area around Stalingrad and all of Army Group B.

The German 371 Infantry Division launched more counterattacks in the Kuporosnoe area and were able to capture more positions held by the 7 Rifle Corps, 422 RD and the 126 RD. These counterattacks finally forced these Russian troops onto the defensive, as they started to dig in wherever they were. However, on the far-left flank, the 36 GRD and a supporting cadet regiment continued their unsuccessful assaults against German positions on hill 145.5.

30 October

Fighting continued on the right flank of the 64th Army. Both sides launched numerous attacks and counterattacks in the area around Kuporosnoe and the nearby Kuporosnoe *Balka*. Units of the 64th Army remained fully engaged by these fluid battles for the rest of the day. At this point, groups of German aircraft, 4-15 strong, intervened in the battle and bombed the rear areas of the 64th Army. These raids were carried out around Beketovka and continued up to the front at Kuporosnoe. General Shumilov must have welcomed the arrival of these Luftwaffe aircraft as a good sign. If the Luftwaffe was bombing 64th Army positions in the south, then they were not bombing the 62nd Army within Stalingrad. These constant assaults out of the Beketovka bridgehead were starting to pay off. During these attacks the 2nd Brigade of the Volga River Flotilla fired more than 3,000 artillery rounds in support, with further fire support coming from artillery on both sides of the river. The Germans were starting to take seriously these efforts by the 64th Army.

Also, on this day, General Shumilov issued Order #0186 to the 64th Army regarding preparation of selected rifle battalions for night actions. Each division or brigade was ordered to start a 10-day training session in night combat with one of its rifle battalions. About 70 percent of their training was to be dedicated to night combat and the remainder to close combat. These troops also were to be issued with and trained in the use of special weapons and equipment like silent weapons, signal pistols, rockets, distinctive badges and compasses.

What Shumilov was planning with this order is unknown. Perhaps he was preparing a surprise nighttime assault against the German 371 Infantry Division: a quick penetration into the Germans' main defense line during the night with a small force and then support any success with more units during the day. That would be a risky move, but it might offer a chance to crack this tough German defense line. More likely, this order was really related to the upcoming general offensive. This is plausible for several reasons. First, the order was issued to all the rifle divisions and brigades within the Army so that each unit would have its own nighttime trained assault force. This implies the use of these specially trained troops all along the front line and not at a single location. Second, the scheduling called for 10 days of training with units reporting back to Shumilov by 12 November, which was close to the original Operation Uranus start date. Third, these assault troops were supposed to be issued with signal pistols, signal rockets, and compasses, which implied a breakthrough operation followed with an advance in a specific direction, with the use of signaling equipment to mark their progress. This type of effort also would lend itself to controlling prearranged artillery bombardments. Once the first objective had been reached, the troops would launch signal flairs or rockets, telling the supporting artillery to shift their fire to the next set of targets.

Thus, this second possibility might be the correct one. This could be significant because General Shumilov would have been informed by higher headquarters about the need for these training efforts but not necessarily the reason why. He might not have been directly informed about the upcoming offensive by this date, but he certainly would have understood the implications.

31 October
General Shumilov launched further attacks toward the north against the German 371 Division and against German positions on hill 145.5. These assaults achieved only minor gains during the day. The 36 GRD, 422 RD, 126 RD as well as formations from the 7 Rifle Corps and one cadet regiment, took part. [273] [274] The continuation of these unsuccessful attacks were launched mostly as a diversion, just like the efforts of the 62nd Army within Stalingrad. The intent was to keep the German High Command focused upon the area directly adjacent to Stalingrad and to not look too closely at activities further away.

The 64th Army continued to do its best to keep German forces distracted by conducting further offensive operations well into the next month. With overall preparations for the winter offensive nearing completion, the Red Army required only a few more weeks for the buildup. But these few weeks were vital; the Germans could still slip away if they learned of the impending Operation Uranus attack too soon.

1 November 1942	
64th Army	
Assigned – Stalingrad Front	
Major General M.S. Shumilov Commanding	
Rifle, Airborne and Cavalry Units	7 RC (422 Rifle Division - minus 1334 Regiment, 93, 96, 97 Rifle Brigades) 36 Guards Rifle Division 29, 38, 126, 157, 169, 204 Rifle Divisions 66, 154 Naval Rifle Brigades Two Composite Cadet Regiments
Artillery Units	1104, 1111, 1159 Gun Artillery Regiments 186, 500, 504, 507, 612, 665 Antitank Artillery Regiments 3 Guards Heavy Mortar Regiment (rocket) 2, 4, 18, 47, 76, 90, 91 Guard Mortar Regiments (rocket) 622, 1261 AA Regiments
Armored and Mechanized Units	13, 56, 90 Tank Brigades 28 Separate Armored Train Battalion
Engineer Units	328, 329, 330 Separate Engineer Battalions 175 Separate Sapper Battalion
Reported/Estimated Combat Strength Reported/Estimated Ration Strength	54,351 men [275] (see Appendix 16) 67,603 men

(BSA - 42 and BSSA -42 part 2)

With the start of November, the countdown to the planned winter offensive Operation Uranus had begun. Stalin and Stavka worked closely with front commanders to finish planning and saw to the distribution of critical supplies and combat units. Adjustments to the plan took place until the last minute. The NKVD worked for the success of the offensive as they monitored the rear area, rounded up stragglers and in general kept an eye on the political aspects of the war. The NKVD also continued to support the two army level blocking detachments, #76 and #83, that were still in place behind the 64th Army. These detachments ensured that when the offensive started, all the combat troops of the 64th Army headed in the correct direction. The 77 and 118 Fortified Regions appear to be subordinate to the Stalingrad Front. The 77 FR was deployed on Sarpinskii Island and the 118 FR further south some 18 miles and on the east side of the Volga. Both units were in defensive positions covering the Volga River.

1 November

Weather Report: Sunny, clear, warm, dry, good road conditions.[276] Note: All future weather reports are form this same source.

With the addition of large numbers of artillery units, the 64th Army's strength increased substantially from October. The reported and estimated ration strength increased noticeably to 67,603 men and women as of 1 November, with 54,351 combat troops and 13,252 noncombat troops within the army. The new 622 Anti-Aircraft Regiment boosted air defense against the Luftwaffe, and of special interest were all the new Guards Mortar Katyusha regiments, the 3 Heavy, 2, 4, 18, 47, 76, 90, and 91. They added significant and badly needed firepower to the Army. Also of note, the 169 RD, which was added to the 64th Army sometime in late October but was never used at the front, was not included in these totals.

The 64th Army continued its assault to the north against the German 371 Infantry Division. The 97 RB repelled several German counterattacks in the area of Topor Grove, west of Kuporosnoe. Meanwhile, the 36 GRD, 93 RB, a cadet regiment and tanks from the 90 TB twice launched powerful attacks of their own against German troops near hill 145.5. Strong enemy fire prevented them from gaining any ground.[277][278] Please note that on military maps prominent peaks, hills and other geographic features are routinely referred to by their height, in meters, like hill 145.5. This avoids confusion, allowing troops to understand where they are going, what point they should be trying to capture, or even what enemy position they should bombard with artillery.

Within Stalingrad, the 138 RD fought well since being transferred to the 62nd Army. This division added critically needed strength to the defenses because the German 6th Army wasn't finished yet with its attempt to capture the city.

2 November

Weather Report: Sunny, clear, warm, frost at night, dry, good visibility.

The 64th Army temporarily suspended attacks and shifted the two cadet regiments, the 93 RB, 126 RD and 154 NRB into new defensive positions toward the rear, most likely into a second defensive line. The positions of other units remained unchanged.

The German OKH's Foreign Armies East intelligence report for this date stated, "We must expect there to be continued reinforcement of the enemy confronting the Romanian Third Army and, possibly, even an attack…. We must await further indications."[279]

3 November

Weather Report: Sunny, bright, good visibility.

The 93 RB was back in action in a new location along with the 96 RB; they both joined the attack toward Kuporosnoe to capture the Square Grove area. Once again, strong enemy fire prevented them from gaining any ground.

The German OKH's Foreign Armies East intelligence report for this date stated, "There emerges an increasingly clear picture of preparations for an attack on the Romanian Third Army, though they are in the early stages…We cannot yet be certain whether the objective is an attack designed to lure our forces away from Stalingrad or an operation with much broader objectives."[280] Similar types of reports were dispatched from frontline Axis units all along the line. Italian, Hungarian, and Romanian Armies kept the German command well informed about these Russian activities. In essence, the Germans knew the Soviets planned some type of winter attack. But the real question was, when would it take place, where would they strike, and with how much force?

4 November

Weather Report: Good weather, sunny, bright, warm, good road conditions.

The only action that took place on this day by the 64th Army was aggressive reconnaissance of the enemy's forward positions. Supporting formations engaged the enemy with harassing small arms, artillery and mortar fire.

The German OKH's Foreign Armies East intelligence report stated, "[T]he enemy has by no means given up the fight for the city" and also was reinforcing its forces in the Beketovka bridgehead south of the city.[281] The Germans had noticed the recent reinforcement of the 64th Army.

5 November

Weather Report: Good weather, cool, dry, overcast, good road conditions.

Once again units of the 64th Army conducted reconnaissance activity against enemy forward positions along with the essential supporting harassing fire. Just another normal day on the front line.

A German air attack against the StalGRES power plant located outside Beketovka finally put it out of action for good. The power plant had been operated only at night after a heavy artillery attack on 23 September. But at 08:30, 49 Stuka dive bombers arrived and pounded the power plant for the next 25 minutes. The surviving plant operators finally were evacuated to the east side of the Volga. For the remainder of the battle, there would be no electrical power to run workshops or repair facilities in the surrounding area.[282]

6 November

Weather Report: Foggy, occasional rain, midafternoon heavy rain, very bad visibility, road conditions passable with four-wheel drive.

The 64th Army occupied the same position.

By this date, officers and men within the 64th Army almost certainly were aware that headquarters was planning something big. This information did not come from an official source like the 64th Army newspaper *"For the Motherland"*, but by way of the soldiers' grapevine. Each army and front had a newspaper, and when they were distributed, "People eagerly began to read them, looking for answers to many tormenting questions: about the situation on our front, on other fronts, the country as a whole, but they provided only episodes from the life of the army and the front of exploits of warriors." There also were "articles about the dedicated work of our people in the rear of harvesting and production of military products for the front."[283] But this formal news source was lacking; it didn't tell the soldiers what they wanted to hear because everything was considered a secret! The real information filtered up to the front via rear supply troops or from runners taking messages back and forth. Some scraps of news came from phone operators or even a causal comment by an officer. These bits and pieces of information were secrets for sure, but gossip, even in the Red Army, had a way of making its way through to even the lower ranks.

7 November

Weather Report: Frosty, changing cloudiness, overcast, sharp east wind, 32°F (0°C) during the day, good road conditions.

On this day, most troops of the 64th Army only conducted minor reconnaissance action along the front. But, the 29 RD and 204 RD carry out skirmishes against the 20 Romanian Infantry Division.

In Moscow General Zhukov reported to the GKO (State Defense Committee) on the status of preparations for the offensive. He said the operations were proceeding according to plan.

8 November

Weather Report: Clear, light frost, dry, good visibility.

The 64th Army occupied the same position with no reported action.

General Zhukov sent a message to General Rokossovsky, the Don Front commander, and to General Eremenko, the Stalingrad Front commander. This message stated that the "Resettlement" (i.e., a code name for Operation Uranus, which in itself was a code name for the winter offensive) was set to start on 13 November. Army level commanders had not yet been informed about the impending offensive.[284] But it is hard to believe that Generals Chuikov and Shumilov knew nothing about the offensive. Chuikov stated that after defeating the last German attack in the city on 12 November, "Everyone was convinced that the next attack, a powerful and irresistible one, would be made by our armies. The progress of the fighting since the second half of July had created the conditions necessary for this."[285] To further expand on this issue, General Eremenko said, "Long time army commanders knew nothing about the counteroffensive, but, of course, guessed that something important is preparing."[286] So, individual army commanders did not officially know about the upcoming offensive, but they knew all the same.

9 November

Weather Report: Sunny, bright, dry, cold, night 5°F (-15°C), sharp wind, frosty, good visibility.

No ground action took place, but the 64th Army was hit by short artillery and mortar attacks all along the front.

The small section of the front that had been held by the German 29 Motorized Division was taken over by the 20 Romanian Infantry Division. These released German troops rejoined the 29 Division in its assembly area just to the west of the 20 Romanian Infantry Division.

The German 6th Army, within the general area of Stalingrad, had no reserve units at all.

10 November

Weather Report: Sunny, clear, light frost, good visibility, good road conditions.

The 64th Army saw little action along the front and only suffered limited, short barrages of artillery and mortars on its defensive positions. But on the other side of the line, the German 29 Motorized Division received a warning order "to be prepared to assume positions behind the Romanians on the shortest notice."[287] This order referred to the area of the front adjacent to and south of the 64th Army. Jason Mark, in his book *Panzer Krieg Vol 1*, stated that by early November "the 29 Motorized Division was the only rested and full-strength division in the Stalingrad region." This fact was a powerful indication of just how depleted German forces were in and around Stalingrad. Constant combat in the city and on the flanks had drained most German formations of their offensive power. This inherent weakness and lack of reserves would not bode well for the German 6th Army if the situation at the front were to suddenly change.

The German OKH's Foreign Armies East intelligence report stated, "The appearance of the Soviet headquarters for the Southwestern Front somewhere to the northwest of Serafimovich [this was near the center of the Romanian Third Army's front, 100 miles northwest of Stalingrad] indicates that a major enemy attack is in sight."[288]

At this point, Russian planning for the winter offensive reached a stage that it became necessary to inform the actual commanders of the attacking forces. So, on 10 November, a meeting was scheduled for army commanders, commanders of corps, and divisions of the Stalingrad Front. This meeting was held in the vicinity of the command post of the 57th Army. Attending were Comrade Khrushchev, General Zhukov, General Eremenko, the commanders of the 57th and 51st Armies and a host of other senior officers, both political and military. These assembled officers were informed of the plan and the scope of the attack. Incidentally, General Eremenko gave the same briefing separately to General Shumilov and his staff at the command post of the 64th Army, most likely on the same day.[289] The Southwestern and Don Front commanders also would have conducted similar briefings for their army, corps, and divisional commanders on or near this date.

Somewhere around this time, General Eremenko made the statement, "After a final decision [and] until the beginning of the counteroffensive all our attention was focused on the implementation of a number of urgent preparations. The most important of these were regrouping of troops and guaranteeing secrecy of all preparations for a counteroffensive." He went on to say, "Redeployment of forces to create strike groups [was] in these days especially intense, but in strict accordance with the plan of the front staff." This redeployment involved the movement of large numbers of troops along the front line, which if not carried out secretly might alert the Germans to the impending attack. So, from the 64th Army two divisions, namely the 169th and 36th Guards Rifle, were transferred to the 57th Army, and the 126th Rifle Division to the 51st Army. General Eremenko was talking only about the southern Stalingrad Front arm of the offensive; the northern arm was handled by the Southwestern and Don Fronts. The 57th Army also was ordered to transfer its 15 GRD to the 51st Army.[290]

In order to carry out the overall plan for the southern group of forces, the left flank of the attack had to be reinforced. Therefore, three rifle divisions were transferred out of the 64th Army into the main assault areas further south, and one additional division was moved from the 57th to the 51st Army. At first this does not make sense to move troops away from Stalingrad, but the attack plan, in the south, was in essence a turning movement of the southern three armies. The 64th Army would undertake only a moderate advance to the west and then quickly wheel to the north, securing that section of the planned pocket around the German forces in and near Stalingrad. Next, the 57th Army would thrust a bit further westward, and then in turn, also wheel to the north, keeping pace with the 64th Army. Finally, the 51st Army, taking the longest route, would swing around the outside flank of the southern attack group and linkup with the northern attack group, somewhere in the region of Kalach, on the Don River. Additionally, the 51st Army also would dispatch a few formations far to the south along the rail line toward Kotelnikovo, to act as a left flank guard and to drive any enemy units away from Stalingrad. (see Map #23)

To secretly withdraw three rifle divisions from the 64th Army's front would take some effort. Two of them, the 169 and 36 Guards Rifle, needed to march south 7 miles to the 57th Army's assembly positions. The 126 RD need to move south about 34 miles to the 51st Army's attack area. Overall, these transfers would require at least 3-5 nights to complete. This time frame also included the 15 GRD transferred from the 57th Army to the 51st Army. Note: Records show the 169 RD was with the 57th Army by 1 November, the 36 GRD by 10 November, and the 126 RD within the 51st Army by 10 November.[291] So the 169 RD never really was part of the 64th Army for very long and quickly was transferred to the 57th Army for the offensive. The other two divisions were already redeployed by 10 November. This implies the transfer of these units out of the 64th Army was ordered earlier, perhaps around 5 November, long before General Eremenko's official briefing.

To concentrate mobile forces for the offensive, the 13th Mechanized Corps, (formerly the 13 TC), transferred out of the 64th Army's area to the 57th Army and was assigned as its main mobile assault force. The 4th Mechanized and 4th Cavalry Corps also were assigned as the main mobile striking force for the 51st Army. The 64th Army only had its remaining rifle divisions and a few tank and rifle brigades with which to conduct this major attack.

11 November

Weather Report: Sunny, bright, frosty, good visibility, good road conditions.

The 64th Army occupied the same positions with little or no reported action. But this was only the outward appearance. During the night, three of its rifle divisions, their supporting artillery, supplies, and sub units continued to head south to take up their appointed places for the assault.

The German 6th Army launched a heavy attack against the 62nd Army within Stalingrad. The situation in the city had not changed.

General Zhukov sent a message to Stalin and the appropriate front commanders, stating that Operation Uranus would not be ready as scheduled due to logistical problems. The new date was set for 15 November, but that date depended upon delivery of critical munitions and other supplies. He went on to say that 27,000 trucks were at work transporting troops and freight, along with 1,300 rail cars (about 29 full trains) that were arriving daily to supply the offensive. This level of railroad activity alone amounted to 20,000-26,000 tons of supplies that arrived daily to the Southwestern, Don and Stalingrad Fronts. River and road transport added to this tonnage and still there were shortages.[292] Additionally, due to the winter weather, the water level of the Volga had dropped over 12 feet by mid-November. Before the river froze over, the falling water level exposed more dry ground along the river banks, leaving docks high and dry which hindered loading and unloading of supply ships and barges operating along the river.

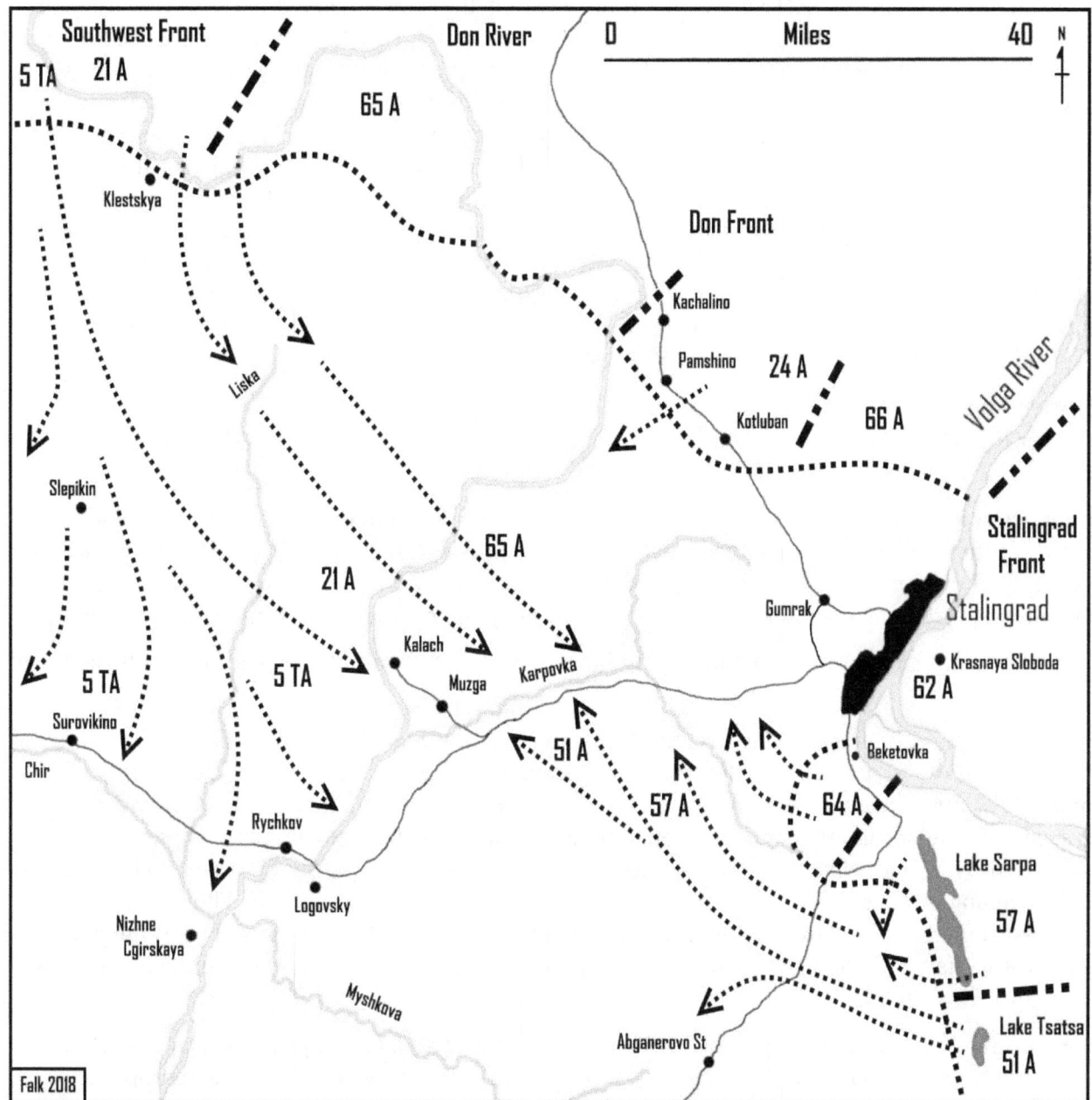

12 November

Weather Report: Overcast, light frost, dry, clearing afternoon, good weather, good road conditions.

Units of the 64th Army undertook only minor reconnaissance action along with skirmishes and small arms fire against the enemy. This type of activity was just to keep the enemy under observation and to harass them.

German units continued their attack against the 62nd Army. They gained little ground and by the end of the day, the German forces strength was spent.

On this date, the German OKH's Foreign Armies East intelligence issued a large report; it stated in part, "In front of the Army Group, the enemy's attack intentions, which we have long suspected, are gradually becoming clearly defined. In addition to establishing two main groups of forces, which we have detected opposite the two wings of the Romanian Third Army – where the enemy can now be said to be ready to attack." The report went on to say, "While it is not possible to make any overall assessment of the enemy situation with the picture as uncertain as it is at present, we must expect an early attack against the Romanian Third Army, with interruption of our railroad to Stalingrad as its objective so as to endanger all German forces further to the east and compel our forces in Stalingrad to withdraw." This report acknowledged concern about other possible attacks further to the west against the Italian 8th Army and the Hungarian 2nd army.[293] Note: These ongoing German intelligence reports make little mention of the large Russian forces gathering south of Stalingrad, a clear failure to determine the true scope of the planned Russian offensive.

13 November

Weather Report: Sunny, clear, bright, dry, frosty, below 32°F (0°C), visibility and road conditions good.

Minor activities took place all along the front, from the 64th Army in the north to the 57th and 51st Armies further south. Units were busy strengthening their positions, keeping warm, and conducting any required reconnaissance missions. All of these armies also were stockpiling supplies, conserving ammunition, arranging positions, and secretly assembling their forces for the impending offensive.

Within Stalingrad, only minor amounts of fighting occurred as worn-out German forces adjusted their front lines.

Deputy Supreme Commander Zhukov and Red Army Chief of Staff General Vasilevsky met with Stalin in Moscow for two days to discuss the impending operation. They reported the following points: "The mission assigned to the various Fronts, armies, and formations had been finalized. The air armies would apparently complete preparations not earlier than 15 November. Delivery of ammunition, fuel and winter clothing was somewhat tardy, but there was every reason to believe that by the evening of either 16-17 November, the troops would have these supplies. The counter-offensive could be started by the Southwestern and Don Fronts on 19 November. The Stalingrad Front was to start 24 hours later. After a brief discussion, the plan for the counter-offensive was given full approval by Stalin."[294]

After several earlier delays, all parties finally agreed they were ready; the plan for the destruction of the enemy forces in and around Stalingrad was set. At this time, Stavka transmitted appropriate orders to the front commanders. All along and behind the fronts, the Russians needed to complete only a final few preparations. They still needed to keep the secret from the enemy, and there was hope for some assistance from the weather. Frozen ground would allow for a rapid advance across the open steppes, and speed was essential for this operation.

The German OKW's Daily Communiqué reported the Romanian forces south of Stalingrad repelled several enemy attacks of local significance. Also, similar attacks occurred along the Don front area.[295] These attacks were probing efforts by Russian units to locate the enemy forward positions and artillery placement. The Russians also wanted to determine if they could find any weakness in the defenses or establish if any changes had been made along the Axis line.

14 November

Weather Report: Sunny, bright, windy, light frost, good visibility, good road conditions.

In the morning, the 96 RB attacked into German defenses just south of Kuporosnoe and were able to reach the center of the city, capturing the leather factory. In order to achieve this success, this attack must have been supported by considerable amounts of artillery, both from the rear of the 64th Army and from across the Volga. This small attack and the ground that they gained must have worried the German 371 Infantry Division, because soon the counterattacks began.

In Stalingrad, German forces reported a quiet night apart from some harassing artillery fire. Light frost and good road conditions were reported. Minor and localized combat took place within Stalingrad.

15 November

Weather Report: Overcast, under 32°F (0°C), good road conditions.

The 96 RB held onto its gains within Kuporosnoe for the entire day.

Once again, German forces reported a quiet night in the city with only harassing artillery fire and no enemy bombers during the night.[296] During the day, both sides undertook small unit actions while most units consolidated any recent gains.

By this date the three rifle divisions that transferred out of the 64th Army were fully settled into their new positions within the 57th and 51st Armies. With the transfer of these divisions, the estimated ration strength of the 64th Army had understandably fallen to approximately 50,233 men and some 40 tanks. This estimate was based upon 36 GRD having 4,144 men (reported), 126 RD having 4,224 men (reported), and the 169 RD having 9,000 men (estimated), at or near full strength at the time of their transfer. This amounted to a loss/transfer of about 17,368 men from the 64th Army since the beginning of the month. Planning for the offensive required such shifting of forces to reinforce some attack zones while reducing troop strength in other areas.

Also, on this date, Stalin sent an incredible message to General Zhukov at Stalingrad, "You can set the day for the resettlement [the offensive] of Fedorov [General Vatutin, commander of the SW Front] and Ivanov [General Eremenko, commander of the Stalingrad Front] at your discretion and subsequently report back to me upon arrival in Moscow. If the thought occurs to you for either of the two to begin the resettlement one or two days earlier or later, I authorize you to decide that matter at your own discretion."[297] This message is unbelievable because Stalin was permitting General Zhukov to shift the date of the great offensive at his own discretion! This would have been unthinkable even as little as six months before. Stalin clearly was much more at ease in allowing senior Red Army commanders to perform their duties without undue scrutiny.

During the planning phase of the offensive, security was strict. Nothing on paper regarding the offensive was allowed to leave the Kremlin in Moscow. Security officers even assigned code names to each of the leading commanders. In this way, even if the enemy intercepted and then broke a coded message, they would not easily understand for whom the message was intended. In this regard Stalin became Vasiliev, Zhukov – Konstantinov, Vasilievsky – Mikhailov, Rokossovsky – Dontsov, Eremenko - Ivanov,[298] and Vatutin – Fedorov. These code names were confusing, but that was the intent!

16 November

Weather Report: Sunny, bright, dry, light frost, light snow within Stalingrad, 28°F (-2°C), good visibility, good road conditions.

German forces within Stalingrad reported another quiet night with no bomber activity. During the day they conducted limited attacks within the city.

The German 6th Army HQ reported to higher headquarters (Army Group B) that between 13 September and 14 November (three months) the 6th Army had suffered the following casualties:

- 7,487 men killed
- 30,313 wounded
- 1,081 missing in action

For a total of 38,881 casualties.[299]

Within the Beketovka bridgehead, the 96 RB repelled all German attacks and held onto the center of the Kuporosnoe. The 64th Army proved to be an on-going and dangerous annoyance to German forces covering the southern approaches to Stalingrad.

17 November

Weather Report: Heavy frost, very overcast, thick fog, heavy ice on the Volga River, afternoon freezing rain, snowstorm north of Stalingrad, good road conditions.

The heavy fog and bad visibility were just what Operation Uranus needed. German reconnaissance planes were grounded just when Russian units were moving into their final assault positions. But the heavy ice on the Volga was another thing. Heavy ice moving down river made it difficult to move supplies over the Volga to the 62nd, 64th, 57th and 51st Armies and was causing some delay in preparations for the offensive.

Once again German forces in Stalingrad reported a quiet night with no night bomber or ground activity.

Hitler sent a message to General Paulus stating, "I am aware of the difficulties of fighting in Stalingrad and the decline in combat strengths." He went on to say, "I expect, therefore, that the leadership and troops will once more, as they often have in the past, devote the same energy and spirit that they have shown in the past to fight their way through to the Volga at the gun factory and the metallurgical plant and to take those sections of the city."[300] That the 6th Army had the strength to reach the goals set by Hitler is doubtful. In any case, time was running out. General Paulus and the 6th Army already had launched, and failed, in their last major attack to capture the last few bits of Stalingrad. At this point, the 6th Army was so weakened that they undertook only scattered combat for minor local positions.

In the south, German commanders shifted some forces of their own. Troops from the 371 Infantry Division, along with the 221 Infantry regiment, on loan from the 71 Infantry Division and supported by 10 tanks borrowed from the 24 Panzer Division, were formed into an ad hoc *Kampfgruppe* (battle group) to deal with the latest incursion by the 64th Army. German troops were skilled in forming these temporary battle groups to achieve a set objective. In this case, the objective was to remove Russian forces from Kuporosnoe. The *Kampfgruppe* struck the 96 RB in the Kuporosnoe city center, where the Russians were holding the leather factory. By the end of the day, after heavy fighting, the 96 RB was forced out of the leather factory; they retreated to a gully to the south of the city. The 371 Infantry Division had restored its positions and expelled the enemy.

At this point, just two days before launching the offensive, the 64th, 57th and 51st Armies each held special meetings to inform commanders of divisions and regiments about the impending action and their part in the plan. Soon afterward, the common troops also were told. The secret was out to thousands of men all along the front line. Also, on or about this date, General Shumilov started shifting units along the front into their attack locations. Most likely these movements only occurred at night to deceive the enemy.

18 November

Weather Report: Misty, later clearing, 20-32°F (-6 to -0°C) at night, frozen ground, passable roads.

After the previous day's fighting, the entire front-line south of Stalingrad went quiet; the 64th, 57th, and the 51st Armies retained their former front-line positions, limiting themselves to reconnaissance activities and in some areas engaging the enemy with small arms fire.

The VVS once again was not active in the Stalingrad area, and the German troops were starting to notice, "Why are the Russians so quiet?" This lack of VVS activity was planned; it permitted active air units to rest and to carry out repairs, thus increasing the number of operational aircraft that would be ready for the offensive. This pause also allowed fresh reserve air units to secretly be brought forward to their new operational air bases, adding to the buildup of air power. In contrast, new Russian heavy caliber artillery positions as well as medium artillery batteries started to make themselves known to the Germans. These new batteries needed to start their registration process, to align their weapons in order to fire accurately. They did this by firing off a few rounds from the east side of the Volga and other locations and observing the fall of their shot.[301] This type of activity also should have alerted German forces to some kind of impending Russian action.

19 November - Day One of Operation Uranus

Weather Report: Overcast, foggy, cloudy, 32°F (0°C), clammy fall weather, roads passable.

Once again, the entire front south of Stalingrad was quiet; the 64th, 57th, 51st and 28th Armies held their previous positions, all the while shifting forces to new offensive locations. The 62nd Army within Stalingrad had an estimated fighting strength of only 19,500 men.[302] Floating ice on the Volga severely restricted the movement of men and supplies to the 62nd Army, so until the river froze, General Chuikov had to make do with what he had. The weather report forecasted overcast and foggy, 32° F (0°C) at night, passable roads. While this was not perfect weather from the Russian point of view, it was good enough because northwest of Stalingrad things were about to change. The long-planned offensive, Operation Uranus, was about to begin.

Up until this point, German and Romanian front-line commanders had increasingly reported troop movements and the regrouping of Russian forces in their respective areas. Likewise, air reconnaissance noted an upswing in rail traffic and troop movements behind the Russian lines. But the further away from the front line and the further up the German chain of command, the less creditability those in power gave to these reports. In the end, the German intelligence assessment was that some type of local attack was pending, but they did not expect an attack of the size and power of Operation Uranus. Russian deception efforts were effective in concealing the extent of their offensive preparations.

Early in the morning at 07:30 hours, the Southwestern and Don Fronts went over to the offensive, launching the northern arm of Operation Uranus. The matching southern arm was scheduled to open their attack the next day. The Southwestern and Don Fronts went over to the offensive a day early because they had further to go to reach the planned meeting point near the village of Kalach. Kalach also just happened to be the main German supply line crossing point over the Don River. (see Map #23)

A great deal has been written about this northern part of the offensive, so an outline of events taking place in this area will suffice. Needless to say, this part of the offensive unfolded as a massive surprise attack that caught the defenders off guard and unprepared. The Southwestern Front's attack, launched from a bridgehead south of the Don River, used the 1st Guards Army, 5th Tank Army and the 21st Army against the 3rd Romanian Army holding positions along the Don River, some 120 miles northwest of Stalingrad. The VVS was unable to help in the initial attack as heavy ground fog prevented air operations. This fog also prevented the Luftwaffe from interfering with the initial Russian ground assault.

In the same general area, the Don Front launched a somewhat more limited attack against the junction between the left flank of the German XI Corps and the adjacent Romanian IV Corps with its 65th Army. Surprise was achieved everywhere.

At 10:00 hours, Army Group B Commander General Maximilian von Weichs told General Paulus that "[a]ll offensive operations in Stalingrad are to be halted at once." The German command was starting to respond to the attack, but could they react in time to prevent disaster?

By the end of the day, Russian forces had broken through most of the front-line defenses of the Romanian 3rd Army, and Russian mobile units were advancing into the rear of the Romanian Army, insuring the breakdown of organized defenses. Some Romanian units fought bravely and held their positions, but others gave way. From the first moment of the attack, the battle was an unequal struggle as the Russian forces far outnumbered the Romanian troops in manpower, tanks and artillery. The German XI Corps was able to hold its front line but was under extreme pressure on its left flank from the surprise attack.

20 November - Day Two of Operation Uranus
Weather Report: light snow, 0°F (-17°C), bad visibility, icy roads.

The Southwestern and Don Fronts continued their attacks from the north. The plan called for these forces to reach the vital Kalach Don River crossing area and to come up along the Chir River by the end of the third day. Due to stiff resistance in some areas and bad weather conditions, they already were behind schedule.

In the south, the Stalingrad Front's part in the overall offensive called for the 57th and 51st Armies to start their initial artillery barrage early in the morning. Immediately after the preliminary artillery bombardment, the ground assault would take place against Romanian infantry divisions of the 4th Romanian Army. About 2 hours later, the 64th Army, on the right flank, would follow with their own artillery barrage and ground assault against a single Romanian infantry division and part of a German infantry division. Overall this was an uncomplicated but effective plan, but when did a major offensive ever go according to plan?

As intended, the three assaulting armies of the Stalingrad Front, 64th, 57th and 51st Armies were ready early in the morning to take part in the offensive, but heavy fog hung down to ground level. This fog caused problems for General Eremenko, who was observing conditions right at the front with the 57th Army. The bad weather conditions force Eremenko to delay the start of the offensive for the 57th Army. Weather conditions were somewhat better in the area of the 51st Army, so they started their opening artillery barrage on time at 07:30 hours followed by ground attacks starting at 08:30 hours. In all areas the heavy fog thwarted the use of air support in the opening attack and prevented artillery gunners from directly viewing their targets and correcting their fire. The assigned time for the planned 57th Army attack passed with no change in the weather. Eventually Eremenko received a phone call from Moscow. Stavka was anxious and wanted to know why the attack hadn't been launched. Eremenko explained the local weather situation and assured Moscow the attacking troops were ready. He said he was just waiting for the proper moment to attack. Once again, Stalin accepted a military decision from a senior commander and did not force the issue.

In a slightly different version of this account, Eremenko called Moscow and asked for a delay in the start of his attack. In any case, Eremenko acted like a front commander and took charge. Eventually, the fog lifted just enough that by 09:30 hours General Eremenko ordered the opening artillery barrage for the 57th Army to start at 10:00 hours. After a two-hour delay, the southern arm of Operation Uranus would soon to be fully underway. A simple and effective signal for the start of the 57th Army's ground attack was arranged to be a final 5-minute volley of fire from dozens of Katyusha rocket launchers. The front-line assault troops could hardly fail to see and hear hundreds of rockets flying through the air toward the enemy positions. The ground assault of the 57th Army started at 11:15 hours and went off almost without issue, but the fog still prevented the VVS from flying missions in support.

Initially, only the assault units from the 57th and 51st Armies engaged the Romanian 4th Army while the 64th Army remained silent in front of the German 4th Panzer Army. The 57th and 51st Armies achieved a breakthrough almost at once in several areas, shattering the defending units of the Romanian 4th Army. After the initial barrage phase ended and the breakthrough was underway, numerous batteries of heavy artillery and mortars were pulled out of the 57th Army's main attack area. They moved rapidly north to take up new positions to support the 64th Army's attack. This area of the front simply did not have enough artillery to go around, so the 64th Army was forced to await the arrival of these needed artillery reinforcements before trying to break through the enemy positions along its front. Because of this need to transfer supporting artillery to new firing positions, the initial attack by the 64th Army always had been planned to follow the 57th and 51st Armies attack by two hours. So with the weather delay, the 64th Army finally began its 75-minute artillery barrage at 14:20 hours with the ground assault starting at 15:35 hours (3:35 in the afternoon). (see Map #24)

To recap the opening events of the southern arm of Operation Uranus:

- Bad weather prevented the VVS from supporting the initial attack and hindered supporting artillery fire.
- The 51st Army launched its attack on time with its opening artillery barrage starting at 07:30 hours followed by ground attacks starting at 08:30 hours.
- Due to heavy fog, the 57th Army began its opening artillery barrage two hours late at 10:00 hours followed by the ground assault at 11:15 hours.
- Because of the delay in the 57th Army's attack, the 64th Army only started its artillery barrage at 14:20 hours with the ground assault at 15:35 hours. Note: On this day, sunset was at 16:18 hours, so the ground troops of the 64th Army only had about one hour of full daylight left to conduct their initial assaults.

At this point it would be useful to make note of the forces involved in the 64th Army's attack. At first only the 64th Army's left flank was involved in the offensive with units in the center and on the right flank holding fast. The initial assault force was composed of the 38, 157, and 204 RD supported by the 13 and 56 TB, totaling some 20,000 men and 40 tanks. This entire force was directed against one German regiment of the 297 Infantry Division defending the area around the town of Elkhi and two regiments of the Romanian 20 Infantry Division which only had some 7,700 men on hand. Both of these defending units were assigned to the German 4th Panzer Army.

Map #24 64th Army Deployment and Attacks - 20 November 1942

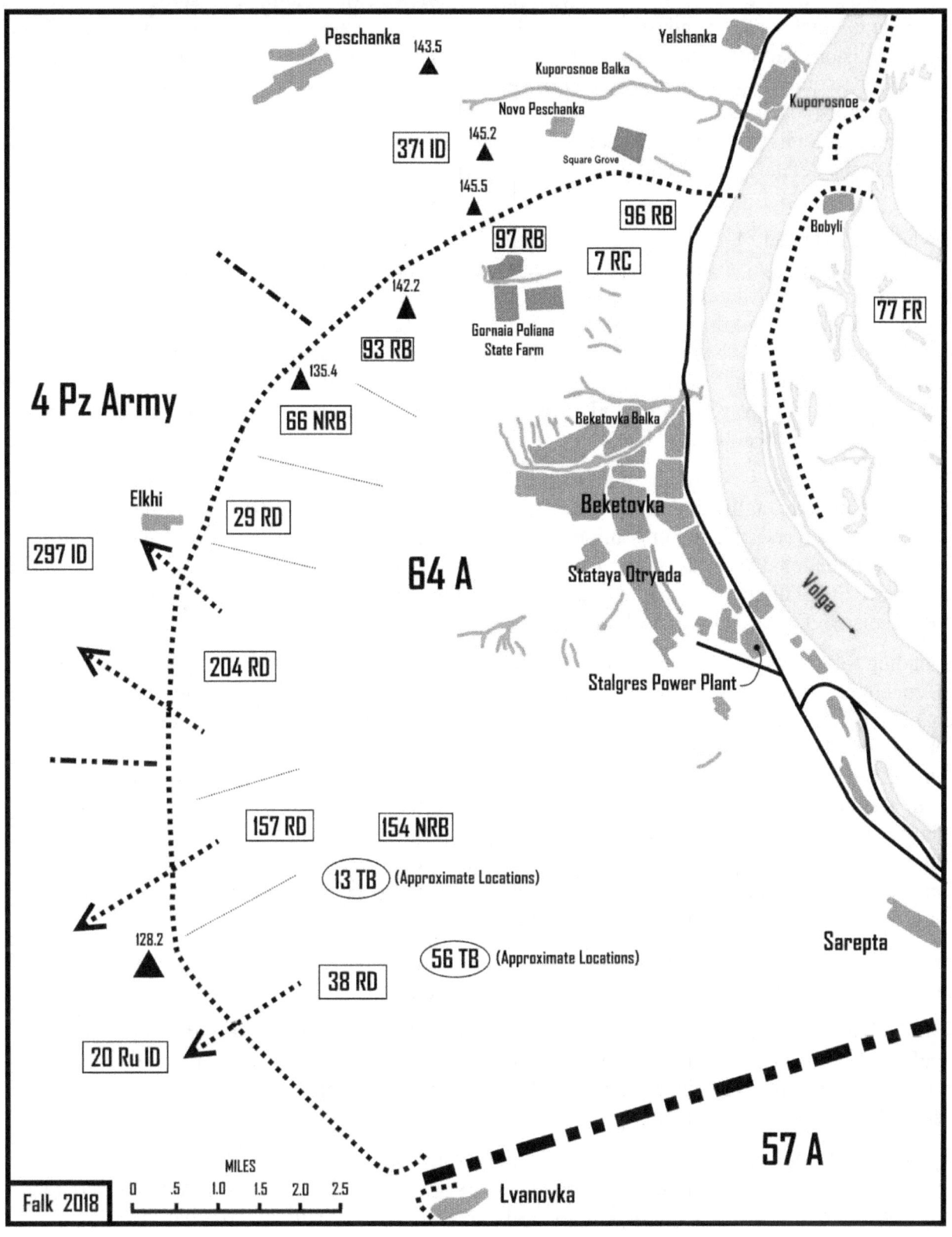

In the 64th Army sector, the 204 RD was able to only make minor penetrations of positions held by the German 523 Regiment/297 Infantry Division. Even after making multiple assaults, the 204 RD was unable to break through and capture the heavily fortified town of Elkhi. To the south, the 157 and 38 Rifle Divisions had better luck. Each of these two divisions attacked a single Romanian regiment and was able to quickly overwhelm the Romanian defensive positions. The third Romanian infantry regiment that also occupied the front line was attacked by the 169 RD of the 57th Army. So, these three Russian rifle divisions each attacked a single Romanian infantry regiment. No wonder they were able to push the defenders back and to quickly advance deeply into Romanian defense positions by nightfall.

Likewise, the 57th and 51st Armies made breakthroughs in their areas. By nightfall the 57th and 51st Armies infantry units advanced through the main defensive positions of the 2 and 18 Romanian Infantry Divisions, surrounding both divisions. The 13th Mechanized Corps, attached to the 57th Army, exploited the breakthrough and penetrated deeply into the rear areas. The 13th Mechanized Corps was stopped only when the German 29 Motorized Division counterattacked that evening, causing a fierce battle that lasted most of the night. The 51st Army made deep penetrations in the area of the 1 Romanian Infantry Division and released the attached 4 Mechanized Corps and 4 Cavalry Corps to exploit the breakthrough.

By the end of the day, the German 29 Motorized Division and the Romanian 20 Infantry Division were still fighting and falling back to the northwest. The Romanian 2 and 18 Infantry Divisions were largely surrounded and offering only minor resistance to Russian forces. The 1 Romanian Infantry Division was shattered and was offering little resistance to the mobile units of the 51st Army. Despite the late start, the Stalingrad Front had largely achieved its mission of breaking through the German and Romanian defenses south of Stalingrad.

Responding to the twin offensives, Hitler ordered the creation of a new army group, Army Group Don, encompassing the Romanian 3rd Army, the German 6th and 4th Panzer Armies, and the Romanian 4th Army. Searching for a commander for this new Army Group Don, Hitler eventually ordered Field Marshal Eric von Manstein, who was currently in the Leningrad area, to form and then command this new force. "Very poor weather prevented [von] Manstein from flying south so he and his staff were compelled to travel all the way to Novocherkassk [his new HQ location] in his armored command train [leaving on 21 November], entailing a lengthy journey of nearly 2,000 Km [over 1,200 miles] strung out over five days."[303] Despite the urgency of the situation, the German High command responded slowly to the crisis. Precious days were lost as the Russian offensive continued to unfold.

21 November - Day Three of Operation Uranus

Weather Report: Morning fog, heavy snow in the afternoon, -15°F (-26°C) at night.

Overnight Hitler ordered the 6th Army to hold its ground "regardless of the danger of a temporary encirclement." The German IV Corps from the 4th Panzer Army was transferred to the German 6th Army. This order effectively deprived the 4th Panzer Army of any combat troops and temporarily added to the confusion in the south.[304]

At the start of the day, attacking armies of the Southwestern and Don Fronts were about one-third of the way to their objectives, although 36 hours behind schedule. Stalin and Stavka were concerned at the rate of progress and urged that mobile forces be pushed forward with all possible speed.[305] In response, the Southwestern and Don Fronts continued their attacks, shattering the remains of the Romanian 3rd Army and surrounding several major units. Left flank divisions of the German 6th

Army fell back to the southeast and occupied new defensive positions. German and Romanian counterattack efforts failed to defeat the Russian offensive in this area. By the end of the day Russian mobile units were approaching the Don River.

In the south, the Stalingrad Front's mobile units of the 57th and 51st Armies struck deeply into enemy territory making good progress, advancing some 25-37 miles and scattering Romanian rear area troops. For all practical purposes, no real defensive line blocked the advance of the 57th and 51st Armies. Further north, the 64th Army dealt mostly with dug in German units that did not budge.

In the 64th Army sector, the same three assault divisions, 38, 157 and 204 RDs, once again were sent against the German 523 Regiment defending the town of Elkhi and two regiments of the Romanian 20 Infantry Division. What Shumilov did not know at the time was that overnight units of the German 29 Motorized Division had been ordered to support the Romanian 20 Infantry Division. Also, the German 297 ID had shifted another regiment south to boost the defense of Elkhi. Because of these reinforcements, the Russian attacks were met with immediate, powerful counterattacks backed up with tanks. This unexpected strength stopped both the 157 and 204 RDs from making any progress and pushed back the 38 RD somewhat with heavy losses.[306]

22 November - Day Four of Operation Uranus

Weather Report: Unknown, most likely clear skies with a full moon, cold at night and during the day.

The battle was not going well for Axis forces in the north. The start of the day saw parts of two Romanian Infantry Corps of the Romanian 3rd Army surrounded but still fighting near the Don River. This pocket was called Group Lascar after its commander. These surrounded troops were forced to hold this hopeless position by Hitler's refusal to allow them to break out of the pocket and to reach safety. Meanwhile, the rest of the Romanian 3rd Army was in full retreat to the southwest and south toward the Chir River. In turn, units of the German XI Corps withdrew toward Stalingrad under heavy pressure from attacking Russian forces. They tried desperately to avoid being encircled on the west side of the Don River. For them, safety meant crossing to the east side of the Don and rejoining the main body of the 6th Army.

Taking advantage of the failing defensive efforts of German and Romanian forces, Russian troops surged forward toward their objectives. By day's end, mobile units of the Southwestern and Don Fronts finally started to capture key locations, including the vital bridge over the Don River at Kalach. As gratifying as these achievements were, the mobile units were still two days behind schedule, thus allowing Axis forces the time they needed to regroup and to build new credible defensive positions. Ultimately, the German High Command had to act quickly and decisively, or the loss of the Kalach position would seal the fate of the entire German 6th Army and all Axis forces in and around Stalingrad. In response to Russian exertions, General Paulus decided to withdraw the entire XI Corps over the Don as soon as possible. This in effect consolidated 6th Army forces between the Don and Volga Rivers and made it possible to form and then defend a compact pocket of resistance around Stalingrad.

Meanwhile, the Stalingrad Front also made progress. After assessing the situation, Eremenko ordered his three attacking armies to continue their attacks to the northwest. The 51st Army also sent forces to the southwest to pursue the Romanian units that were withdrawing toward the Myshkova and Aksay Rivers.

After resupplying overnight, the 4 Mechanized Corps from the 51st Army swung wide around defending Axis units and reached and occupied the town of Sovetskii and the nearby train station of

Krivomuzginskaia. This town was situated on the rail line between Stalingrad and Kalach. This flanking movement effectively cut the main German supply line to Stalingrad.[307] Not to be outdone, the 13 Tank Corps of the 57th Army with supporting infantry divisions helped force back toward Stalingrad the German 29 Motorized Division and the 20 Romanian Infantry Division from their blocking positions.

The 64th Army in turn attacked with its left flank and held fast with its center and right flank units. Due to previous heavy losses, Shumilov withdrew the 38 RD from the front line and replaced it with the 36 GRD, which had recently returned from the 57th Army. At 13:00 hours, these fresh troops, supported by the 56 TB, the 1104 Artillery Regiment, and the 4 Guards Mortar Regiment, assaulted the 20 Romanian Infantry Division and supporting units of the German 297 Infantry Division. The 157 and 204 RDs, supported by the 13 TB, also joined in the attack. After heavy fighting all day, the 36 GRD advanced some 3 miles. The 157 RD made almost a 2-mile advance with the 204 RD pushing forward about 2.5 miles. Due to flanking advances made by the 57th Army during the day and the relentless attacks mounted by the 64th Army, the German command was forced to make a tactical withdraw. It ordered the 20 Romanian Infantry Division and the German 297 Infantry Division to withdraw northward some 6.5 miles to a much stronger defensive position. This new position was along the north side of the Karavatka *Balka* ravine, which covered about 7 miles from the village of Tsybenko until it reached the fortified town of Elkhi. From Elkhi to the Volga, the front line had not moved with parts of the German 297 and the entire 371 Infantry Divisions holding the line. This new, very strong defensive line proved to be hard to breach, allowing the Germans to hold out there for many weeks. This new German defensive line occupied part of the old Russian K and S defensive positions that were constructed before the battle for the city of Stalingrad began.

At the end of the day, units from the Stalingrad Front's 4 Mechanized Corps from the 51st Army were located along the Stalingrad and Kalach rail line. To the north, the Southwestern Front's 26 Tank Corps from the 5th Tank Army were positioned in and around Kalach on the Don River. Only about 13 miles separated these two spearhead forces. The Russian commanders thought one more push would be enough to close the gap and to achieve the main objective of Operation Uranus, the encirclement of all Axis forces in and around Stalingrad.

Later during the night, General Paulus asked Army Group B for "freedom of action," that in effect implied permission to abandon Stalingrad if necessary, and to break out of the forming pocket. Hitler, of course, denied Paulus's request, thus giving up the best chance to save his army from destruction.

23 November - Day Five of Operation Uranus
Weather Report: Unknown, but conditions were most likely similar to 22 November's weather: clear skies and cold with a waning moon.

Stavka pressed for the Southwestern and Don Fronts to finish the encirclement from the north "at all costs." Generals Vatutin and Rokossovsky were only too glad to comply.[308]

The Don Front continued to pressure the German XI Corps as the Corps slowly withdrew over the Don. Despite constant attacks, heavy losses in men and equipment, and trying weather conditions, the German XI Corps maintained its combat effectiveness.

The Southwestern Front's advance continued to the southwest, with cavalry and rifle forces finally reaching the Chir and Krivaia River areas. Further north, the surrounded Romanian troops of Group Lascar were compelled to surrender. By the next day, 27,000 Axis officers and men had given themselves up. Only a small force was able to break out of the pocket and to reach the safety of the newly forming Chir River defensive line. Toward the southeast, the 26 Tank Corps finally captured the town of Kalach, forcing defending German troops to fall back to the east. While this was taking place, units of the 4 Tank Corps, coming from the north, moved southeast toward the town of Sovetskii where they found units of Stalingrad's Front 4 Mechanized Corps. The northern and southern arms of the offensive joined hands at last. The entire German 6th Army, units from the 4th Panzer Army, and the 4th Romanian Army now were truly surrounded.

Overall, this change in fortunes produced a very unusual military situation. Within Stalingrad, the Russian 62nd Army was surrounded by the German 6th Army. At the same instant, the 6th Army found itself surrounded by six intact and powerful Russian armies, the 66th, 24th, 65th, 51st, 57th and 64th Armies. Elements of several more armies, the 21st, 51st and, of course, the 62nd Army within Stalingrad itself supported this effort. The besieger had now become the besieged. (see Appendix 17 for historical context)

During the night, the 64th and 57th Armies were ordered to continue their attacks to the north against the Karavatka *Balka* and Elkhi positions. Units of the 51st Army were directed to the north in order to reinforce the linkup with forces of the Southwestern Front, and at the same time to send units to the southwest, reaching the Aksay River area in some force.

On the far-right flank of the 64th Army's front, the 93 RB launched a probing attack against the German 371 Infantry Division. This was most likely done to keep units of the 371 Division in place, preventing them from being used elsewhere. Shumilov then committed his main force on the left. The 29 RD and the 204 RD moved against the strong positions around Elkhi, the 157 RD against the center of the 20 Romanian Infantry Division and the 36 GRD on the far-left flank against the Karavatka Balka position east of Tsybenko. These attacks advanced the front line several miles, bringing units of the 64th Army into close contact with the new German and Romanian fortified positions along Karavatka *Balka* from Tsybenko to Elkhi.

So, after sustaining substantial losses while achieving its goals during the offensive, major advances by the 64th Army were now over for the next several weeks. Due to the very tough enemy defensive positions the Army was currently facing, the "64 Army would simply tie down as many German forces as possible in the south of Stalingrad by conducting local attacks, raids, and feints."[309]

While all the dramatic action took place far outside Stalingrad, General Chuikov was still in the fight by using his 62nd Army to pin down as many German units within the city as possible. As a result, General Paulus was prevented from freely transferring forces out of the city to cover the threatened western sector of the pocket. General Paulus also had other concerns. North of Stalingrad, along the Volga, German General Seydlitz ordered the two divisions of his LI Corps to withdraw southward to a new defensive line without orders from Paulus. This act threatened to destabilize the entire northern section of the pocket.

By accomplishing the linkup between the northern and southern assault groups, Soviet forces achieved the first major objective of Operation Uranus; Axis forces in and around Stalingrad were surrounded. We may never know the exact number of Axis troops that were caught within the pocket, but an informed estimate is possible. Quoting David Glantz and Jonathan House from their book, *Endgame at Stalingrad Vol. 3*, table #27 shows the following values:

- Estimates range from 250,000 to 330,000 men, with the most likely figure being 284,000 men within the pocket.
- This includes some 12,000 to 15,000 Luftwaffe troops
- About 20,000 Romanians

- And some 60,000 *Hiwis* (Russian volunteer auxiliaries working for the Germans)

At the same time, Stavka produced a rather low estimate of the surrounded Axis forces at around 80,000 to 90,000 enemy troops. This low estimate led Stavka into thinking the pocket could be reduced quickly by relatively few Russian forces. Eventually they were proven wrong in this regard.

Because the Red Army took five days to complete the encirclement instead of three days as originally scheduled, problems began to appear at the front. Logistically speaking, the attacking troops started to run out of supplies. Before the start of the offensive, the Southwestern, Don and Stalingrad Fronts had amassed a reserve of 4 to 5 days' worth of food, fuel and ammunition for their armies. This was thought to be sufficient to sustain the offensive through the first phase of Operation Uranus and then some, but the armies had now consumed the bulk of these supplies. From this point forward, the victorious Russian armies were supplied directly from the rear over newly won territory and an unforgiving winter landscape. In turn, the Axis forces within the newly formed Stalingrad pocket were, of course, cut off from their normal source of supply. The logistical battle confronting both sides now intensified and reached new levels of urgency over the next few weeks. The outcome of this battle for supplies ultimately would decide the fate of entire armies and the battle for Stalingrad.

24 November - Day Six of Operation Uranus
Weather Report: Heavy night fog.

The Luftwaffe transported the first 84 tons of supplies to the encircled forces using 42 aircraft.[310] This was the official start of the German airlift operations. Staff officer calculations revealed the 6th Army required some 750 tons per day to be combat effective. At the start of the airlift, a bare minimum amount of 400 tons per day worth of fuel and ammunition were needed to keep the army fighting. Once the on-hand food stocks ran out (this included normal rations and then slaughtered horses) this bare minimum amount rose to about 550 tons per day because food would then need to be supplied from outside the pocket.[311] Hitler announced the surrounded 6th Army now would be called "Fortress Stalingrad."[312] Hitler decided that the 6th Army would be supplied by air. There would be no breakout attempt; instead a relief force would be sent to linkup with the pocket…In essence Hitler said, hold on, help is coming! and General Paulus believed him. In support of these statements by Hitler, within the Stalingrad pocket were five primary Airfields that could be used by the Luftwaffe to deliver supplies. The two main ones at Pitomnik and Basargino were fully operational and even were capable of nighttime use. Three smaller Airfields were located at Gumrak, Stalingradskii, and Karpovka. These smaller Airfields were undeveloped and dangerous to use, which resulted in many inevitable crashes.[313]

Field Marshal von Manstein arrived at the Headquarters of Army Group B and was briefed on the Stalingrad situation.[314] He then sent his first official message to the 6th Army; basically, it said he would take over command of Army Group Don on 26 November. Until then, the 6th Army was to hold on and to wait for strong outside forces to open a supply route to the pocket from the outside. General Paulus accepted this message and took appropriate measures. So, both Hitler and von Manstein told Paulus the same message, hold on and we will get you out.

For various reasons, Stalin and Stavka wanted to reduce and/or destroy the Stalingrad pocket as soon as possible. Therefore, Stavka gave the forces of the Southwestern, Don and Stalingrad Fronts the following mission for 24 November: "[D]ismember the encircled enemy Stalingrad grouping by launching converging blows to Gumrak [a secondary Airfield to the west of Stalingrad], and destroy it in piecemeal fashion." Stavka also ordered the outer ring of forces to be strengthened and defensive positions to be pushed further out away from Stalingrad.[315]

Despite resuming attacks all along the front, the Southwestern and Don Fronts achieved little more than pushing German forces back a few miles. The armies of the Stalingrad Front, in turn, even after heavy fighting were unable to gain ground against dug in German and Romanian forces. The 64th Army itself launched attacks on its left flank with four rifle divisions, the 36 GRD, 157, 204 and the 29 RDs. All these attacks failed after gaining little ground and suffering heavy losses. The only bright spot was when the 36 GRD captured a small bridgehead on the north bank of the Karavatka *Balka*, near the village of Tsybenko. (see Map #25) On the right flank, the 7 Rifle Corps, with the 93 and 97 Rifle Brigades, attacked positions of the 371 Division. These attacks also failed to produce any usable results. General Shumilov shifted some small units around in rear areas to reinforce his defensive positions. He apparently was concerned the Germans might choose his sector of the front to attempt a breakout from the pocket.

On the north side of the pocket, the German XI Corps still held a small bridgehead on the west side of the Don. The situation in this area was going to change in the near future.

On the south side of the pocket, the 6th Army HQ finally decided to consolidate the remaining troops of the 20 Romanian Infantry Division, containing some 5,000 men, into the German 297 Infantry Division.[316]

25 November – Day Seven of Operation Uranus
Weather Report: Grey morning, light frost, snow.

The Luftwaffe transported 66 tons of supplies into the pocket and flew out 487 men to safety using 33 aircraft, all Ju-52s. This equals 14.7 men taken out per aircraft.

The three fronts encircling Stalingrad were issued virtually the same instructions from the day before: attack and crush the Axis forces within the pocket. Once again, the Southwestern and Don Fronts achieved little progress with these attacks. Units of the 51st Army were being diverted more toward the southwest in order to build up the Aksay River line. This was a prudent move because the Germans were almost certain to attempt a relieving attack, and Stavka did not know where or when the attack would take place.

Within the Stalingrad Front's area of responsibility, the 64th Army was ordered to transfer the 36 GRD to the 57th Army along with its defensive sector at the front. With this transfer, the 57th Army was ordered to continue attacking toward the village of Tsybenko and the Karavatka *Balka*. In turn, the 64th Army was to support these efforts with minor attacks on its right flank near the town of Peschanka, with two battalions from the 97 RB. Little or nothing was achieved by these efforts. The Germans were dug in too well to dislodge with these small forces.

On the north side of the pocket, the German XI Corps continued to hold a small bridgehead on the west side of the Don as it slowly withdrew its troops to the east side of the river.

26 November – Day Eight of Operation Uranus
Weather Report: Overcast, cold, light snowfall, icy wind.

The Luftwaffe transported 72 tons of supplies into the pocket using 35 aircraft. The 6th Army HQ ordered rations within the pocket reduced by 50 percent. Returning empty supply planes flew out 290 men, for 8.2 men per aircraft.

On 26 November, Field Marshal Eric von Manstein arrived at his new HQ located in the town of Novocherkask, northeast of Rostov, and took full control of the new Army Group Don on the evening of 27 November.

Attacks continued by all fronts against the Stalingrad pocket with little effect. At this point in the battle, Russian forces needed time to regroup and resupply, but because Stalin was impatient, the attacks continued. The 64th Army dug in at their new positions with no other activity. In the south, scattered units of the 51st Army continued their advance southwest along the rail line toward the supply center of Kotelnikovo. The few German and Romanian troops in this area defended themselves as best they could.

Within the pocket, German forces did what they could to improve their defensive positions and to economize on the use of food, fuel and ammunition. By this date, everyone knew they were surrounded and their only hope was liberation from the outside.

27 November - Day Nine of Operation Uranus
Weather Report: Overcast, afternoon clearing, icy wind, icy snowstorms, frost, 23°F (-5°C) during the day, 19°F (-7°C) at night, very bad visibility, icy roads.

The Luftwaffe transported 28 tons of supplies into the pocket and flew out some 200 men using 14 aircraft, for 14.2 men per aircraft. One must remember that this 28 tons of supplies represents only about four percent of the 750 tons needed to keep the army combat effective.

Once again, the various armies surrounding the German pocket attacked and once again made little progress. On the right flank of the 64th Army, Shumilov sent in units of the 7 Rifle Corps and 97 Rifle Brigade against the German 371 Infantry Division, but they only gained 600-700 feet before being stopped. Other than keeping the Germans occupied and forcing them to expend ammunition, the fighting achieved little. The relentless series of attacks, along with winter weather, stiff German resistance, and the worsening supply situation reduced the speed and power of Russian assaults. Both sides were reaching exhaustion.

On the north side of the pocket, the German XI Corps achieved a victory of sorts by withdrawing its last units over the Don to the east side of the river. Despite vigorous attacks, the Don Front was unable to destroy this isolated corps. After suffering considerable losses in men and equipment, the XI Corps remained a vital fighting force within the 6th Army. The Stalingrad pocket measured some 37 miles east-west by 23 miles north-south.

In the south, advancing units of the 81 and 61 Cavalry Divisions of the 51st Army finally reached the city of Kotelnikovo, a critical Axis rail and supply center. The dismounted cavalry troops from the 81 Cavalry Division infiltrated into Kotelnikovo from the northwest in an attempt to capture the city straight from the march. In support, the 61 Cavalry Division approached from the east. But unknown to the Russians, a major German reinforcement to this section of the front was about to arrive at the train station.

The story started in France. The 6 Panzer Division had been refitting in northern France when on 5 November the Division was alerted to move back to Russia. During preparations for this move, on 8 November, the Western Allies launched Operation Torch, the invasion of North Africa including French Algeria. In responded to this attack, Hitler ordered German Army units stationed in northern France into the unoccupied section of southern France or Vichy France. Due to a lack of locomotives, it was impossible to move all available units to the south. So, the 6 Panzer Division was reassigned back to the Russian Front with the first units of the Division scheduled to start entraining on 14 November. General Raus, the commander of the 6 Panzer Division, insisted his Panzer Division be loaded onto trains in combat formations (i.e. "Combat Trains"), instead of the most efficient packing method, which entraining regulations required. Saving space and reducing the total number of trains and cars was all well and good, but this method also would leave separate trains unable to defend themselves. General Raus knew what it was like fighting in Russia, and he wanted his division to arrive combat-ready. This decision would prove to be the correct one. After leaving France, the 6 Panzer Division was not sent to a specific operational location but instead toward two major cities in Russia: Stalino and Belgorod. When the Russians launched Operation Uranus, the Division was diverted south to Kotelnikovo via the only rail bridge over the Don River at Rostov. After a ten-day, 2,485-mile journey, the first of 78 fully loaded trains of approximately 50 cars each were about to arrive at Kotelnikovo Station.

The two Russian cavalry divisions that were attempting to capture Kotelnikovo were not ready for this type of reception at the train station. As the first German train arrived at Kotelnikovo Station, it suddenly came under small arms and then artillery fire from Russian troops. At this point, German combat veterans literally charged off the train and engaged Russian forces in close combat all around the station. By the end of the day, Kotelnikovo was cleared as Russian troops were forced out of the city. Over the next week or so, the remaining trains carrying the 6 Panzer Division arrived in the same general area and unloaded without undue hindrance.[317] The German Stalingrad relief force was starting to come together.

Ironically, back in August the Russians had sent units of the 208 Rifle Division by rail to the Kotelnikovo Station where they were attacked and destroyed by advancing German units. Strangely, three months later, the same thing almost happened to the 6 Panzer Division as it too arrived at the same station!

To facilitate command and control issues around Stalingrad, Stavka transferred the 21st Army from the Southwestern Front to the Don Front. So, the Don Front located north of Stalingrad would contain four armies, and the Stalingrad Front to the south contained three armies.

28 November - Day Ten of Operation Uranus
Weather Report: Clear, biting wind, frost on ground, 19°F (-7°C), good visibility.

The Luftwaffe transported 101 tons of supplies into the pocket using 55 aircraft and flew out an unknown number of men.

Field Marshal von Manstein completed his first plan for the relief attack toward Stalingrad, using two corps. The LVII Corps would attack from Kotelnikovo northeast toward Stalingrad, while the XLVIII Corps would advance from the Chir River area eastward toward Kalach.[318] Von Manstein still needed quite a few days to gather this assault force, so the 6th Army had to hold until then.

Only the Southwestern and Don Fronts were free to attack vulnerable, less well-developed sections of the Stalingrad pocket's defensive line. Despite putting forth an effort, their attacks only produced minor gains.

For the Stalingrad Front, strong German positions in the south from the village of Tsybenko to the Volga blocked the entire 64th Army and most of the 57th Army from easily advancing. Despite this, both the 64th and 57th Armies conducted major attacks. The 64th assaulted German positions of the 297 Infantry Division with the 38 and 157 Rifle Divisions supported by 50-60 tanks, the 154 NRB and even aircraft from the VVS. At the same time, the 57th Army launched similar attacks to the left of the 64th Army. After gaining some ground, strong German counterattacks forced these Russian troops back to their starting point.[319] The 57th and 64th Armies were not finished yet; these attacks would continue into the next day.

By this date, the 51st Army advanced south of the Aksay River line and along the east bank of the Don River. As German forces started their build up in and around Kotelnikovo and along the Aksay River, Russian commanders recognized a growing need to reinforce defensive positions protecting the southern approaches to the Stalingrad pocket. After the recent battle at Kotelnikovo Station, General Trufanov, the commander of the 51st Army, knew his weak army was no match against German tank units. At Stavka's urging, his army continued to advance, slowly pushing back weak Romanian screening units to the south. General Trufanov did not know how many fresh enemy units awaited his army to the south. It was only a matter of time before one side or the other attacked in force.

Despite all the activity in and around Stalingrad, Stalin and Stavka were deeply involved in planning their next big attack, Operation Saturn. Stalin tried to build on the success of Operation Uranus by means of another great offensive toward the city of Rostov at the mouth of the Don River. Operation Saturn's aim was the complete destruction of German Army Group A which was currently deep in the Caucasus region far to the south of Stalingrad. By this date, the initiative clearly had shifted to the Russians with German planning efforts only responding to Russian actions.

29 November - Day Eleven of Operation Uranus

Weather Report: Gloomy weather, rainy, overcast, cloudy, snowstorms at night, 32°F (0°C), bad ground conditions, especially for tanks.

The Luftwaffe transported 46 tons of supplies into the pocket using 25 aircraft and flew out an unknown number of men.

Despite continuing attacks against German defenses along the north and west side of the pocket, the Southwestern and Don Fronts were too weak and disorganized to make a decisive breakthrough.

The 64th Army was reinforced by the 166 Separate Tank Regiment. This tank regiment was most likely shifted from the 13 Tank Corps of the 57th Army. The 64th and 57th Armies continued their attacks from 28 November against German positions along the Karavatka *Balka* which were manned by the German 29 Motorized Division, 297 Infantry Division and the remnants of the 20 Romanian Infantry Division. German troops reported fierce fighting as they claimed the destruction of 19 Russian tanks with a further 21 damaged. Needless to say, the attacking infantry forces suffered heavy losses. To counter these attacks by the 64th Army, the Germans employed 52 tanks of their own and at least 23 heavy 88 mm Anti-Tank guns.

Once again, the 57th Army launched its own attacks with much the same results. After a series of German counterattacks, the 57th Army was forced back and suffered heavy losses for little or no gain.

30 November - Day Twelve of Operation Uranus

Weather Report: Milder, sometimes sunny, partly cloudy, cold, 32°F (0°C), good visibility.

The Luftwaffe transported 129 tons of supplies into the pocket using 98 aircraft and flew out an unknown number of men.

The end of the month marked the official end of Operation Uranus.

Once again, the 57th and 64th Armies launched separate but coordinated attacks against the southern part of the Stalingrad pocket. These attacks only served to tie down German units in this area and to cause a slow erosion of their offensive strength. All along the rest of the front, various Russian armies conducted sharp but limited attacks against the defenders. However, Russian commanders knew they needed a new approach to reduce the pocket if they wanted to achieve final victory. (see Map #25)

Map #25 Stalingrad Pocket - End of Operation Uranus - 30 November 1942

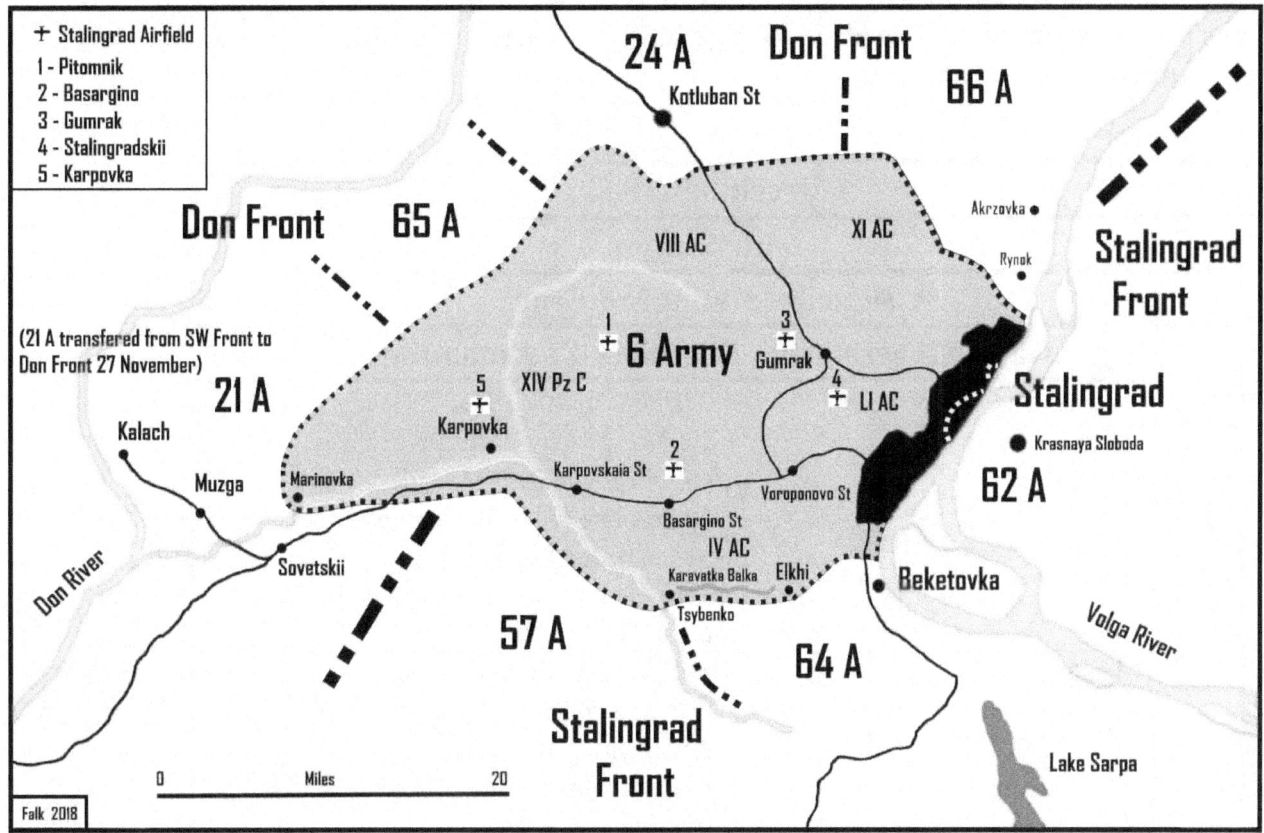

Note: Several additional, very small airfields within the pocked are not displayed.

During November both the VVS and the Luftwaffe greatly increased their sortie rates. At first bad weather hampered these efforts, but with improving conditions most missions were directed against ground targets. Firm numbers are hard to come by, but the reinforced VVS flew some 5,760 sorties from 24-30 November in the Stalingrad area to support Operation Uranus. This was about five times more than the Luftwaffe could produce.[320] Overall the VVS had more operational aircraft and flew vastly more sorties than the Luftwaffe during all of November. At first the Russian ground offensive forced Luftwaffe units away from the Stalingrad area, thus disturbing their airbase operations. Then toward the end of November, the Luftwaffe shifted its focus to supplying Fortress Stalingrad from the air. The VVS, in turn, made a huge effort to support Russian ground forces and to block German attempts to supply the Axis forces trapped at Stalingrad. In general, the struggle in the air would continue with increased intensity during December despite deteriorating weather and strained operational conditions.

Overall, Operation Uranus was a huge military success for the Red Army. Certainly, they made mistakes, and some units suffered heavy losses in men and equipment, but that did not detract from the outcome. The successful conclusion of Operation Uranus signified that Stalin and Stavka devised a workable plan to defeat the Germans, and then they saw the plan through to a successful conclusion. This required long term planning, detailed preparation, the amassing of supplies, and the secret movements of troops and equipment to key locations. All these factors and many more contributed to the defeat of Axis forces over a wide front. By doing so, the Soviet High Command demonstrated to Stalin and to the world that a rebuilt Red Army, commanded by proven leaders, could defeat the

Germans and their allies. With major German forces trapped within the Stalingrad pocket and two entire Romanian armies shattered beyond recovery, no one could deny a resurgent Red Army was a force to be reckoned with. New plans and new offensives were in the works; surely the tide had finally turned against the invaders.

1 December 1942	
64th Army	
Assigned – Stalingrad Front	
Major General M.S. Shumilov Commanding	
Rifle, Airborne and Cavalry Units	7 Rifle Corps (93, 96, 97 Rifle Brigades) 29, 38, 157, 204, Rifle Divisions 66, 154 Naval Rifle Brigades 20 Separate (Tank) Destroyer Brigade 118 Fortified Regions Separate Rifle Regiment (without number, possible Cadet Regiment)
Artillery Units	1111 Gun Artillery Regiment 186, 500, 507, 665 Antitank Artillery Regiments 3 Guards Heavy Mortar Regiment (rocket launchers) 4, 91 Guard Mortar Regiments (rocket) 622, 1261 AA Regiments
Armored and Mechanized Units	56, 235 Tank Brigades 38 Motorized Rifle Brigade 166 Separate Tank Regiment 28 Separate Armored Train Battalion
Engineer Units	328, 329, 330 Separate Engineer Battalions 175 Separate Sapper Battalion
Reported/Estimated Combat Strength Estimated Ration Strength	52,695 men 65,195 men[321]

(BSA part 4 and BSSA part 2)

With the successful conclusion of Operation Uranus, the 64th Army settled down to protracted fighting and holding the southern section of the encirclement next to the Volga. Overall, the Army did not travel far during the offensive, but its left flank attack was vital in allowing the 57th and 51st Armies to do their jobs of completing the southern part of the encirclement.

Even with units being transferred in and out of the 64th Army during Operation Uranus, the estimated ration strength of 65,195 men hardly changed from early November. This was a strong indication that personnel in supply, artillery, engineering and other supporting units suffered few losses during the offensive. Despite major offensive operations and numerous separate attacks during late November, the 64th Army maintained its overall combat strength. Only a steady inflow of replacements into the Army would have made this stability possible. One of the new additions to the 64th Army was the 20 Separate Destroyer Brigade, which most likely contained the following sub-units:

- 1 (Tank) destroyer artillery regiment, 1,000 men
- 1 mortar battalion, 300 men
- 1 separate engineer-mine battalion, 150 men
- 2 battalions of AT guns, 400 men
- 1 separate SMG company, 90 men

Approximate men assigned at full strength = 1,940

The inner encirclement directly around the Stalingrad pocket consisted of the Don Front along the north side with the 21st, 65th, 24th, and 66th Armies. The Stalingrad Front along the south side of the pocket contained the 57th, 64th and 62nd Armies. (see Map #25)

1 December

Weather Report: Unfriendly weather, snowstorms, heavy frost, snow changing to rain, bad road conditions.

The Luftwaffe transported 85 tons of supplies into the pocket using 25 He-111s and 15 Ju-52s. The now empty transport planes flew an unknown number of men out to safety.

The German 6th Army reported 4,000-6,000 wounded within the pocket. With the Luftwaffe flying out a few hundred wounded per day on empty transports, the backlog would not be cleared quickly.[322]

With the success of Operation Uranus, both sides now faced the reality of an enormously changed strategic situation. In their attempt to adjust to this new state of affairs, Stalin and Hitler each were compelled to plan and then to execute several offensive or defensive operations nearly simultaneously.

Stalin and Stavka attempted to plan, support, and then implement three different major operations following Operation Uranus:

1. Operation Ring (*Koltso*), the plan to destroy Axis forces trapped at Stalingrad.
2. Operation Saturn (*Catyph*), the plan to launch a major attack into the Donbass (Donets Basin), thereby cutting off and destroying the bulk of German Army Group South.
3. Finally, the defensive operation (not named) to prevent any Axis relief force from breaking through and freeing the 6th Army trapped at Stalingrad.

Meanwhile, Hitler and the OKH were dealing with their own plans and problems. They needed to plan, implement, and resolve the following issues:

1. Find a way to keep the troops within Fortress Stalingrad supplied and fighting.
2. Launch Operation Winter Tempest (*Wintergewittr*), the plan to break through to Stalingrad from the outside and to save the 6th Army.

3. Launch Operation Thunderclap (*Donnerschlag*) the plan for mobile units of the 6th Army to break out of the pocket and to link up with the relieving troops of Winter Tempest.

4. Prevent the unstable left flank of Army Group B from collapsing. This defensive line ran along the Chir River, then north to the Don River and then along the Don all the way to the city of Voronezh.

5. Finally, decide whether to withdraw Army Group A from the Caucasus in a timely manner.

The new commander of Army Group Don, Field Marshal von Manstein, issued Directive #1: Operation Winter Tempest would be launched on 8 December provided the required forces were available. The 6th Army was required to gather all possible mobile forces in the southern part of the pocket to be used in Operation Thunderclap.

At Stalingrad, the seven encircling Russian armies awaited a plan of action that would lead to the destruction of the German forces within the pocket. The 62nd Army within Stalingrad was in no position to do more than launch minor supporting attacks against the trapped Germans. Delays were inevitable as Stalin, Stavka, and the Front Commanders attempted to craft a workable plan after they slowly realized the trapped German forces were much larger than estimated. After some debate, they clearly understood the need to reinforce the inner encirclement directly around Stalingrad if Russian forces were to clear the pocket in a timely manner. This would be in support of Operation Ring. Furthermore, they also realized that Operation Saturn required additional powerful forces to successfully complete the planned attack against enemy units west of the Don. Finally, they recognized the need for additional troops to block and to defeat any German relief efforts directed against the outer encirclement.

As the Russian High Command tried to organize these different operations, the logistical situation made matters worse. Most Russian armies were fighting far from their established supply bases, out on the open steppe, during the winter. This would complicate any logistical planning. The tangled logistical situation was due to the advancement of the front, constant fighting, the winter weather, and limited rail capacity. In essence no working rail lines were in the newly captured lands south of the Don or west of the Volga. Everything the fighting armies needed had to be moved by truck or wagon over the frozen rivers and landscape from the nearest rail heads to the new front lines. Shortages and delays everywhere behind the lines helped to slow the tempo of Russian operations.

On the inner encirclement, the 57th and 64th Armies exchanged fire with enemy troops with no change in positions. General Shumilov and other army commanders followed orders to launch nearly constant attacks to weaken and to tie down German troops within the pocket. More attacks would follow.[323]

2 December

Weather Report: Heavy snow at night, light frost, clear, cold, overcast, then sunny, two to four inches of snow on the ground, 16°F (-8°C) day, 9°F (-12°C) night.

The Luftwaffe transported 120 tons of supplies into the pocket using 70 aircraft and then flew out an unknown number of men.

On the inner encirclement, the Don and Stalingrad Fronts conducted coordinated attacks against the southern and western sides of the pocket. The 57th and 64th Armies captured little or no ground, but the 64th Army falsely reported (to the Front HQ) that its 29 RD captured the fortified village of Elkhi. Meanwhile, the 204 and 157 RDs actually did capture some enemy frontline foxholes in the area of the German 297 Infantry Division before being stopped by heavy enemy fire.

3 December

Weather Report: Cold, wet weather, foggy, light rain, sleet, thick fog, blizzard, 32°F (0°C), visibility under 328 feet (100 meters).

The Luftwaffe transported 0 tons of supplies into the pocket and no men out of the pocket.

The armies of the Don Front launched new attacks or used the time to dig in. In the south, the Stalingrad Front armies, the 57th and 64th Armies, also carried out their harassing attacks into the face of heavy enemy fire. No progress was made. At this time, Operation Ring had low priority with Stavka because they thought that once started, the attack would easily dispose of German forces encircled at Stalingrad. Also, the armies holding the inner ring needed time to gather reserves and supplies in order to support a major attack.

Stalin signed an order approving Operation Saturn, which was to be launched on 10 December.

4 December

Weather Report: Cold, wet weather, foggy, misty, light snow at dusk, 32°F (0°C), bad visibility.

The Luftwaffe transported 143 tons of supplies into the pocket on 74 aircraft and then flew out the wounded, 356 Germans and 56 Romanians, for 5.5 men per aircraft.[324]

The Don Front resumed its attacks, using all four of its armies; they encountered heavy enemy resistance.

To make new forces available, on 4 December Stalin finally agreed to move the 2nd Guards Army to the Don Front, to spearhead the reduction of the Stalingrad pocket. But due to logistical confusion behind the lines, the trains transporting units of the army were slowed, further delaying the start of Operation Ring.

During previous weeks, Russian forces of the Southwestern Front tried to breach the thin German defense line along the Chir River. To counter these powerful Russian attacks and to hold the vital Chir River line, Field Marshal von Manstein reluctantly agreed to release the XLVIII Corp from his Winter Tempest force for use in this area. That was a hard decision to make because with this loss, Winter Tempest was reduced to only the LVII Corps poised to attack from the Kotelnikovo area.

5 December

Weather Report: Cold, wet weather, 32°F (0°C), foggy, ground fog, eight inches of snow, poor visibility, very icy roads.

The Luftwaffe transported 61 tons of supplies into the pocket using 29 Ju-52 aircraft and then brought out an unknown number of men. Note: The Ju-52 was capable of transporting 12 stretcher cases or 18 healthy men under normal conditions. So, these twenty-nine aircraft could have theoretically carried 348 to 522 men to safety, but these were not normal conditions.

The Don Front continued to attack, while units of the 64th Army remained in their current positions and exchanged fire with the enemy. In general, the 57th and 64th Armies fortified their positions and shifted some forces around. The 64th Army gained the 19, 147 and 198 Separate MG Battalions, most likely from the Stalingrad Front's reserves. The 57th Army also transferred to the 64th Army the following units: 38 MRB, 143 RB, 36 GRD, 169 RD and the 422 RD.[325] This transfer almost certainly was due to the moving of the boundary line between the 57th and 64th Armies' areas of responsibility. In this case, the 64th Army gained control of a new section of the front and the units covering that area. This shifting of responsibilities between armies and other units occurred all the time and would occur again as the situation demanded. (see Map #26)

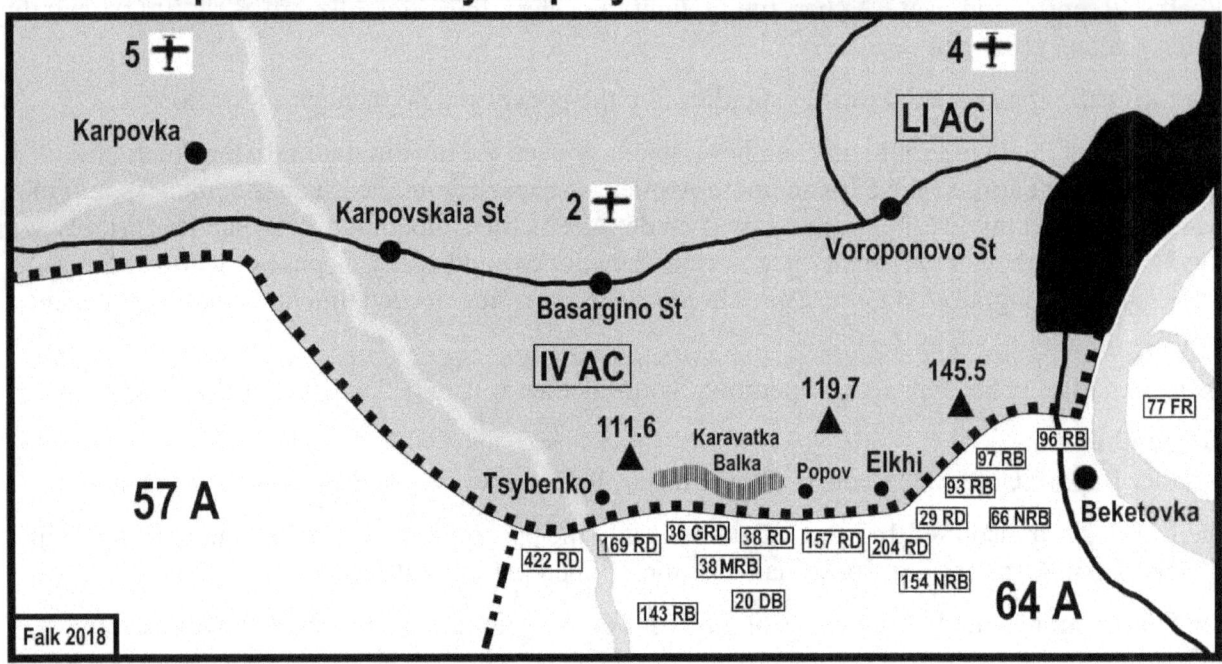

6 December

Weather Report: Cold, wet weather, light frost, ground fog, drizzling rain, 32°F (0°C), bad visibility, light icing on roads.

The Luftwaffe transported 72 tons of supplies into the pocket using 44 aircraft. The first men of the command staff from the disbanded 384 Infantry Division were flown out. That left some 243,700 Germans, approximately 11,000 Romanians, and an estimated 20,300 Russian auxiliaries (*Hiwis*) within the pocket.[326] The 57th and 64th Armies remained in their current positions with little or no fighting.

7 December

Weather Report: Very cold, foul weather, light frost, afternoon snow, 23°F (-5°C), bad visibility, icy roads.

The Luftwaffe transported 362 tons of supplies into the pocket using 135 aircraft. This was the largest single day amount of supplies so far. An unknown number of men were flown out.

Units of the 64th Army remained in their current positions while a few units from the 57th Army, holding sections of the outer encirclement, engaged the enemy along the Don River.

Also on this date, the 422 RD with 5,664 men once again appears on the order of battle for the 64th Army despite being transferred on 5 December. This illustrates the type of delay that can be found in the historical record when documenting transfers.

8 December

Weather Report: Frost, icy winds, clear, later partly cloudy, drizzling rain, light afternoon snow, bitter cold, 32°F (0°C).

The Luftwaffe transported 209 tons of supplies into the pocket using 107 aircraft and flew out an unknown number of men.

The 64th Army attacked on its right flank with the 7 Rifle Corps toward hill 145.5 at 09:30 hours, with armored trains of the 28 Separate Armored Train Battalion providing fire support. The attacking 93rd Rifle Brigade made little progress against these German positions. The 57th Army held its previous positions.

On 8 December Stalin issued Stavka Order #170699 about forming the new 5th Shock Army for the Stalingrad Front, to be used to defeat the expected German relief efforts launched toward Stalingrad.[327] Operation Saturn and Ring would have to wait for further developments as Stalin and Stavka once again shifted the focus of Russian efforts and tried to anticipate German actions.

The German Winter Tempest relief attack was originally planned to start on this day but was delayed due to lack of forces. Field Marshal von Manstein still tried to gather enough troops to make a credible attack possible. Meanwhile, Hitler continued to hold Army Group A in place, far to the south. Army Group A continued to defend its positions against nearly constant Russian attacks, but if Hitler had allowed a limited strategic withdraw, a significant number of these forces could have been released for used at Stalingrad.

9 December

Weather Report: Overcast, light snow, 32°F (0°C), poor visibility, roads passable.

The Luftwaffe transported 0 tons of supplies into the pocket and 0 men out. So far, the Luftwaffe had flowen at least 6,441 people from the pocket.

After shifting units around once more, the 64th Army attacked on its right flank with its 38 Rifle Division, this time toward the village of Peschanka. Overcoming stubborn resistance, they advanced 250-300 feet.

10 December

Weather Report: Clear, frosty, light clouds, sunny, falling temperatures, good visibility, roads icy.

The Luftwaffe transported 156 tons of supplies into the pocket using 74 aircraft and then flew out 467 wounded. That worked out to 6.3 men per aircraft.

The 64th Army fortified its positions but also dealt with several German counterattacks. On the far-right flank, the 93 RB along with the composite student rifle regiment attacked and captured several German trenches. Meanwhile the 422 RD repelled six counterattacks against its positions. The 38 RD also repelled two enemy counterattacks. Obviously, German combat troops were not passive despite being surrounded. This should have been a clear indication to Russian commanders of the tough fighting to come.

11 December

Weather Report: Cold, clear, sunny, frost, 23°F (-5°C) day, 14 F (-10°C) night.

The Luftwaffe transported 266 tons of supplies into the pocket using 117 aircraft and then flew out 449 wounded. That worked out to 3.8 men per aircraft. The ratio fell because more He-111 bombers were used as transports this day, and they could not carry as many troops as Ju-52s.

The 64th Army launched a major attack on its right flank with the 93 RB, 422 RD, 38 RD and the 38 MRB against the German 371 Infantry Division. Stubborn enemy resistance prevented any success.

12 December - Day One of Winter Tempest
Weather Report: Sunny, clear, frost, somewhat overcast, 23°F (-5°C), good visibility.

The Luftwaffe transported 114 tons of supplies into the pocket using 56 aircraft and flew out 196 wounded. That was only 3.5 men per aircraft. The VVS continued to be active, shooting down transports flying to and from Stalingrad. The VVS also bombed the Pitomnik Airfield within the pocket 42 times between 6-12 December. All this took a toll on Luftwaffe effectiveness.

The 64th Army held fast on both flanks but launched attacks in the center with the 93 RB, 422 RD and 169 RD with minor success. They captured a few trenches but achieved no breakthrough.

Almost on a daily basis General Shumilov was deliberately shifting his forces, and the focus of his attacks, back and forth along the front. He probed the enemy positions and looked for weakness. These constant attacks also served to keep the enemy pinned down and forced them to expend their limited amounts of ammunition. This rapid shifting of combat strength was possible because the Army's entire front line at this time was only about 20 miles long. So, a rifle division could launch an attack on one flank, and a few days later attack on the opposite flank. The Germans never knew when or where the next series of attacks would come from, thus greatly complicating their defensive efforts.

Field Marshal von Manstein could wait no longer. Operation Winter Tempest began; the German relief attack had to somehow fight its way through some 80 miles of defending Russian troops to reach Stalingrad. The main attack force of the LVII Corps was composed of the 6 and 23 Panzer Divisions, supported by several fragile German infantry divisions and the remnants of several more equally weak Romanian infantry divisions. This relatively weak force offered the only remaining chance to free the German 6th Army trapped at Stalingrad. Field Marshal von Manstein was not optimistic about its chances of success, but on the first day combat units advanced some 18 miles. Operation Winter Tempest had a good start.[328] Standing in their way was the depleted 51st Army, which was now in full retreat, but help was on the way. Stalin and Stavka had other forces available that they could use to defeat this new German attack.

In response to the German relief attack, the 57th Army basically went onto the defensive and held its current positions. It also concentrated the 235 and 234 Tank Regiments toward the south, but still within the 57st Army's sector, just in case they were needed. On his left flank, General Shumilov likewise decided to "create a stronger general reserve (two infantry divisions) and … (three anti-tank artillery regiments) and also at the ready, two guards mortar regiments "Katyusha".[329] The two rifle divisions that were moved to the left flank were almost certainly the 29 and 38 Rifle Divisions. They were deployed in a way to be ready to attack or to defend in either direction, to the north or south.

13 December - Day Two of Winter Tempest
Weather Report: Frost, clear, sunny, cloudless, 32°F (0°C).

The Luftwaffe transported 133 tons of supplies into the pocket using 73 aircraft and then removed 531 men from the pocket, or 7.2 men per aircraft.

Operation Winter Tempest continued, reaching the Aksay River as German units seized several bridgeheads over the river. This advance triggered a large, daylong tank battle that swirled around the village of Verkhne-Kumskii, located between the Aksay and Muschkova Rivers. At the end of the day, German tank units withdrew to the Aksay River to resupply fuel and ammunition. The badly damaged Russian 4 and 13 Mechanized Corps of the 51st Army held their positions.[330]

The 64th Army also held its positions with little or no activity. On or about this date, the Don Front ordered the 64th Army to transfer its 38 Rifle Division to the 51st Army. Ultimately the 38 Rifle Division only stayed under the command of the 51st Army until the Russians halted Winter Tempest. They eventually returned to the 64th Army in February 1943.

The 57th Army continued to hold its current positions and also transferred the 235 and 234 Tank Regiments south to the 51st Army. The Stalingrad Front started to respond to the long-expected German relief attack. The 57th and 64th Armies prepared for potential attacks by the German relief force as well as breakout attempts from the German 6th Army. Needless to say, the 57th and 64th Armies closely coordinated their efforts to contain the 6th Army within the pocket and, if necessary, to prevent a breakthrough into the pocket from the outside.

The 64th Army HQ reported their combat strength as 41,877 men. With combat operations approaching from the south and a possible breakout from the north, the 64th Army assembled and deployed all the strength it could possibly find.

After repeated debate, Hitler once again made no decision about the withdrawal of Army Group A from the Caucasus.[331] Due to the changed strategic situation, these units clearly were committed in the wrong place. They could have been redeployed to help the breakthrough to Stalingrad, but by 13 December it was too late to move them north in time.

Because of the German relief attack, logistical difficulties, and limited available forces, Stalin and Stavka decided to change Operation Saturn into Operation Little Saturn (*Malyy Saturn*). Little Saturn was a less daring but still potentially decisive offensive against Axis forces in southern Russia. Its ultimate aim was to destroy or to capture all Axis forces operating within the Don River bend. Little Saturn was to attack southward from the Don River and crush the left flank of German Army Group B. This area was defended by the Italian 8th Army, then a bit further south by Army Group Hollidt, and finally by the remnants of the 3rd Romanian Army along the northern Chir River.

Stalin also ordered the transfer of the powerful 2nd Guards Army from the Don Front to the Stalingrad Front, to counter the German relief attack. This transfer would go into effect on 15 December, which meant Operation Ring now was postponed until the Red Army defeated the German relief attack.[332]

14 December - Day Three of Winter Tempest

Weather Report: Overcast, foggy, windy, 32°F (0°C) day, lower at night.

The Luftwaffe transported 135 tons of supplies into the pocket using 85 aircraft and then flew out 233 wounded, or 2.7 men per aircraft.

Units of the 51st Army launched fierce attacks against German positions south of Verkhne Kumskii but were blocked from crossing the Aksay River. German tank units were resupplying and repairing damaged tanks. Winter Tempest could not advance until the Germans had defeated Russian mobile units in the area. At this point the VVS and Luftwaffe intervened in force, starting equally fierce battles in the air. All the while, bombers and ground attack aircraft struck at whatever targets they could find along the front and in the rear areas.

The 64th and 57th Armies held their current positions. The 64th Army transferred the 20 Separate (Tank) Destroyer Brigade to the 57th Army. This transfer was another boundary line shift as the Destroyer Brigade doesn't actually change its position on the front line.

Due to the power of the German relief attack, Stavka issued Order #170708, officially postponing Operation Ring. The order also directed the Don and Stalingrad Fronts to "continue systemically destroying the encircled enemy forces from the air and with ground forces, deny the enemy rest both by day and night, compress the encirclement ring, and halt attempts to break out from the encirclement at the onset."[333] That the Don and Stalingrad Fronts lacked the required strength to achieve these aims was irrelevant. Attacks against the pocket would continue with whatever forces were at hand.

The 2nd Guards Army continued to shift toward the south in order to block the German relief attack.

15 December - Day Four of Winter Tempest

Weather Report: Gloomy, overcast, 32°F (0°C).

The Luftwaffe transported 91 tons of supplies into the pocket using 50 aircraft and then carried out 549 men from the pocket to safety. That number was almost 11 men per aircraft. However, not all these German transports flying to and from Stalingrad made it safely back to German airbases in the rear. Fighter aircraft of the VVS and anti-aircraft guns on the ground steadily inflicted losses on these almost unarmed transports. To complicate matters, the dreadful winter weather accounted for a number of aircraft lost on both sides.

Infantry and tank units of the 51st Army continued to attack the German bridgehead positions over the Aksay River. Meanwhile, the Germans prepared for further offensive action the next day.

The 64th and 57th Armies held their current positions while exchanging harassing fire with the enemy. On this date, the 77 FR apparently was returned to the control of the 64th Army; however, it remained in the same position covering the islands in the Volga.

16 December - Day Five of Winter Tempest

Weather Report: Clear, later cloudy, rain showers, strong northeast winds, sinking temperatures, 32°F (0°C).

The Luftwaffe transported 93 tons of supplies into the pocket using 94 aircraft and flew out 612 men, or 6.5 men per aircraft.

The weak 17 Panzer Division, which had only 5-10 operational tanks,[334] was added to the Winter Tempest force. The 17 Panzer Division crossed the Aksay and advanced on the left flank of the offensive moving northward along the Don. Meanwhile strong German units once again attacked in the Verkhne Kumskii area.

The 64th and 57th Armies held their current positions, exchanging fire with the enemy and conducting reconnaissance.

To make the overall situation even more complicated for the Germans, Stavka launched Operation Little Saturn against Army Group B's left flank. Defensive positions of the Italian 8th Army were shattered by this powerful attack on the first day.

17 December - Day Six of Winter Tempest

Weather Report: Clear, cloudless, -4°F (-20°C) night, good visibility.

The Luftwaffe transported 129 tons of supplies into the pocket with 47 aircraft and then carried 461 men from the pocket, for 9.8 men per aircraft.

German and Russian forces battled in and around Verkhne-Kumskii. The area between the Aksay and Muschkova Rivers had become the decisive battleground for Operation Winter Tempest. Late in the day, German forces capture Verkhne-Kumskii.

The 64th and 57th Armies held their current positions and conduct reconnaissance against enemy positions.

18 December - Day Seven of Winter Tempest
Weather Report: Frost, cloudless, then snow, 14°F (-10°C) day, 1°F (-17°C) night, good visibility.

The Luftwaffe transported 85 tons of supplies into the pocket and removed 284 men using 31 aircraft. That worked out to 9.1 men being saved per aircraft.

The VVS was active against ground troops. Between 17-18 December, VVS aircraft of the 8th AA flew 499 combat sorties against units taking part in Winter Tempest. Everything possible was being done to stop the relief attack from reaching the Stalingrad pocket.

The 6th Army reported to OKH that its ration strength stood at "249,000 men, including 13,000 Romanians, 19,300 auxiliary volunteers *Hiwis* and attached *Zugeteilte* [attached personnel] and some 6,000 wounded."[335]

Once again Russian and German units attacked and counterattacked each other in and around Verkhne Kumskii. During the night German mobile units broke off combat with Russian forces and thrust toward the north and the Muschkova River. By dawn, a surprise German attack captured a bridge over the Muschkova River. The 6th Army was waiting and time and the strength of the relief force was running out fast.[336] Further to the north, the Russian 2nd Guards Army finally moved into position south of the Stalingrad pocket along with elements of the newly formed 5 Shock Army.

Operation Little Saturn continued to advance toward the southeast, destroying Axis forces along the middle Don River. For the German High Command, the success of this new Russian offensive placed all Axis forces in the south in an impossible position, a position they could not win. This of course was what Stalin and Stavka had intended.

Once again, the 64th and 57th Armies held their current positions while conducting combat reconnaissance against a reported inactive enemy. General Shumilov must have been concerned as the relief battle raged to the rear of the 64th Army. Powerful German forces approached from behind as the equally powerful units of the 6th Army might be gathering to launch an attack from the front; Shumilov knew that was not the best position to be in.

19 December - Day Eight of Winter Tempest
Weather Report: Overcast, frost, later sunny, 25°F (-3°C), good visibility.

The Luftwaffe transported 273 tons of supplies into the pocket and flew out 1,007 men on 146 aircraft, for 6.8 men per aircraft. Even as wounded were leaving the pocket, some men actually were trying, against orders, to get into the pocket to rejoin their units. These brave souls would rather share the fate of their comrades than to remain outside the pocket.[337]

During the day, German Winter Tempest units consolidated their positions along the Muschkova River. With only 25-30 miles separating the relief force from the 6th Army, they had reached a decisive point in the battle. Largely unknown to the German command, powerful formations of the Russian 2nd Guards Army and the newly formed 5th Shock Army already were moving to block any further German advance to the north.

The 64th Army held its position, exchanging harassing fire with the enemy. The 57th Army made minor changes to its frontline positions, almost certainly due to the German relief force approaching from the south.

On this day, a confident Stavka ordered General N.N. Voronov to create a new plan for Operation Ring. Working quickly, General Voronov presented his new plan to Stalin and Stavka in Moscow on 27 December. In essence, the plan consisted of a three-stage offensive, mostly against the west side of the pocket. The surrounding Russian Armies would force the 6th Army out of its defensive positions and push it into Stalingrad where they could destroy it. General Voronov's plan also specified the operation would require only seven days of combat to finish off the 6th Army. After making revisions mandated by Stavka and covered under Directive #171718, Stalin and Stavka finally accepted the modified plan in late December.[338] If all went as planned, Operation Ring would be launched in January 1943 and the German 6th Army would be swiftly subdued…in just seven days.

20 December - Day Nine of Winter Tempest
Weather Report: High clouds, frost, clear, light snowfall, 32°F (0°C).

The Luftwaffe transported 215 tons of supplies into the pocket with 114 aircraft and removed 1,211 men on the way back. That worked out to 10.6 men per aircraft.

With strong aircover, the 51st Army launched attacks against the German positions along the Muschkova River while the Germans consolidated their positions and held on. At the same time these attacks were taking place, Field Marshal von Manstein tried to get Hitler's approval to launch Operation Thunderclap, the breakout of the 6th Army from the Stalingrad pocket, but Hitler refused to allow it.

Once again, the 64th and 57th Armies held their current positions, conducting reconnaissance against an inactive enemy. This inactivity seemed like strange behavior for the Germans, but they were in no shape in terms of men or ammunition to perform unnecessary combat actions. In turn, the 64th and 57th Armies certainly were amassing men, supplies, ammunition and other equipment that would be needed for the final offensive against the Stalingrad pocket. The 64th and 57th Armies also were tasked with repelling any breakout attempt from within the pocket. Constant reconnaissance of the enemy's forward positions, using both active and passive methods, was necessary to spot preparations for such an attack.

21 December - Day Ten of Winter Tempest
Weather Report: Foggy, 25°F (-3°C), good road conditions, ice in fighting positions, very bad visibility.

The Luftwaffe transported 362 tons of supplies into the pocket using 144 aircraft. These same aircraft then flew out 857 men, for 5.9 men per aircraft.

German positions along the Muschkova River were once again attacked by the 51st Army with units of the 2nd Guards Army joining in. The weakened divisions of the German LVII Panzer Corps were barley holding on.

The 64th and 57th Armies fortified their positions and conducted reconnaissance activities. Officially, this was all that occurred for the entire day. However other military events were indeed taking place within the Army, but mostly behind the lines. Things like regrouping of forces, logistical provisioning, maintenance, and other necessary day-to-day upkeep activities continued endlessly within the army. All the while, thousands of men were out in the cold holding the front line as they awaited further orders.

22 December - Day Eleven of Winter Tempest

Weather Report: Light frost, foggy, surface ice, rain, later snow, 32°F (0°C).

The Luftwaffe transported 142 tons of supplies into the pocket and removed 204 men from the pocket. They used 114 aircraft for this effort, resulting in only 1.7 men per aircraft on the way out. That was an appallingly low number of rescued men.

Russian forces launched a series of wave attacks against German positions along the Muschkova River. In turn, German counterattacks regained any lost ground. Both sides were caught in a death struggle for this slice of land; Stalingrad was so close.

The 64th and 57th Armies continued to reconnoiter enemy outlying positions and planned future attacks.

23 December - Day Twelve of Winter Tempest

Weather Report: Light snow cover, bad visibility.

The Luftwaffe transported 83 tons of supplies into the pocket using 32 aircraft (all He-111) and then flew 361 men out, for 11.2 men per aircraft. That was a great improvement over the previous day's effort.

In the morning, German Winter Tempest troops received orders (presumably from von Manstein) outlining a final, massive all-out attack to the north. Starting on 24 December, the entire 6 Panzer Division would be used as an iron fist to break through to the Stalingrad pocket.[339] But later in the day, under great pressure from Russian attacks all along the southern front (Operation Little Saturn) including the Chir and Muschkova River lines, Field Marshal von Manstein ordered the abandonment of Operation Winter Tempest.[340] [341] Instead of attacking toward Stalingrad, General Hoth, commander of the 4th Panzer Army, was ordered to transfer the 6 Panzer Division to the Chir River line to counter Russian attacks in that area. At this point, the fate of the German 6th Army was sealed. It was now only a matter of time before all the encircled troops were either killed, captured, or evacuated by air if they were lucky. Alas, this final mass attack by the 6 Panzer Division actually had no chance of reaching Stalingrad. Mostly unknown to Field Marshal von Manstein, the entire 2nd Guards Army with some 120,000 men and 600 tanks stood ready behind the Muschkova River line to block any further German advance toward Stalingrad. The 5 Shock Army also stood ready to lend support. Operation Winter Tempest had ended in failure. (see Map #27)

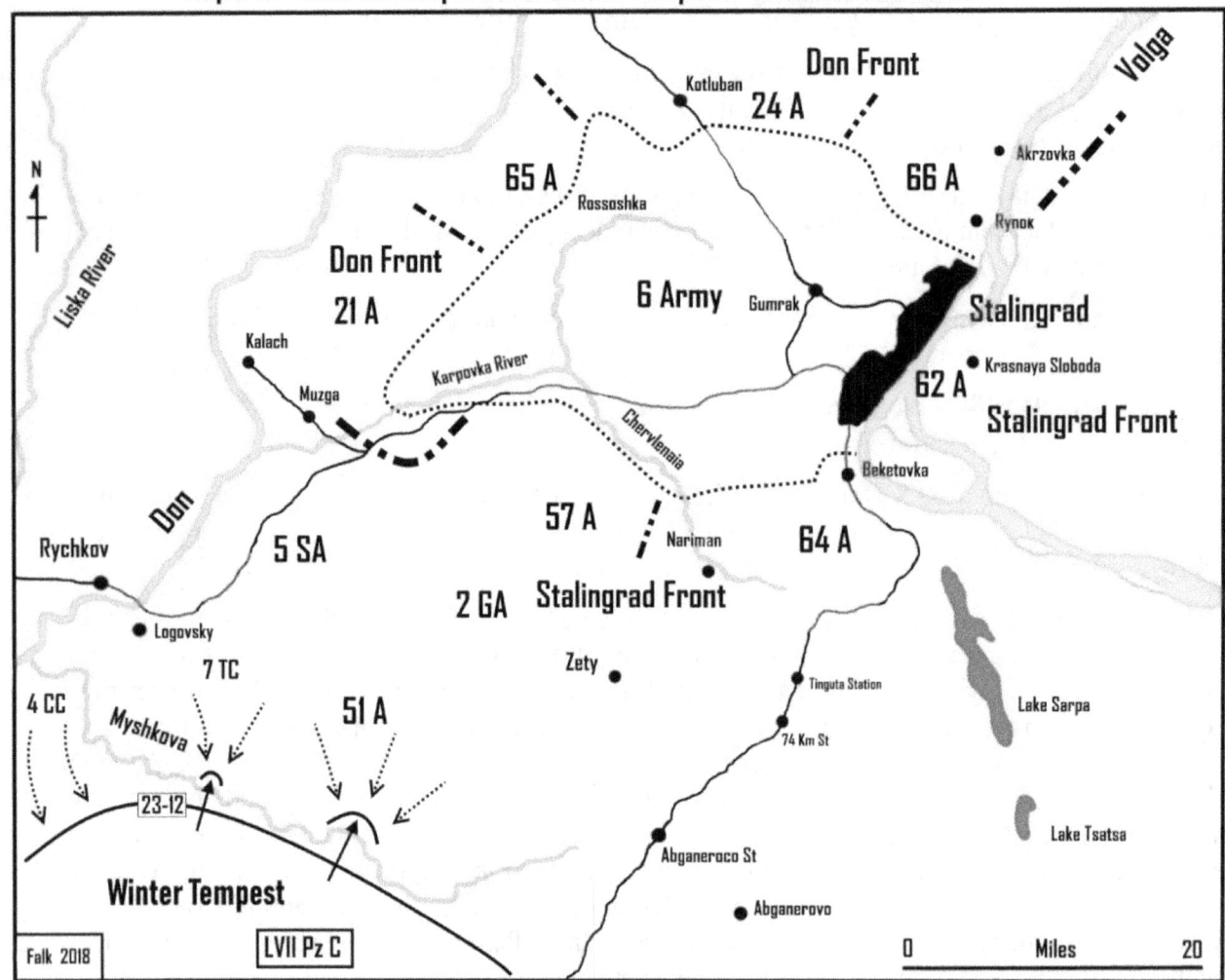

Map #27 The End of Operation Winter Tempest - 23 December 1942

The impact of Operation Little Saturn now started to dictate German responses and options. As Little Saturn continued to unhinge the Axis front, Army Group A, with some 1.5 million men, was at real risk of being destroyed.[342] As a result, Field Marshal von Manstein was forced to shift his thinking to the future and what might occur over the next several months. The trapped 6th Army had become the least of his problems. He now had to find some way to save not just an army but also an entire army group.

The 64th and 57th Armies continued to do their job of occupying their previous positions and pinning down the German 6th Army. Of note, sometime after 23 December, the 38 RD was transferred from the 64th to the 57th Army as a reinforcement for the upcoming Operation Ring.

24 December

Weather Report: Frost, fog, snowstorm.

The Luftwaffe transported 0 tons of supplies into the pocket and naturally no one flew out. The 6th Army reported only 354 casualties for the day. This was also the day the Russian 24 Tank Corps reached the main Luftwaffe supply base at Tatsinskaya. This was the so-called "Tatsinskaya Airfield Raid," which totally disrupted the German Stalingrad air supply efforts for several days.

Of course, Christmas Eve found most of the German troops contemplating their future and the holiday at the same time. Troops in the Red October plant had "carved a Christmas tree out of wooden crates. Their Christmas Eve feast was a slice of horse-meat, a piece of bread, and extra cigarettes. Someone produced some rum and a little wine. Two of the officers argued the merits and demerits of suicide."[343] Throughout the pocket, groups of men gathered together what they could to create decorations and shared what little food they had for the special day. Most knew this would be their last Christmas. On the shortwave radio, some men heard a Christmas broadcast staged by Dr. Joseph Goebbels in far off Berlin. Most were not fooled when soldiers "from Stalingrad" joined in on the broadcast, "A joyful chorus of soldiers' voices filled the airways."[344] They were singing "Silent Night Holy Night from the Volga Front" for the folks back home.[345] The broadcast was in fact pure propaganda.

The 64th and 57th Armies fortified their positions and reconnoitered against a passive enemy.

The Stalingrad and Southwestern Fronts, using the 5th Tank, 5th Shock, 2nd Guards and the 51st Army, started to attack all along the front from the Chir River to the Muschkova River. German units were forced to retreat toward the Kotelnikovo area. Operation Winter Tempest was put into reverse by the application of overwhelming Russian force.

25 December

Weather Report: Heavy frost, icy east wind, snowstorm, -22°F (-30°C) at night, early good visibility, roads bad for wheeled vehicles.

The Luftwaffe transported 7 tons of supplies into the pocket, using a reported 9 aircraft. This was a low tonnage per aircraft amount. These same aircraft should have been able to carry out 108 men, at 12 men per aircraft. But the number flown out is unknown. This poor day's effort was partly due to the ongoing Russian ground attack against the Luftwaffe's Tatsinskaya Airfield and the surrounding area. At this time, Tatsinskaya was the largest Airfield supporting the Stalingrad supply effort. Early in the morning on 24 December, advancing units of the Russian 24 Tank Corps conducted a tank raid against the Airfield, severely disrupting flight operations and causing significant losses to the Luftwaffe.[346] To further complicate evacuation efforts from Stalingrad, the Surgeon-General of the 6th Army, Dr. Otto Renoldi, barred frostbite cases from being flown out of the pocket, as they might be self-inflicted casualties.[347] What a nice Christmas gift to the wounded.

The 64th and 57th Armies held their positions. With the defeat of the German relief attack, the armies surrounding the Stalingrad pocket started to prepare, in peace, to crush the encircled German force.

26 December

Weather Report: Very heavy frost, clear, snowstorm in north, -22°F (-30°C).

The Luftwaffe transported 78 tons of supplies into the pocket with 37 aircraft and rescued 250 wounded, for 6.7 men per aircraft. Russian troops still occupied the Tatsinskaya Airfield area.

The 64th and 57th Armies held their positions and conducted reconnaissance. They too were fighting the cold and conserving strength for the final attack against the Stalingrad pocket.

27 December

Weather Report: Clear, light frost, -6°F (-21°C), good visibility.

The Luftwaffe transported 127 tons of supplies into the pocket using 97 aircraft and took 580 men out of the pocket, for 7.3 men per aircraft. Because Russian forces continued to occupy the main supply Airfield of Tatsinskaya, the Luftwaffe was forced to use secondary airfields further away from Stalingrad reducing the overall supply effort.

The 64th and 57th Armies held and improved their current positions.

28 December

Weather Report: Clear, frosty, -2°F (-18°C), good visibility.

The Luftwaffe transported 35 tons of supplies into the pocket using 10 aircraft. and flew out about 50 men, for 5 men per aircraft. German troops finally retook the Tatsinskaya Airfield, but the damage to the Airfield and supporting equipment was major. Because of this, the air supply effort was badly disrupted.

The 64th and 57th Armies held their previous positions.

Hitler issued Operations Order #2 that in part directed Army Group A to gradually withdraw to a shorter line. This was the first realistic order from Hitler concerning saving Army Group A from being surrounded and destroyed in the far south.[348] In reality, this order was hopelessly late and should have been issued weeks before. Army Group A might have been used to help free the 6th Army from the Stalingrad pocket, but now it would be considered a victory if Army Group A could reach safety itself.

29 December

Weather Report: Clear, frosty, -4°F (-20°C).

The Luftwaffe transported 124 tons of supplies into the pocket using 96 aircraft and then removed 830 men from Stalingrad, for 8.6 men per aircraft.

Becoming active once again, the 64th Army attacked on its far-left flank with its 36 GRD against the German 297 Infantry Division near Tsybenko. A small amount of progress was made, with an advance of some 600-800 feet into the enemy's forward positions with heavy fighting. The 57th Army continued to hold its positions.

30 December

Weather Report: Light frost, windy, overcast, -2°F (-18°C).

The Luftwaffe transported 224 tons of supplies into the pocket using 81 aircraft and removed 980 wounded for 12 men per aircraft. This was close to a max load for the transport aircraft.

The 64th and 57th Armies held their previous positions and prepared to participate in Operation Ring.

Stavka issued Directive #170720, abolishing the Stalingrad Front and creating a new Southern Front to improve command and control to the south of Stalingrad. At the same time, the 57th, 62nd, and 64th Armies were transferred to the Don Front's control to conduct Operation Ring. This order would go into effect 1 January 1943. So, General Rokossovsky finally would control all the armies surrounding the Stalingrad pocket. This made it much easier for Rokossovsky to launch and to coordinate Operation Ring.[349]

Also, on this date, the Soviet leadership finally started to respond to the almost unique situation of dealing with large numbers of enemy prisoners. The Deputy of the NKVD (Peoples Commissariat for Internal Affairs) A.I. Serov sent a memo to his superior L.P. Beria, head of the NKVD, regarding the treatment of POWs. Serov pointed out the high mortality rate among prisoners taken in and around Stalingrad. This memo covered four main points:

1. Romanian and Italian prisoners had not been fed by the Germans for 6-7 (and even 10 days) before being captured.

2. Once captured, prisoners were marched 125-190 miles to reach a railroad, without their own supplies or food for 2-3 days. The rear part of the Red Army was not organized to handle this POW effort.

3. The Red Army did not feed or clothe the POWs during their forced march and upon reaching the trains, they were given flour to eat instead of bread.

4. The rail wagons provided to transport the prisoners to the rear were not fitted with bunks or stoves, and each car was loaded with 50-60 prisoners. (Note: An enclosed rail car [box car] would normally carry 40 men.)

He summed up his memo by saying, contrary to orders, wounded and sick prisoners of war were not taken to frontline hospitals but were sent to the regular collection points for POWs. He then went on to say, "All these circumstances lead to physical exhaustion of war, as a result they die before sending to the rear, as well as in transit."[350] Beria responded by ordering comprehensive changes in how POWs were to be handled. Orders implementing these changes were issued by 2 January, but few changes actually were made. The mistreatment of prisoners continued, and the death rate rose even further.

31 December

Weather Report: Light frost, sharp east wind, light snowfall, moderate visibility, 12°F (-11°C).

The Luftwaffe transported 310 tons of supplies into the pocket using 158 aircraft and flew out 982 men, for 6.2 men per aircraft. Mostly wounded troops were being flown out, but healthy men were ordered out also. These healthy individuals might be stranded Luftwaffe aircrew, redundant command staff, or even specialist troops like radio operators and engineers.

The 64th and 57th Armies held their previous positions.

During the day, General Rokossovsky was on an inspection tour of the new armies that had been added to his Don Front. He first visited the 57th Army, then the 64th Army and finally the 62nd Army still surrounded within Stalingrad. Upon arriving at the HQ of the 64th Army, the following comments were made: "At the command post of the 64th Army were waiting for us the army commander Major General M.S. Shumilov, a member of the Military Council Major General K.K. Abramov, [and] Chief of Staff Major-General I. Laskin…. In the [HQ] dugout [which] was spacious, clean and warm."[351] They had previously heard about the comforts the commander of the 64th army provided himself and his staff and noted the thoroughness in which the dugout was equipped, clean and tidy, with carpets on the floor and rugs on the wall, challenging the customary idea of uncomfortable military accommodations. Before his guests left the 64th Army, General Shumilov invited them to a hearty dinner, which was appreciated by all.

Also, on this day, for his hard work over the last few months, Major General Shumilov was promoted to Lieutenant-General. As the year ended, things were looking up for the new Lieutenant-General, the 64th Army as a whole, and the Red Army overall. This promotion must have come directly from Stavka because the visiting General Rokossovsky made no mention of it.

With the defeat of Operation Winter Tempest and the retreat of Axis forces all along the southern front, Russian armies were free to exploit their victories. But to do this successfully, they had to first launch Operation Ring to finally destroy the trapped German forces in and around Stalingrad. This would not only free up troops and equipment, but it also would ease logistical problems by opening the rail lines through the Stalingrad area. These rail lines were desperately needed to support further attacks during the winter and to keep the victorious Red Army advancing! Second, they needed to expand Operation Little Saturn toward Rostov at the mouth of the Don River. If the Soviets could successfully carry out this expansion of Little Saturn (along with two additional Operations, Star and Gallop), then the Red Army could achieve further great victories that winter, namely the destruction of German Army Group A. Hitler was firmly to blame for putting Army Group A at risk by keeping it isolated from the decisive battlefield at Stalingrad. Stalin and Stavka were eager to take advantage of Hitler's strategic errors. To fully seize this opportunity, the Red Army needed to complete these two pressing operations: finish off the Axis forces trapped at Stalingrad, and concurrently capture the city of Rostov at the mouth of the Don.

Later that night, the 64th Army greeted the New Year, 1943, with a pounding display of artillery fire, followed by mortars, machine guns, and colorful rockets. Of course, all these lethal fireworks were directed toward the enemy lines, where they might do some good. From the German positions, "only star shells were fired. High-explosive rounds could not be wasted."[352] So ended a very demanding year for all those involved.

1 January 1943	
64th Army	
Assigned – Don Front	
Lieutenant General M.S. Shumilov Commanding	
Rifle, Airborne and Cavalry Units	7 Rifle Corps (93, 96, 97 Rifle Brigades) 36 Guards Rifle Division 29, 157, 169, 204 Rifle Divisions 143 Rifle Brigade 66, 154 Naval Rifle Brigades 77, 118 Fortified Regions Composite Cadet Regiment
Artillery Units	70 Guard Artillery Regiment (from 19 Artillery Division) 85 Guards Gun Artillery Regiment 1104, 1111 Gun Artillery Regiments 186, 500 Antitank Artillery Regiments 1261 AA Regiment
Armored and Mechanized Units	90 Tank Brigade 38 Motorized Rifle Brigade 35, 166 Separate Tank Regiments 28 Separate Armored Train Battalion
Engineer Units	47, 175, 328, 329, 330 Separate Engineer Battalions
Reported Combat Strength Reported/Estimated Ration Strength	51,844 men 65,243 men (see Appendix 18)[353]

BSSA - 43, and BSA - 43 (see Appendix 1)

The 64th Army was now part of the Don Front under the command of Lieutenant General Rokossovsky. The old Stalingrad Front was renamed as the new Southern Front.

By the start of January, the composition of the 64th Army had changed yet again. The last Composite Cadet Regiment was still in the line, fighting with the 93 Rifle Brigade, 7 Rifle Corps. At some point in the near future, combat units of the Army would absorb any remaining Cadets, but until that time came, they continued to do their duty. A notable shift in firepower was the removal of all the Guards mortar Katyusha rocket launcher units from the Army; they were most likely moved into Don Front reserve. Eventually, most of the Katyusha launchers would move to the western side of the pocket to support the main assault of Operation Ring in that area. Other interesting changes to note are the reintroduction of two Fortified Regions, 77 and 118, to the Army. These Fortified Regions had been in and out of army control for weeks and also were being used to hold inactive sections of the front or critical defensive areas, thus freeing up rifle units to assault the pocket. Also on this date, the 64th Army had the following sub-units attached:

- 82, 83, 84, 85, 86 Anti-Profiteer Detachments (also known as Barrier Units)
- 838 Radio Battalion
- 172 Radio Company
- 22 and 103 Officer Companies
- 62, 63, 64, 65, 66 Separate Penal Companies
- Plus one or more Field Water-Supply Companies

Mixes of sub-units like these routinely were attached to the 64th Army; however, these small units are seldom listed in the historical record, so they are difficult to identify.

Overall, the ration strength of the 64th Army had hardly changed since 1 December. This was somewhat deceptive because the composition of the Army had changed. Firepower was added to the existing strength, and that is what was needed for the upcoming fight to win Stalingrad back from the enemy.

1 January
Weather Report: Overcast, cloudy, heavy morning fog, light frost, cold, 14°F (-10°C).[354]

The Luftwaffe transported 205 tons of supplies into the pocket with 78 aircraft. The now empty aircraft flew out 863 men to safety for 11 men per aircraft.[355]

All the armies of the Don Front were busy getting ready for the upcoming Operation Ring offensive. The newly approved offensive plan called for the reduction of the pocket by means of powerful attacks primarily against the far western side of the pocket. The general idea was to break through the German defenses in the west, thought to be the weakest area, and to collapse the pocket in toward Stalingrad. The Soviets hoped this approach would bypass the strongest enemy defensive positions directly to the north and south of the city. This new plan also put an emphasis on the use of firepower to overcome German defenders and not mass infantry attacks by the weakened encircling armies of the Don Front. The 64th Army continued to perform reconnaissance and artillery fire missions against German units along their section of the front line. The 64th Army also conducted a company-size reconnaissance attack against the 371 Infantry Division. Additionally, the 57th and 64th Armies were chosen to launch major but largely supporting attacks along the southern part of the pocket during Operation Ring.

General Rokossovsky stated that GHQ warned him not to count on reinforcements in infantry or tanks for the offensive. He went on to say that by the beginning of Operation Ring, the Don Front had received some 20,000 replacements from GHQ reserves, but he said "this was a drop in the ocean" for the depleted armies. He was able to gather another 10,000 men by "cutting down logistical personnel, screening hospitals and medical battalions and placing everyone capable of carrying arms in the line." He clearly was concerned about amassing sufficient troops to reduce the pocket quickly. He went on to say that GHQ did assign to the Don Front a few artillery regiments and three guards armor regiments, but "[t]hat was all we could count upon. For the rest we had to rely on the means at our disposal to carry out our mission."[356] Of course the trapped German formations could count on no such reinforcement. They had to keep fighting with whatever they had on hand, which became more difficult every day.

2 January

Weather Report: Icy, wet weather, fog, drizzling rain, 32°F (0°C), visibility 300 feet (91.44 meters).

The Luftwaffe transported 0 tons of supplies into the pocket and no one came out. This result was due to terrible weather conditions.

On the southern part of the pocket, the 57th Army conducted a strong reconnaissance attack, this time against the German 297 Infantry Division. The 57th and 64th Armies took turns probing German defenses.

Also on this day, orders from General Eremenko reassigned the 28 Separate Armored Train Battalion from the 64th Army to the newly established Southern Front. During its stay with the 64th Army (July-Dec), the 28 Separate Armored Train Battalion performed over 89 artillery barrage attacks against German positions, expending some 13 thousand rounds of 76-mm ammunition. The Battalion also endured at least 15 enemy air raids and was able to shoot down several German aircraft in the process. The 64th Army had lost a good and loyal support unit.[357]

3 January

Weather Report: Light icing, windy, light frost, broken clouds, 32°F (0°C), 23°F (-5°C), at night, good visibility, black ice on roads.

The Luftwaffe transported 168 tons of supplies into the pocket using 97 aircraft and then evacuated 554 men, for 5.7 men per aircraft.

The 57th and 64th Armies performed reconnaissance and artillery fire missions against German units along their section of the line. This was mostly harassing fire, with a few small, quick ground actions.

4 January

Weather Report: Sunny, clear, light frost, sharp wind, 18°F (-8°C), good visibility.

The Luftwaffe transported 270 tons of supplies into the pocket using 145 aircraft and then flew out 1,220 men, for 8.4 men per aircraft.

Armies of the Don Front maintained their harassing attacks against the 6th Army as preparations continued for Operation Ring.

The NKVD's 21 Rifle Brigade (People's Commissariat for Internal Affairs) arrived in the Stalingrad area and began to secure the rear area behind the 57th, 62nd, 64th and 66th Armies. Eventually these NKVD troops would be charged with securing all prisoners of war and searching for enemies of the state.

5 January

Weather Report: Clear, later cloudy, heavy frost, sharp east wind, 3°F (-16°C), good visibility.

The Luftwaffe transported 161 tons of supplies into the pocket and flew out 613 men using 53 aircraft, for 11.5 men per aircraft.

While the Don Front prepared to assault the Stalingrad pocket and the expanded version of Operation Little Saturn continued to advance, a review of the medical consequences of these victories might be in order. Simply put, how was the Red Army and the Soviet Military Medical Service dealing with the casualties generated by all this fighting?

By this stage of the war, the Red Army had a fairly well worked out doctrine covering medical care for combat and non-combat casualties. For instance, within the 64th Army this doctrine stated that medical care started with:

- The wounded themselves. They "used their own means (first aid kit, or other means) for first aid treatment 5.9 percent of the time."

- If they were unable to help themselves, starting right behind the front, medical orderlies and stretcher-bearers gave immediate first aid to wounded troops, some 53 percent of the time, during their transport to the rear.

- Next, the unit and battalion doctors gave a total of some 9 percent of the aid.

- The remaining 32 percent of all medical care was given at division level – army level – front level hospitals – or at deep rear area hospitals. (Percent values are averages for the entire war.)[358]

This arrangement was typical for most armies in World War II. The first medical care started during and immediately after combat, and then continued and expanded as the wounded were evacuated to larger medical facilities located in the rear.

This basis for combat medical support was modern, but the Russian execution of the process was lacking throughout the war.

For instance, during Operation Ring, Don Front ambulances were not available for use by the 64th Army despite a stated need for 60-80 ambulances to move seriously wounded to the rear, (that is the Front's rear, outside the 64th Army's zone of control). At the time, the entire 64th Army only had 48 ambulances available, and of these, only 24-26 vehicles were operational. This problem was only partially overcome when General Shumilov finally ordered empty supply trucks to be used to transport the seriously wounded to the rear. Compounding this problem, and due to the remoteness of the Don Front's hospitals, the 64th Army was forced to transport their badly wounded to hospitals within the old Stalingrad Front/new Southern Front. But this arrangement ended when the 64th Army became part of the Don Front and therefore an outsider to the Southern Front. So, the Southern Front hospitals, which were themselves overflowing with wounded, stopped accepting even more wounded from the 64th Army. Therefore the 64th Army had to largely care for its badly wounded using only internal resources.[359] Needless to say, these makeshift arrangements caused much suffering and many unnecessary deaths.

But the 64th Army did appear to have, by Soviet standards, adequate medical resources to call upon. At the time of Operation Ring, the 64th Army most likely had at its disposal the following:

- One Field Evacuation Point, #24 (to transfer wounded to the deep rear).

- Ten Mobile Surgical Hospitals (or PPGs as they were known) #34, 58, 869, 2204, 2207, 2208, 2209, 2294, 2409, and 2712.

- Two Evacuation Hospitals, #3247 and 3255.

- One Infectious Hospital, #224.

- Two Vehicle Medical Companies, Auto Medical Company #97 (ambulances) and Horse Medical Company #251 (wagons).

- And various medical support facilities like medical depots, laundry detachments, medical labs, bath houses, and other sanitation services.

Because of the restricted nature of the Beketovka bridgehead area during the earlier part of the battle for Stalingrad, only some 1,400 hospital beds were on the right, west side of the Volga. The remaining medical units of the 64th Army containing some 2,700 beds were located on the far side, the left side of the Volga River. This arrangement required the wounded to make a long and hazardous trip over the river to reach medical help. Only after the launch of Operation Uranus were some of these distant medical units brought over to the right side of the Volga to directly support the army during Operation Ring. This effort increased the number of available beds directly behind the fighting; even so, it was not adequate support.

Overall, during this period, the 64th Army level medical facilities provided some 3,600 - 4,100 beds for wounded of all types. The chief medical officer for the Don Front A. I. Barabanov considered it necessary that each field army have about 8,000 beds to ensure a successful major offensive.[360] Operation Ring would show just how inadequate the 64th Army's medical resources really were.

6 January

Weather Report: Overcast, heavy frost, sharp east wind, ground fog, 7°F (-14°C), visibility moderate.

The Luftwaffe transported 49 tons of supplies into the pocket and flew out 196 wounded men on 29 aircraft, for 6.7 men per aircraft. Wounded troops were given preference to be flown out of the pocket, but others were given clearance to leave also. The OKH and 6th Army HQ considered just how to save the most men and the best officers from certain death or capture.

On the section of the front covered by the 57th and 64th Armies, all was quiet. All parties were taking a break awaiting the next big action.

Far to the south, German Army Group A at last was withdrawing in an orderly fashion. Despite Hitler's meddling, Field Marshal von Manstein was able to keep the army group moving north, out of the trap. The 6th Army at Stalingrad had to hold out a bit longer, in order to engage large Russian forces that might interfere with von Manstein's efforts to restore the southern front.

7 January

Weather Report: Clear, later cloudy, moderate frost, ground fog, afternoon fog, 16°F (-9°C), poor visibility.

The Luftwaffe transported 125 tons of supplies into the pocket and flew out 480 men on 63 aircraft, for 7.6 men per airplane.

The 64th Army made several probing attacks against the 297 Infantry Division near Elkhi. The German 297 Infantry Division reported to higher headquarters that enemy troops were amassing to its front.

8 January

Weather Report: Clear, morning snow, scattered cloud coverage, 16°F (-9°C).

The Volga River was completely frozen. Supplying the 62nd Army within the ruins of Stalingrad would be much easier now over a frozen Volga.

The Luftwaffe transported 117 tons of supplies into the pocket and flew out 308 men using 76 aircraft, for 4 men per aircraft. The ammunition supply within the pocket was reaching dangerously low levels. Restrictions were placed on the use of all munitions.

The 64th Army launched heavy attacks against the 297 Infantry Division using the 36 Guard, 157, 169, and the 204 Rifle Divisions. The 57th Army supported this attack with the 15 Guards, 38, and 422 Rifle Divisions.[361] The Germans repelled all these attacks.

With the full knowledge of Stalin and Stavka, Lieutenant General Rokossovsky, Don Front Commander, sent official emissaries to the 6th Army and Colonel General Paulus. Under a white flag, two Russian officers approached German positions with terms of surrender and were allowed to deliver their ultimatum.[362] General Paulus informed Hitler of the ultimatum and asked for freedom of action. Hitler refused and insisted the 6th Army fight to the last man and bullet. At the end of the surrender document, it stated, "Your reply is to be given in writing by ten o'clock, the 9th of January 1943." Paulus knew that with the setting of this date and time, the final assault upon the Stalingrad pocket was approaching quickly.

9 January

Weather Report: Morning clear, later complete cloud cover, sharp frost, light snow, 30°F (-1°C).

The Luftwaffe transported 349 tons of supplies into the pocket and flew out 864 men using 102 aircraft, for 8.4 men per plane. Up to this point, the Luftwaffe had flown out of the pocket approximately "29,000 wounded and 7,000 specialists." [363] Even to this day, there is no firm agreement on the total number of men flown out of the pocket.

Underscoring all these facts, Colonel French L. Maclean in his book, *Stalingrad The Death of the German Sixth Army on the Volga 1942-1943, Vol 2,* stated on this date, "The Sixth Army reported to the OKH the following daily food allowance for soldiers in the encirclement: 2.6 ounces of bread, 0.85 ounces of vegetables, 7.05 ounces of horsemeat, 0.42 ounces of fat, 0.31 ounces of coffee and one cigarette." In many different ways the end for the 6th Army rapidly approached.

The 64th Army launched harassing attacks against the 371 Infantry Division near Kuporosnoe.

The Soviet High Command waited for a reply to their ultimatum from General Paulus, but by the end of the day, they knew General Paulus would not or could not surrender the 6th Army. Operation Ring was set to start the next day. That evening, German troops reported mostly a quiet night, nothing special. That would change in the morning.

10 January - Day One of Operation Ring

Weather Report: Very beautiful winter day, moderate frost, light overcast, good visibility, 21°F(-6°C).

During the day, the Luftwaffe transported 161 tons of supplies into the pocket and flew out 942 men using 102 aircraft, for 9.2 men per plane. In turn, to support Operation Ring, the VVS launched 642 sorties against the Stalingrad pocket.

At the start of Operation Ring, the 64th Army reported a combat strength of 53,748 men.[364]

The Army also had 51 tanks assigned, in three armored units.

- The 90 Tank Brigade had 20 tanks: 1 KV-1, 12 T-34, 3 T-60, and 4 T-70.
- The 35 Separate Tank Regiment had 11 tanks: 7 T-34, and 4 T-70.
- The 166 Separate Tank Regiment had 20 tanks: 11 T-34, and 9 T-70.[365]

The Army's artillery totaled 291 guns, (45 of them 152 mm), and 1,071 mortars.

Overall, the seven armies of the Don Front had overwhelming forces available to crush the remaining Germans within the pocket:

- 281,158 Soviet men vs. roughly 212,000 Axis troops. Many Axis troops were starving, sick men within the pocket.
- 264 Soviet tanks vs. 125 Axis tanks. Some Axis tanks were without fuel and most were low on ammunition.
- 1,702 Soviet artillery and 6,247 mortars vs. 1,222 Axis artillery and mortars. Most Axis guns were low on ammunition.
- 400 VVS Aircraft vs. approximately 100 Luftwaffe aircraft (not counting transports).[366]

Operation Ring began. The plan was to use firepower to overwhelm the defenders instead of relying upon masses of tanks and troops to do the job, and that is just what General Rokossovsky delivered. Over the next 23 days, the duration of Operation Ring, the armies of the "Don Front reportedly fired some 24 million rifle and machine gun rounds, 911,000 artillery shells and 990,000 mortar shells."[367] Ammunition expenditures at these levels led to only one conclusion: the upcoming battle was going to be a slaughter.

At 06:50 hours, German positions in the western section of the pocket were pounded by a 55-minute artillery barrage followed by heavy attacks by the 24th, 65th, and 21st Armies. Fighting was at times hand-to-hand. Late in the day, Russian units broke through German defenses and advanced toward the Rossoshka River. The VVS added to the destruction as hundreds of Russian aircraft roamed overhead, attacking at will all over the pocket and acting in support of the ground attack.

At 09:00 hours, the 57th and 64th Armies joined in, launching their own massive barrage and assault against the south side of the pocket. The 64th Army concentrated almost all of its artillery on the far-left flank of the army to support the selected breakthrough site. Directly west of this area, the 57th Army attacked with its right flank in tandem with the 64th Army, and they deployed their artillery accordingly. Once again, the army boundary between the 57th and 64th Armies had been shifted, and with this shift, some units also had been transferred.

In all the 64th Army launched its initial assault with three rifle divisions, 36 GR, 204, 157 RD and one rifle brigade, 143 RB, supported by the 90 TB and the 35 and 166 TR. Additionally, the Don Front allocated four guard mortar rocket regiments (M-13 launchers) and one heavy guard mortar rocket brigade (M-30) to support this southern attack. These rocket launchers were very mobile, so their appearance in support could have occurred quickly, just before the opening barrage.[368]

For the southern attack, the main objective for the 57th and 64th Armies during the first phase of Operation Ring was to break through German defenses, to thrust north to reach the main east-west rail line and to capture the rail stations of Basargino and Voroponovo. (see Map #28)

The 36 GR and 204 RD attacked toward hill 116.6 and advanced 1.5 miles on their right flank against opposition from the German 297 Infantry Division. The 143 RB and 157 RD reportedly captured hill 119.7 and enveloped the village of Popov. On the far-right flank of the Army, the 96 and 97 RB of the 7 RC launched supporting attacks against the German 371 Infantry Division and seized some trenches. The 57th Army, supported by 55 of their own tanks, captured ground around Tsybenko and forced back the German defenders. By the end of the day the 57th and 64th Armies had advanced some 2 miles into German defensive positions, and the Germans were too weak to force the attackers back.[369] The advance would continue the next day.

Map #28 Start of Operation Ring - 10 January 1943

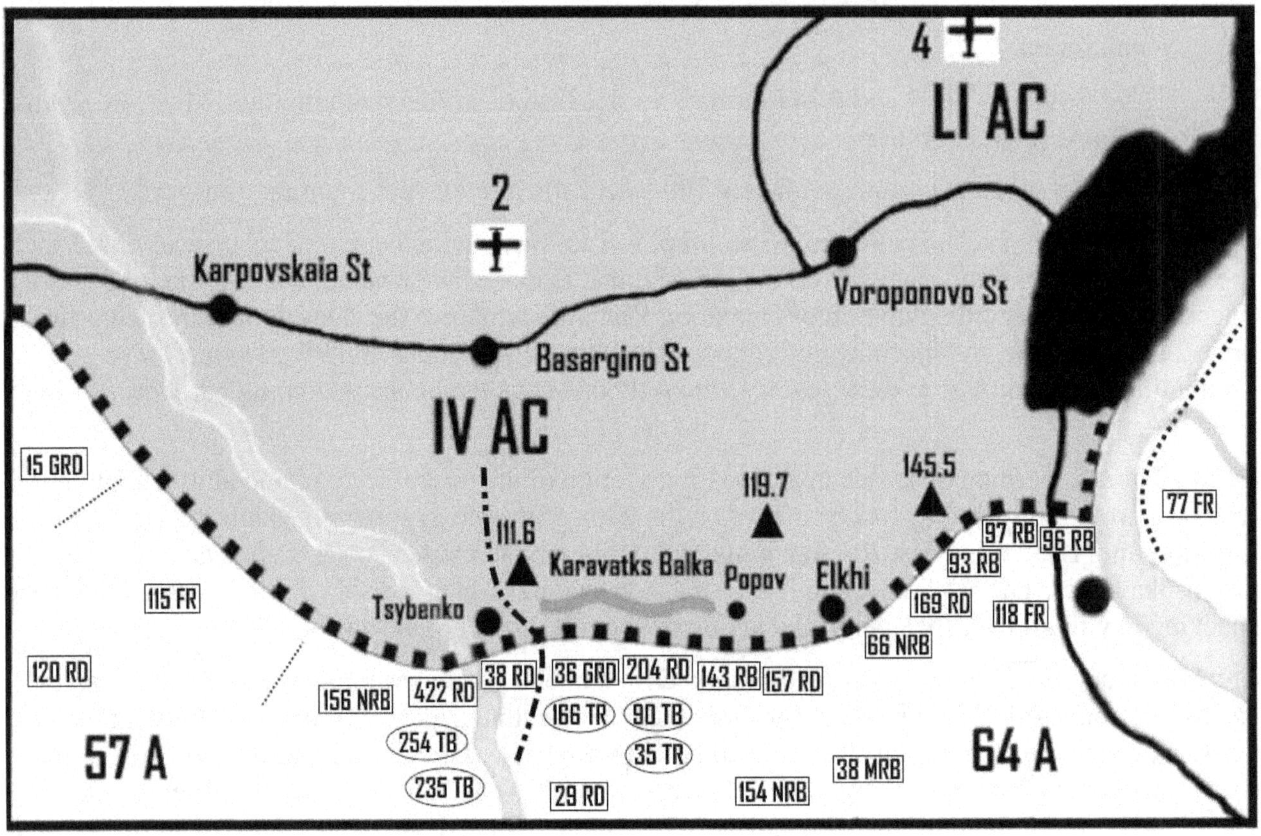

11 January - Day Two of Operation Ring

Weather Report: Light frost, sharp northeast wind, sinking temperatures, 19°F (-7°C), poor visibility.

The Luftwaffe transported 189 tons of supplies into the pocket and flew out 460 men using 95 aircraft, for 4.8 men per aircraft.

The VVS launched 73 combat sorties against the Stalingrad pocket.

The attacking 24th, 65th, and 21st Armies on the western side of the pocket continued their advance toward the Rossoshka River. The German defenders were unable to hold them back and were forced to give up ground. During this retreat, the lack of fuel forced the abandonment of much heavy equipment and vehicles, severely reducing German firepower during the remainder of the battle.

In the south, the 57th Army held fast to its gains, but the 64th Army pushed forward, assaulting the badly weakened German 297 Infantry Division. Their goal for the day was to capture more ground, inflict casualties on the Germans and pin down any reserve units in the area, thus making the main attack on the western side of the pocket easier. On the right flank of the Army, the 7 Rifle Corps repelled several enemy attacks. On the left flank, the 29 Rifle Division and the 154 Naval Rifle Brigade were committed to the battle, which helped to expand the penetration into enemy defenses. Units advanced slowly in the face of stubborn resistance.

12 January - Day Three of Operation Ring
Weather Report: Heavy night snow, heavy frost.

The Luftwaffe transported 61 tons of supplies into the pocket and flew out just 10 wounded men. A total of 51 aircraft were used for this effort. The reason for this dismal evacuation performance is unknown, but the Luftwaffe could do more to fly men out of the pocket.

The VVS launched 433 sorties against the Stalingrad pocket.

The armies on the western side of the pocket continued their advance toward the Rossoshka River. German forces in this area desperately tried to build a new defensive line to stop this attack.

On the southern section of the pocket, after heavy fighting and using an intensive artillery bombardment, the 57th Army captured the town of Tsybenko using the 422 Rifle Division with the support of the 154 Naval Rifle Brigade from the 64th Army. In the 64th Army sector, the 157 Rifle Division once again assaulted hill 119.7 from its southern slopes, despite reporting that it had captured the hill the previous day. The supporting 143 Rifle Brigade was lodged on the southwestern slopes of hill 119.7, and the 204 Rifle Division was on the western side of hill 119.7. Needless to say, confused fighting took place throughout the area. On the left flank, the 36 Guards Rifle Division finally reached and captured hill 111.6. The 36 Guards Rifle Division then pushed about a mile further north supported by the 29 Rifle Division and the 204 Rifle Division. Despite these efforts, and with only 13 tanks remaining in support out of the original 51 tanks, the 64th Army only achieved minor gains during the day.

13 January - Day Four of Operation Ring
Weather Report: Clear, frost, 14°F (-10°C) day, -4°F (-20°C) night.

The Luftwaffe transported 224 tons of supplies into the pocket using 69 aircraft and flew out an unknown number of men. But one of these lucky men was the commander of the 20 Romanian Infantry Division, Major General Nicolae Tataranu. His division had been disbanded, so he was redundant and ordered out of the pocket. By this date, the Germans had lost the small Airfield north of Karpovka, so only four airfields continued to operate.

The VVS launched 541 sorties against the Stalingrad pocket. In turn, the Luftwaffe only flew supply transport missions to Stalingrad. Any available fighter or bomber aircraft were fully engaged elsewhere.

On its right flank, the 57th Army continued to develop its attack toward the north with the 38 Guards Rifle and the 422 Rifle Divisions. On the far-left flank of the 57th Army, the 15 Guards Rifle Division launched attacks against German positions. This effort supported the advancement of the neighboring 24th, 65th, and 21st Armies.

In the 64th Army sector, the 157 and 204 Rifle Divisions advanced around hill 119.7 with support from the 143 Rifle Brigade and the newly introduced 154 Navel Rifle Brigade from the army reserves. Clearly General Shumilov was trying to force a breakthrough in this area. On the far-left flank of the army, the 36 Guards Rifle Division and the 29 Rifle Division slowly advanced to the north. The German 297 Infantry Division suffered under these heavy attacks but vigorously defended every inch of ground.

In general, the Don Front armies on the western side of the pocket used the next two days to regroup forces, bring up supplies, and prepare for the next phase of the battle. But fighting at relatively low levels still went on in other locations.

14 January - Day Five of Operation Ring

Weather Report: Ground fog, fog, cloudless, frost, later clear, -4°F (-20°C), good visibility.

The Luftwaffe transported 65 tons of supplies into the pocket using 74 aircraft and flew out an unknown number of men, but they included 11 German officers who were no longer needed within the pocket. The 6th Army was evacuating staff officers and experts from the pocket who would be useful for the war effort in the future.

The VVS launched 147 sorties against the Stalingrad pocket.

The attacking Russian forces on the western side of the pocket were delayed temporary by German units defending behind the Rokossovsky River line.

In the south, the German 297 Infantry Division reported its right flank was open, not supported by other friendly units. The 57th and 64th Armies exploited this gap in the German defensive line. Both the 57th and 64th Armies made progress toward the north, with the 57th Army crossing the Chervlenaia River in force and advancing the furthest. The western and southern sections of the German pocket started to crumble, mostly due to intense fighting and to low Axis ammunition stocks.[370]

Other Don Front armies continued to regroup even while the 57th and 64th Armies kept pressure on the Germans along the southern part of the pocket. (see Map #29)

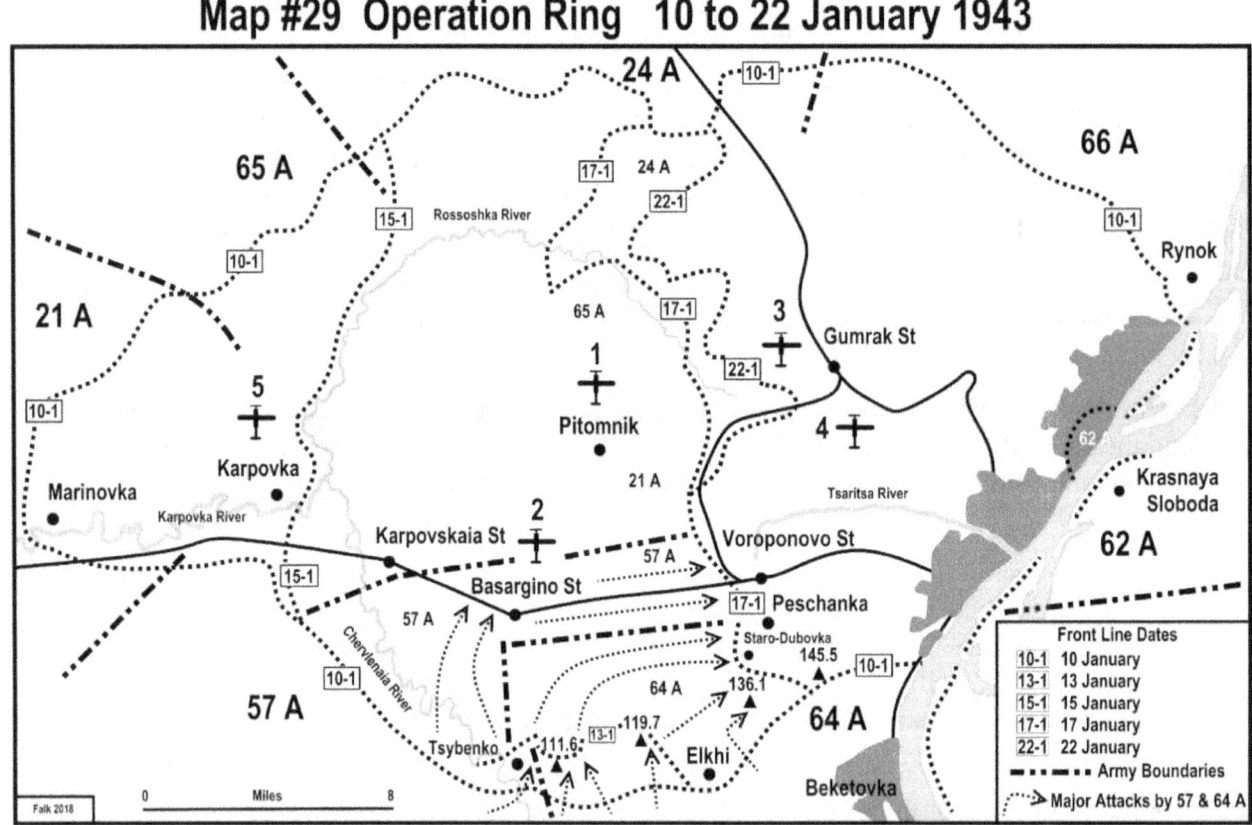

15 January - Day Six of Operation Ring

Weather Report: Clear, frost, cloudless, -4°F (-20°C) day, -31°F (-35°C) night, good visibility.

The Luftwaffe transported 105 tons of supplies into the pocket and flew out an unknown number of men using 56 aircraft. The VVS launched 158 sorties against the Stalingrad pocket.

Stavka must have liked what Lieutenant General Rokossovsky was doing with the Don Front because on this day Stavka promoted him to Colonel-General.

Having completed their regrouping and resupplying activities, the 24th, 65th, and 21st Armies resumed their advance eastward over the Rokossovsky River line. Defending German units were driven back; lack of fuel forced them to leave large amounts of equipment behind. The pocket was starting to collapse faster now.

The 57th Army captured Basargino Railway Station with the 38 Guards Rifle Division. The supporting 15 Guards Rifle Division advanced to the north of Basargino and captured the major Airfield there. With this loss, the Stalingrad pocket had only three remaining Airfields in operation: Pitomnik, along with the small Gumrak and Stalingradskii Airfields.

Reported losses for the 57th Army - 10 to 15 January were 2,175 casualties in 6 days of combat.[371]

- 669 killed
- 1,385 injured
- 10 missing
- 111 other losses (sick, frostbite, etc.)

Unfortunately, during this period the losses for the 64th Army were not recorded by the same source that was used for the 57th Army, but they were recorded by another source.[372] Due to the 64th Army having more units actively in combat, (5 rifle divisions, 2 rifle brigades and 3 tank brigades vs. 3 rifle divisions, 1 rifle brigade, and 2 tank brigades for the 57th Army), the losses for the 64th Army were much larger. The total reported losses suffered by the 64th Army from 10 to 15 January were 4,337 total men for all casualties.

My estimated breakdown of these losses, into the main categories, are.as follows:

- 1,084 killed (at a rate of 25 percent of total losses)
- 3,059 injured (at a rate of 70 percent of total losses)
- 43 missing (at a rate of 1 percent of total losses)
- 151 other losses (at a rate of 4 percent of total losses)

Note: These breakdown values and percentages are slightly rounded off to match the total reported losses of 4,337.

The German 297 Infantry Division attempted to stop penetrations by units of the 64th Army. However, the 204, 169 Rifle Divisions, 143 Rifle Brigade and the 90 Tank Brigade overwhelmed the central and western positions of the 297 Infantry Division, causing such heavy losses this German division become combat ineffective.[373] On the right flank of the army, the 7 Rifle Corps attacked with its 93 and 97 Rifle Brigades, penetrating into positions held by the German 371 Infantry Division. These two weakened German divisions were unable to resist the advancing 64th Army. The 36 Guards and the 29 Rifle Divisions fought behind the main German defensive line, in the Gornaia Poliana area north of hill 119.7. The 154 Naval Rifle Brigade fully captured hill 119.7 after fighting in this area for several days. The 157 Rifle Division pushed forward on the right and fought in the area northeast of hill 119.7. Adding to these successes, the 66 Naval Rifle Brigade achieved a long overdue victory by finally capturing the fortified village of Elkhi. This vital village position originally had been scheduled to be taken during the early days of Operation Uranus, some seven weeks before. With this village firmly in Russian hands, the entire German southern defensive position was outflanked and greatly weakened. Note: So as not to block the advance of other attacking armies, both the 57th and 64th now were forced to turn their attacking units 90⁰ to the right and head directly toward southern Stalingrad. Remarkably, this was the same maneuver the 4th Panzer Army undertook back in early September when it too was advancing upon Stalingrad.

16 January - Day Seven of Operation Ring

Weather Report: Snow, cloudless, 1°F (-17°C) day, -13°F (-25°C) night, good road conditions.

According to the revised plan for Operation Ring, 16 January was scheduled to be the final day of combat against the Stalingrad pocket. Apparently, General Voronov had underestimated German tenacity. Despite suffering tremendous losses and knowing they were fighting a hopeless battle, the German determination to keep resisting had not yet been broken! The struggle continued mostly because the average German soldier had a deep sense of duty; they simply followed orders and kept fighting. But at a more basic level was a great fear of being captured by the Russians. They knew the enemy did not treat Prisoners of War (POWs) very well. Both sides had committed too many atrocities during earlier fighting for either side to trust the other now. The battle would go on until the defenders could no longer resist.

The Luftwaffe transported 68 tons of supplies into the pocket using 39 aircraft and flew out an unknown number of men.

The VVS launched 612 sorties against the Stalingrad pocket. With VVS sortie rates like these, the Luftwaffe was hard pressed to do little more than send in a trickle of transports to supply the pocket.

Due to the lack of fuel, the 6th Army no longer was able to deliver supplies to the defending forces fighting on the western side of the pocket.

The attacking 21st, 24th and 65th Armies on the western side of the pocket continued their advance toward the east and approached the large Airfield at Pitomnik.

The 57th and 64th Armies advanced to the north and east. The 57th Army held its position on the left flank of the 64th Army as both armies wheeled to the right and headed for the southern end of Stalingrad.

Units of the 7 Rifle Corps reached positions near hill 136.1. The 169 Rifle Division advanced to the area just south of Peschanka with the 29 Rifle and 36 Guards fighting nearby. The 66 Naval Rifle Brigade and 157 Rifle Division fought in and around the village of Staro-Dubovka. All other units were holding their previous positions, moving up, and/or policing the rear areas. The German 297 and 371 Infantry Divisions attempted but failed to block the advancing Russian units in their area. Fighting continued.

17 January - Day Eight of Operation Ring

Weather Report: Cloudless, icy fog, 0°F (-18°C) day, -13°F (-25°C) night.

The Luftwaffe transported 52 tons of supplies into the pocket using 57 aircraft and flew out an unknown number of men.

Pitomnik, the only major Airfield still operational within the pocket, fell to Russian troops. Because of this, Luftwaffe supply and evacuation efforts relied upon the last two small Airfields that remained operational, Gumrak and Stalingradskii.

The VVS launched 960 sorties against the Stalingrad pocket. The Luftwaffe was powerless to stop these attacks.

By this date the pocket had been reduced by about two-thirds of its original size. The attacking Russian armies took turns periodically stopping their advance to replenish ammunition and supplies and also to regroup their forces. Meanwhile the defenders attempted to form a new, hasty defensive line out in the open steppes. The ground was frozen solid, so digging trenches was out of the question.

With the 57th and 64th Armies approaching Stalingrad proper, the frontages for each army had been greatly reduced as the pocket was compressed. At this time, the 57th Army attacked on a front of about 3 miles, and the 64th Army attacked only along about 5 miles of its total 8-mile front.

The 169 Rifle Division fought southwest of hill 145.5 with the 157 Rifle Division nearby. Hill 145.5 (also called Bald Mountain) was the site of much fighting during the previous months. The 204 Rifle Division attacked Staro-Dubovka, enveloping it from the north and south with the support of the 143 Rifle Brigade. The 29 Rifle and the 36 Guards Rifle Division fought on the approaches to Peschanka. Meanwhile, units of the 7 Rifle Corps held the line next to the Volga.

18 January - Day Nine of Operation Ring

Weather Report: Icy fog, sharp east wind, snow storm, 5°F (-15°C).

The Luftwaffe transported 24 tons of supplies into the pocket using 42 aircraft and flew out an unknown number of men. The Gumrak Airfield was still in full operation despite the reduced size of the pocket and the proximity of Russian troops.

The VVS launched 758 sorties against the shrinking Stalingrad pocket. With nearly 8.5 hours of daylight at this time of the year, that meant about 89 aircraft attacked each hour. However, some of these sorties must have been flown at night. The Luftwaffe could do little to stem this crushing wave of Russian airpower.

Due to the unexpected fierce resistance from German forces, General Voronov, Stavka representative at Stalingrad, and General Rokossovsky submitted a new plan to Moscow for the destruction of the remaining pocket. This plan called for a regrouping of forces to launch a powerful blow, using the 21st, 24th, and 65th Armies against the German defenders in selected areas on the western side of the pocket. "To organize such a blow, two to three days are necessary for the concentration, deployment, re-supply of ammunition, and also the necessary respite for the infantry. The intended start date was 20 January." [374] Stalin quickly approved the plan without amendments or reservations that same day. The 57th and 64th Armies would support this powerful attack by assaulting German positions south of the main rail line and into the southern part of Stalingrad.

With the approval of this new plan, the entire Don Front rested and resupplied its forces from 18-21 January. General Paulus also took advantage of this pause in operations to regroup his remaining forces as much as possible.

Paulus must have known the next series of Russian attacks would be overwhelming and would most likely mean the end of the 6th Army. The few available German combat troops formed into a new, thin defensive line, covering the approaches to the city. Because the ground was frozen rock hard, these exhausted troops were forced to scrape at the snow and ice to build up some type of protection. These hardships did not really matter anymore, because ammunition and food were almost gone. For the next four or five days, these desperate men waited out in the cold for the next crushing attack; they knew the end was near. On the other side of the line Russian forces also suffered from the cold, but they at least had warm clothing and food to reduce the impact of winter. Meanwhile, behind the Russian lines, artillery and tanks regrouped, supplies were brought forward, and the troops rested as best they could. General Rokossovsky was in a hurry to finish off the Germans, so preparations for the final offensive moved along swiftly. Rokossovsky also knew that Stalin too was waiting for his final victory in the city of Stalin!

The 57th and 64th Armies also rested and resupplied. But even during such a respite, the action continued with reconnaissance efforts and low-level skirmishes taking place all along the front line.

19 January - Day Ten of Operation Ring

Weather Report: Light snow, sharp east wind, 23°F (-5°C), poor visibility.

The Luftwaffe transported 68 tons of supplies into the pocket and flew out an unknown number of men using 66 aircraft. The VVS continued to pound the pocket, launching 466 sorties against German positions.

By this date the pocket was greatly reduced, with the Red Army overrunning enemy positions and coming upon the results of battle. Mikhail Alexeyev, an infantryman with the 64th Army, was looking for a place to rest one night.

> I was looking for a place to hide until morning, to get some sleep. There were many German positions [that were] safe, but nothing was available. I couldn't use them, because the bodies of the German soldiers took up all the ground. They were everywhere, piled high on the fields and in the dugouts. I could not find one piece of open field nor any unoccupied dugout. There were also so many maggots because of the dead bodies. And oh, the lice, the lice. I saw a dugout in the snow. I had warm clothing so I went in there and bumped against something very stiff. It was dark, so I couldn't see what it was. I thought they were sacks of something. So I made myself comfortable lying on these sacks. In the light of the morning, I saw that I was sleeping on the bodies of killed German soldiers.[375]

The 57th and 64th Armies continued to skirmish with enemy forces. Besides fighting the enemy, winter conditions added to the difficulties for both armies. "Other losses" added to the totals of dead and wounded as cold and disease took their toll.

During the preceding four days of moderate and light combat, 16-19 January, the estimated casualties for the 64th Army were about 1,600 men.

The estimated breakdown of these losses into the main categories were

- 400 killed
- 1,120 injured
- 16 missing
- 64 other losses

20 January - Day Eleven of Operation Ring

Weather Report: Snow, cold, 14°F (-10°C).

The Luftwaffe transported 52 tons of supplies into the pocket and flew out 130 men using 54 aircraft, for 2.4 men per aircraft. One of the few lucky men to leave the pocket was the commander of the 9 Flak Division, Major General Wolfgang Picket.

The VVS was unable to launch any sorties against the Stalingrad pocket due to bad weather conditions.

The 57th and 64th Armies continued their reconnaissance efforts and used artillery to pound enemy positions.

The 64th Army reported a ration strength of 40,112 men and women. This represented a reduction of over 18,000 from the start of January.

21 January - Day Twelve of Operation Ring

Weather Report: Ground fog, light snow, sharp frost, afternoon clearing, 18°F (-8°C).

The Luftwaffe transported 99 tons of supplies into the pocket and flew out 390 wounded men on 108 aircraft, for 3.6 men per aircraft.

The VVS launched 93 sorties against Stalingrad.

Fighting flared up along the northern and northwestern sides of the pocket with Red Army units advancing less than two miles. The 65th and 24th Armies pushed forward as German units offered firm resistance.

On the southern aide of the pocket, the 57th and 64th Armies continued their low-level activities against enemy positions while they gathered strength for the next round of combat.

22 January - Day Thirteen of Operation Ring

Weather Report: Foggy, light snow, overcast, 14°F (-10°C), good visibility.

The Luftwaffe transported 93 tons of supplies into the pocket and flew out 19 wounded men using 69 aircraft. Army Group Don (General von Manstein) also ordered the evacuation of five outstanding young general staff officers and five outstanding ordnance officers. Russian troops were within two miles of the Gumrak Airfield. The 6th Army HQ moved from the Gumrak Airfield area to the Univermag Department Store in central Stalingrad. General Paulus once again asked Hitler for permission to surrender the 6th Army. Hitler denied his request.

The VVS launches 105 sortied against Stalingrad.

The Don Front again launched major attacks against the pocket, gaining minor amounts of ground against determined resistance. The Front's tank strength had been reduced to about 110 machines, with most of them supporting the attack against the western and southern sides of the pocket.

Along the southern section of the pocket, the 21st, 57th and 64th Armies went back into action at 10:00 hours after a 70-minute artillery bombardment by 4,100 guns.[376] The 57th Army captured Voroponovo and Alekseevka train Stations. In turn the 64th Army's 204 Rifle Division captured the Staro-Dubovka area, but heavy enemy fire prevented advancement by other units of the army. Many German noncombat units fled into the southern outskirts of Stalingrad and sought shelter from the cold and safety from the fighting.

23 January - Day Fourteen of Operation Ring

Weather Report: Night snow, heavy day fog, 28°F (-2°C), unchanged road conditions.

Seven days behind the original schedule for the destruction of the German 6th Army, and still the fighting continued!

The Luftwaffe transported 80 tons of supplies into the pocket and flew out an unknown number of men using 78 aircraft. Note: Breaker McCoy in his 2006 book, *Secrets of Stalingrad,* states that 18 wounded Croatian soldiers were flown out of Stalingrad on this day. They were the last Croatian soldiers to leave the pocket by air.

The last aircraft flew out of the pocket during the day as Russian troops captured both Gumrak and Stalingradskii Airfields. Five hundred wounded men were at the Gumrak Airfield hospital when the Russians overran the area; the Russians troops mostly ignored the wounded; they would soon die of starvation and cold. Within the tunnels of Tsaritsa Balka, behind German lines, were 3,000 badly wounded soldiers, and in the city center at the Gorki Theater laid another 600 wounded. Thousands more wounded and dying men were scattered around Stalingrad, reportedly some 12,000 in total by this date.[377] The armies of the Don Front advanced 1-5 miles in the face of decreasing enemy resistance. Don Front reported the liberation of some 15,000 starving Russian POWs and the capture of over 1,000 Germans. Every day more Germans surrendered. The battle neared its end.

The VVS launched 340 sorties against Stalingrad.

The 57th Army continued to advance toward Stalingrad. The 64th Army captured the Zelenaia Poliana and Peschanka regions and advanced some 2.5 miles to the east, pushing remnants of the German 297 and 371 Infantry Divisions into the southern part of the city.

According to historical documents, the 64th Army reported a ration strength of 38,477 men and women.

24 January - Day Fifteen of Operation Ring

Weather Report: Snow, heavy cloud cover, rain, 14°F (-10°C), icy roads, black ice.

Without an operational airfield within the pocket, the Luftwaffe was forced to airdrop, via parachute, 12 tons of supplies into the pocket using 9 aircraft. The 6th Army reported to Army Group Don that 20,000 wounded remained with no medical support within Stalingrad. Due to unsanitary conditions, German doctors also reported an increase in the reported number of Typhus cases within the army. Typhus was spread by lice and would kill a large percentage of those infected if not treated. Unsanitary conditions bred millions of lice that lived on and fed upon the bodies of their victims. In doing so, the lice infected them with the disease. After the battle, some German POWs, protesting the horrid conditions, scraped handfuls of lice off their bodies and threw them at Russian guards. The guard's response was prompt execution of the offending prisoners.[378] German troops were hungry, tired, cold, dirty and increasingly sick, and still they fought on. General Paulus and General von Manstein each independently asked Hitler for permission to allow the 6th Army to surrender. Hitler refused both of their requests.

The VVS launched a total of 1,494 sorties on this day, but an unknown number of these were directed against the Stalingrad pocket.

The 57th and 64th Armies seized Sadovaia Station. Both armies now fought in the outskirts of southern Stalingrad, with the 64th Army mainly in the Minina suburb area.[379] Their combat frontages were further reduced with the 57th Army attacking on a 1.8-mile front and the 64th Army on a 1.5-mile front.

With the Volga frozen over, General Shumilov prudently moved the 93 Rifle Brigade over to Golodnyi Island, joining the 77 Fortified Region covering that section of the front. Some desperate German troops actually tried to use the frozen Volga as an escape route out of the city. (see Map #30)

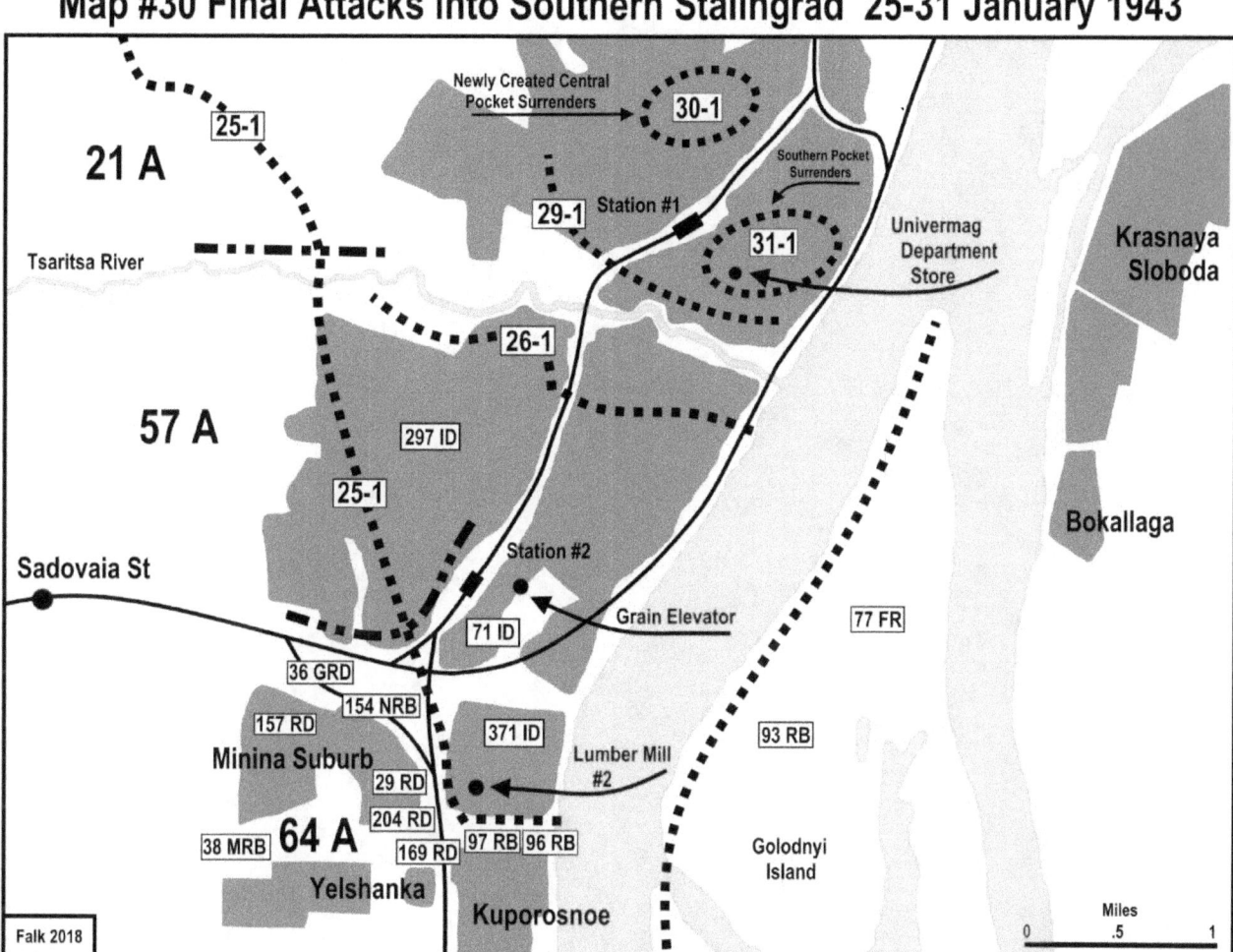

25 January - Day Sixteen of Operation Ring

Weather Report: Cloudless, clear frost, -4°F (-20°C) day, -13°F (-25°C) night, good visibility.

The Luftwaffe used 21 aircraft to airdrop 13 tons of supplies into the areas remaining under German control; resistance continued.

The VVS launched 255 sorties against the much-reduced pocket.

Russian emissaries once again approached the 6th Army front lines asking for their surrender. General Paulus refused to talk with them. Hitler sent a message to the 6th Army, "The army will hold its positions to the last man and bullet!"

Acting together, the 57th and 64th Armies seized Stalingrad's southern suburbs. The 64th Army finished capturing Kuporosnoe, Yelshanka, and the rest of the Minina suburb area as it moved north along the Volga, reaching the Lumber Mill #2 area by the end of the day. In response, the German 371 and 71 Infantry Divisions offered varying levels of resistance in the southern part of the city. German forces found it increasingly difficult to mount a credible defense.

Between 20-25 January, the 64th Army had the following reported losses: (Unfortunately, the totals covering a longer period of time were not recorded for the 64th Army.)

- 392 men killed
- 1,142 injured
- 11 missing
- 54 other losses

Those figures total 1,599 casualties during six days of combat.[380]

In comparison, during 11 days of combat, between 20-30 January, the 57th Army had the following reported losses:

- 731 men killed
- 3,176 injured
- 24 missing
- 282 other losses

Those figures total 4,213 casualties during 11 days of combat. This was more than double the losses suffered by the 64th Army during almost twice the number of days. Because of this, a confident estimate of losses for the 64th Army covering the entire 20-30 January period, are the following:

- 784 killed
- 2,838 injured
- 25 missing
- 110 other losses

Those figures total 3,757 casualties during 11 days of combat. Overall these figures are a conservative but reasonable estimate. (see Appendix 19) Note: Casualty totals for combat actions taking place on 31 January are not available. Undoubtedly units of the 64th Army would have suffered some losses during this day, probably between 100-200 men. These losses should be added to the monthly totals.

On this same day, General von Manstein, commander of Army Group Don, issued his so called "Gag Order." Order #23079 stated in part, "I demand that those officers who by virtue of their positions have access to conversations, telecommunications, etc., concerning large-scale command and logistics, will observe their duty to maintain silence, even vis-à-vis their fellow officers, about the destruction of Sixth Army."[381] Von Manstein realized, very clearly, the end was near for the 6th Army, and he wanted to reduce, as much as possible, the impact of such a loss on the officers and units under his command.

Joseph Stalin also issued a communiqué; it was a congratulatory message to Soviet forces about their victories over Axis armies during the previous two months. "Our forces have gained a serious victory.

The offensive by our forces is continuing."[382] The offensive Stalin referred to was a series of major attacks that had pushed German forces back all along the southern part of the front, from Voronezh in the north to Rostov and the Caucasus in the south. The main front line was quickly moving to the west, in the direction of Germany, and toward the end of the war!

26 January - Day Seventeen of Operation Ring

Weather Report: Foggy, -4°F (-20°C), bad visibility.

The Luftwaffe airdropped 46 tons of supplies from 52 aircraft into the pocket.

The VVS continued to support Russian ground units by flying 182 sorties against German positions in and around Stalingrad, dropping over 400 tons of bombs on the city.

On this day, units of several Russian armies broke through German resistance and linked up with troops of the 62nd Army. After months of isolation, the 62nd Army finally had direct land contact with the rest of the Red Army.[383] Not to be outdone, General Chuikov also did his part in the battle by urging his very tired troops to attack and to capture key buildings to keep up the pressure on the defending Germans.

The 57th and 64th Armies advanced to the north, reached the Tsaritsa River gully area, and captured Train Station #2 along the way. The 57th Army was positioned to the left of the main rail line and the 64th Army was wedged in on the right between the rail line and the Volga. Notably, the 157 Rifle Division captured the famous grain elevator from German defenders. This was the same grain elevator that was so heroically defended by a handful of Russian troops of the 62nd Army in September 1942. The fighting front for both armies now was reduced to just over half a mile wide. Under constant attack, the survivors of the German 297 Infantry Division, some 1,880 men strong, surrendered to units of the adjacent 21st Army. The command staff of the 64th Army certainly must have been wondering when the remaining Germans finally would give up.

Historical documents show the 64th Army had a reported combat strength of 26,526 men and women.

The other armies of the Don Front continued to push into Stalingrad from the north and west. By the end of the day, the Stalingrad pocket was split firmly into two major pockets. Conflicting reports make it difficult to clarify the total number of pockets formed during these last few days. Apparently the northern pocket contained mainly the remnants from the German XI Corps, while the central pocket contained elements of the IV, VIII, XIV Panzer, and the LI Corps, along with the 6th Army HQ.[384] These units were not really Corps but a mixture of combat and noncombat troops desperately trying to stay alive. General Paulus and the 6th Army HQ were located in the basement of the Univermag Department Store which was close to Stalingrad Train Station #1. From this location, the 6th Army HQ remained isolated within the central pocket and had little communication with the northern pocket. Due to this, General Strecker, the commander of the XI Corps, assumed overall command of the northern pocket. The 6th Army HQ reported to Army Group Don that 30,000 - 40,000 wounded troops within the city were totally unsupplied.

In keeping with their lack of preparation for dealing with the POW issue, the Don Front finally acted at this late date. The growing numbers of POWs falling into their hands had become a problem so the Deputy Commander of the Don Front rear, Lieutenant General I.G. Councillors, ordered the Chief of Logistics of the 64th Army, Major General G.V. Alexandrov, to organize a camp for prisoners of war from Stalingrad in the area around Beketovka, to be operational by 31 January.[385] General Laskin was present when this order was delivered to General Alexandrov, and he remembered how

Alexandrov hurriedly ran to the phone and called the chief of the rear of the Red Army General A.V. Khrulev, wanting to know why he was being stabbed in the back this way! General Alexandrov shouted into the phone, "Why are you assigning me to feed ninety thousand prisoners? Where are we going to get the food and how are we going to cook it? I have an army of my own people. The rear of the army [is] already exhausted by providing [for] six months under enemy fire a huge army. Consider the situation and that on the orders of the commander [Shumilov] we collect weapons, equipment and ammunition of the enemy scattered over a wide expanse of steppe near Stalingrad. Please entrusted the catering of prisoners to another army." [386]

General Khrulev told Alexandrov that he had the "skill to come out of any difficult situation" and that was why he was given this task. He also stated, "We'll help you" (that is, the Don Front will help). Despite promises, General Alexandrov knew it was not possible to carry out this order under current conditions. In the end, the order was not changed and promised food and medical support from the Don Front really never arrived. So, the 64th Army would do what it could to handle the growing flood of prisoners with its limited resources. After all, these POWs were the hated enemy; who really cared if they starved, froze, or died of disease?

27 January - Day Eighteen of Operation Ring
Weather Report: Overcast, light snow, 1°F (-17°C).

The Luftwaffe airdropped 103 tons of supplies into Stalingrad using 124 aircraft. The hope was that some of the supplies would reach German troops that were isolated in small pockets around the city.

The VVS continued to support the ground attack by flying 52 combat sorties in the Stalingrad area.

Due to the Red Army's good progress at Stalingrad, and the greatly narrowing of combat frontages, Stavka withdrew the 24th Army from combat and reassigned it to Stavka reserves. This army was the first major Russian formation to be released from Operation Ring.

The 6th Army had done its duty tying down significant Russian forces in the Stalingrad area.[387] By this date, the two armies of German Army Group A, 1st Panzer and 17th, were clear of the Caucasus and were safely withdrawing to new positions. This was mostly due to von Manstein's unwavering efforts to save these two armies, despite Hitler's irrational orders to do otherwise. This also meant any further resistance by the 6th Army would have little effect on the greater battle being fought further to the west. From this point on, Hitler's instance that the 6th Army fight to the last was only increasing the army's suffering, all for no real military value.

The 57th and 64th Armies advanced together, shoulder to shoulder, toward the north. They reached and crossed in a few places the Tsaritsa River gully. During the previous few days, the 64th Army had started to take many more prisoners as it advanced through the ruins of the city. An end to the fighting was in sight.

The 64th Army's reported combat strength was 25,994 men and women.

28 January - Day Nineteen of Operation Ring
Weather Report: Heavy cloud cover, clear, powerful northeast wind, 5°F (-15°C) day, -8°F (-22°C) night, icy road conditions.

From 87 aircraft, the Luftwaffe airdropped 83 tons of supplies into Stalingrad. That action sounds so simple. These Luftwaffe aircrews not only had to fly long distances in bad weather, but they also needed to evade patrolling VVS fighters along the way. After arriving in the Stalingrad area, they struggled to identify the different German pockets still holding out in the ruins of the city, all the while avoiding anti-aircraft fire from the ground. After all this effort, estimates were that German troops recovered only 50 percent of these supplies. The Russians got their share too. Therefore, it should come as no surprise that on this day the 6th Army officially stopped issuing food rations to wounded troops. Since the loss of the last airfield on 23 January, almost no hope remained for anyone suffering from a serious, or even a modest wound. By stopping food distributions to the wounded, the 6th Army HQ acknowledged that most of their wounded soldiers were certainly going to die within the pocket very soon. The living needed the food to fight and survive; the dead and dying did not. General Paulus was under no illusions the Russians were going to share their limited food and medical supplies with the despised enemy. Adding to the confusion, further Russian attacks during the day split the central pocket in two. The reduced central pocket held mostly units from the VIII, and LI Corps, and the new southernmost pocket contained the IV Corps, the XIV Panzer Corps, as well as General Paulus and the 6th Army HQ.

VVS units flew 235 combat sorties over Stalingrad, pounding the rubble in the city center.

The 57th Army continued its attack into the southern part of Stalingrad. Meeting unexpected strong enemy resistance, the 57th Army made only minor progress. In turn, the 64th Army launched its attack at 12:00 and also was met by tough enemy resistance all along the Krasnoznamenskaia Street line. The 64th Army made no forward progress at all during the day.

29 January - Day Twenty of Operation Ring
Weather Report: Sunny, bright, cold, considerable snow, 14°F (-10°C) day, -4°F (-20°C) night.

The Luftwaffe airdropped 108 tons of supplies from 111 aircraft into the city, while the VVS flew 380 combat sorties against German positions.

General Paulus sent Hitler a message, "To the Fuhrer! The 6th Army greets their Fuhrer on the anniversary of your taking over power. The Swastika flag still flies over Stalingrad. May our struggle be an example to present and future generations never to surrender in hopeless situations, so that Germany will be victorious in the end. Heil, mein Fuhrer!"[388]

During the day the 57th Army advanced about half a mile into the city in sustained fighting. The 64th Army made slightly less progress, about four-tenths of a mile, overcoming strong enemy resistance along the way. The 64th captured several buildings and reached positions along Uritskaia, Oktiabrskaia, and Gogolia Streets. Over the previous three days, the 64th Army had taken over 15,000 prisoners. This was a good sign! [389]

Later during the evening, Stavka ordered the 57th Army and its supporting units reassigned to Stavka reserve. This order would take effect on 31 January, and the Army was to prepare for transfer to other fronts.

30 January - Day Twenty-One of Operation Ring
Weather Report: Clear, icy, good road conditions, falling temperatures of 18°F to 5°F (-8°C to -15°C) day, -4°F (-20°C) night.

Two weeks beyond the originally scheduled end date the battle continued because German forces were still resisting within the ruins.

The Luftwaffe airdropped 128 tons of supplies into the city using 130 aircraft. Most of these supplies never would be found in the deep snow. In turn the VVS used 241 aircraft to drop a far less friendly cargo into areas of German resistance.

During the night, officers from the 20 Romanian Infantry Division approached the front lines in the area of the 204 RD with an offer to start surrender negotiations. After being informed of this, General Shumilov dispatched the 64th Army's deputy chief of staff of political affairs Lieutenant Colonel B.I. Mutovina to take over the negotiations. After a brief talk with the commander of the 20 Romanian Infantry Division, Brigadier General Dmitriesku, talks stalled. At this point Colonel Mutovina said that if the surrender demands were not accepted within 30 minutes that area would be attacked with artillery fire and "Katyusha" rocket launchers. "Dmitriesku quickly said yes and gave his division the order to cease fire, lay down their arms and surrender to Soviet troops." The negotiations (and the war for the 20 Romanian ID) were finished by 21:00 hours. Colonel Mutovina then delivered General Dmitriesku to the headquarters of the 64th Army for interrogation.[390]

From these captured Romanian prisoners, and from earlier captured German officers, General Shumilov learned the location of the 6th Army HQ in the city center Red Square area. In response, he ordered the 38 Motorized Rifle Brigade to the front in order to increase the tempo of the attack. Shumilov also, in essence, aimed the 38 Rifle Brigade at the 6th Army HQ. Obviously General Shumilov was trying to gain prestige by having the 64th Army capture the leadership of the 6th Army.

Also fighting in the city center was the 36 GRD. Their Commander, General M. I Denisenko, reported to General Shumilov that his combat strength was down to 5-8 men in each company. So Shumilov reinforced the 36 GRD with a company of 60 men from the 329 Engineer-Sapper Battalion. With the new support, troops of the 36 GRD were able to advance again and to clear several buildings of enemy troops.[391] The combat strengths of units on both sides of the line were falling to extremely low levels. Only an end to the fighting would stop this slaughter. Everywhere medical facilities were filled to overflowing with wounded.

At about this same time, Russian mortar crew member Mansur Abdulin (293 RD, 21st Army) related how he senselessly lost his best friend Rashid. During an advance, Rashid was shot in the leg, and Mansur put a bandage on the wound of his friend to stop the bleeding. He then "dragged Rashid to the rear…to find a medical unit. Finally, I bring my comrade to the so-called trans-shipment point, where the wounded are being loaded into lorries [trucks], for transportation to the rear. With the aid of an orderly, I help Rashid descend into a gigantic pit, where about 200 men are awaiting their turn to be evacuated. The orderly promises that in half an hour all the boys will be taken away. Saying goodbye to my friend, I run off" …back to his gun crew. Later on, during a pause in the fighting, Mansur received permission "to see how Rashid was doing. When I reached the place, there was not a soul in sight. I felt relieved; it seemed that everyone had been evacuated. But glancing into the pit, I was struck with horror: 200 wounded men were lying silent and motionless, seemingly asleep, their heads covered with fresh snow." They had all frozen to death, and the orderly and his commanding officer were gone. Returning to his unit, Mansur told his friends that everything was fine, but he found it hard to go on "after seeing our soldiers die in such a senseless way."[392]

Manpower totals were so low within attacking Russian infantry units that an ad-hoc assault unit was formed from two artillery regiments. The commander was ordered "to leave four to five people at the guns, and from the others to create a unit that, together with the infantry, was to storm the buildings of the hospital" in downtown Stalingrad. "The combined artillery company of two cannon artillery regiments numbered about 120 people…," and later "the commander of the 422nd Infantry Division sent a machine-gun platoon to reinforce it." The attack started the next morning supported by direct artillery fire. After clearing the assigned buildings, the artillerymen of the joint company returned to their units. [393] That attack was very near the Headquarters of General Friedrich Paulus.

Knowing that the end of the 6th Army was near, Adolf Hitler promoted Colonel General Friedrich Paulus to Field Marshal. Because no German Field Marshal had ever been captured, Hitler hoped Paulus would take the hint and commit suicide instead of surrendering to the Russians. The recently formed German central pocket had surrendered to Russian forces, so only the northern pocket and the southernmost pocket containing the 6th Army HQ remained.

31 January - Day Twenty-Two of Operation Ring

Weather Report: Heavy cloud cover, icy, 14°F (-10°C) day, 5°F (-15°C) night.

The Luftwaffe airdropped 118 tons of supplies into the city from 89 aircraft. The VVS conducted 233 sorties against German positions, if they could be found.

Units of the XI Corps, under the command of General Karl Strecker, continued to offer resistance to the Russians in the northern pocket.

The attacking frontages were so narrow the 64th Army could not deploy all its forces to the front.

The final lineup of units in the front line for the 64th Army were starting from the left flank near Train Station #1, 36 GRD, 29 RD, 38 MRB, 204 RD, 97 RB, and the 96 RB on the Volga River bank. I believe the other rifle brigade of the 7 RC, the 93 RB, probably still was deployed on Golodnyi Island.

During the night, formations of the 38 Motorized Rifle Brigade and the 329th Engineer Battalion of the 64th Army surrounded an area some 300 yards around the 6th Army HQ. By early morning the Germans realized the Russians were very near and tightly surrounding the Red Square area and the Univermag Department Store. At 06:15 hours 6th Army HQ transmitted the following message: "The Russians are at the door. We are preparing to destroy the radio equipment." About an hour later, a second and final message was sent: "We are destroying the equipment."[394] At 07:15, after a brief conversation on the surface, a somewhat impatient 21 year-old Senior Lieutenant Fedor Ilchenko, of the 38 Motorized Rifle Brigade, descended into the basement under the Univermag Department Store and began surrender negotiations with Lieutenant General Arthur Schmidt, chief of staff, 6th Army. This electrifying news was passed up the Russian chain of command, triggering the arrival, at about 08:40 of Colonel Burmakov, commander of the 38 Motorized Rifle Brigade, and Major-General Ivan A. Laskin, chief of staff of the 64th Army (along with several other officers from 64th Army). The appearance of the official surrender delegation from 64th Army HQ, headed by General Laskin, accelerated the negotiations, and instead of suicide, Field Marshal Paulus finally disobeyed Hitler and surrendered himself, his entire HQ staff and the troops still under his direct command within the southern pocket. Appropriate surrender orders soon were sent out to German units within the southern pocket. After that, sounds of fighting in the area gradually settled down with little more than scattered gunfire and an occasional explosion.

While the remaining fragments of the southern pocket were being overrun by the 36 Guards Rifle Division, 204 Rifle Division, 38 Motorized Rifle Brigade, the 96 Rifle Brigade and the 97 Rifle Brigade (7 Rifle Corps) of the 64th Army, along with units of the 21st Army and 62nd Army, the end of the war arrived for the 6th Army HQ. Field Marshal Paulus and key members of the 6th Army Staff were soon on their way south to Beketovka and the HQ of the 64th Army. An impatient General

Shumilov was waiting for the arrival of his new guests. It was going to be a long day for all concerned. Note: At this point in the battle, units of the 57th Army had been blocked from advancing any further by combat formations of the 21st Army to the north and the 64th Army to the south.

The journey into captivity by the command staff of the 6th Army is important enough to describe in detail, so the following are excerpts of the firsthand account of these events by Colonel Wilhelm Adam, ADC or adjutant to Field Marshal Paulus.[395]

> Punctually at 09:00 hours the chief of staff of the Soviet 64th Army [Major General Ivan A. Laskin] appeared to take away the commander-in-chief of the defeated German 6th Army and his staff. We climbed into our ready [German] vehicles. Paulus and Schmidt took seats in the first vehicle, the Soviet general sitting next to the driver. I went in the second one with a Red Army lieutenant. The remaining officers and men followed in the truck.
>
> To me it was like awakening from a frightful nightmare when the vehicle left the city center and drove across open country in a southerly direction. We were moving forwards quickly now. The Volga was on our left. Two hours later we entered Beketovka. We stopped in front of one of the dominating wooden houses. Then our escorting general [Laskin] was already asking us to enter the building. A Soviet general had sat down on the opposite side of the T-shaped table. As it soon turned out, this was Shumilov, the commander-in-chief of the 64th Army. Next to him now sat his chief of staff Major-General Laskin, and an interpreter with the rank of major. Also present were ten other Soviet "military commanders, including the deputy commander of the Don Front, Lieutenant-General KP Trubnikov and the first secretary of the Stalingrad Regional Committee of the CPSU - a member of the Military Council of the front A. Chuyanov.[396]

A brief interview then took place where General Shumilov asked various questions of Field Marshal Paulus. Colonel Adam continued his account.

> Discreetly I considered the Soviet commander-in-chief. Shumilov spoke quietly and pertinently. While I was still in a state of surprise, the general stood up. The interpreter translated his last words: 'Tell the Field Marshal that I am asking him to take a drink with me before I drive off to my headquarters!' Outside, Soviet soldiers helped us into our coats. Excitedly we moved toward the entrance, where Shumilov, wearing a tall fur hat on his head, awaited us. He crossed the road and signed for us to follow him. Shumilov opened a door to a lobby in which an old woman was housekeeping.

Hot water and soap were provided and after washing

> we were invited into the next room. There stood a covered table with various things to eat.

At this point General Shumilov noticed some tension between Schmidt and Adam. He said quietly

> It would be much more pleasant if we had got to know one another under other circumstances, if I could greet you here as my guests and not as prisoners of war.' Vodka was consumed by all from the same bottle. The general invited us to drink with him to the victorious Red Army. But we remained sitting motionlessly. After the interpreter had said a few quiet words, Shumilov laughed: 'I don't want to upset you. Let us drink to the two brave enemies that confronted each other in Stalingrad!' Now Paulus, Schmidt and I raised our glasses.

> We sat together with General Shumilov for more than an hour. Afterwards [Shumilov] got up and escorted us to the vehicle waiting to take us further on. He said goodbye with a handshake. He stood saluting on the edge of the road as the car moved off. He was a truly noble opponent.

This ends Colonel Adams account.

Note: This was only one of several different versions of the same meeting between the victors and the vanquished. Later in captivity, Major Adam became a Communist, so his writings took on a pro-Soviet slant to them. No doubt the command staff of the 6th Army, and the other captured generals, were well treated by the Russians, but the common solider would suffer greatly after the surrender.

In contrast, what follows is a brief description of the same event, this time more from the Russian point of view. The captured command staff of the 6th Army arrived in Beketovka at about 12:00, where photographers were on hand to record the event. General Shumilov asked several basic questions of Paulus and the other German officers, like their rank, position within 6th Army, and so on. Finally, Shumilov informed the German officers, "The Soviet command guarantees you life and safety, and the safety of his uniform and medals." This statement pleases Paulus and the other officers, as they had feared a harsh reception or even a quick execution at the hands of their captors. Shumilov also attempted to get Paulus to order the surrender of the northern pocket; this Paulus refused to do, citing that as soon as he had surrendered, he was no longer in command of the army.

Upon completing his superficial interrogation of the captured generals, Shumilov passed the relevant information on to the Don Front commander, General Rokossovsky. Rokossovsky then ordered the prisoners to be sent to his front headquarters that afternoon. With his official duties finished, Shumilov invited the German officers to lunch. Paulus and Schmidt hardly touched their food, only Colonel Adam ate his fill.

Everyone up the Red Army chain of command wanted to see and to be seen with the captured German officers, especially Field Marshal Paulus. So, after lunch, Paulus, Schmidt and Adam were sent on to the headquarters of the Don Front, located some 55 miles away at the village of Zavarykin. This village was north of Stalingrad along the middle of the Llovlia River. General Shumilov had his moment of fame with the commander of the defeated 6th Army, and now Shumilov stood watching as Paulus drove away into the fading afternoon light. With this great achievement, General Shumilov knew the fighting would end soon with just a few more holdouts to capture or eliminate.

Years later, General Shumilov was asked which day (during the war) was the most joyful for him? Shumilov answered this question as follows: "January 31, 1942, when I was sitting in front of Paulus, the first Field Marshal of Hitler's army, taken prisoner by the Red Army - or rather, the 64th Army."[397]

Returning to the course of battle, the 57th Army largely was blocked from advancing any further into the city by units of the 21st Army approaching from the northwest and the 64th Army advancing from the south. The 57th Army was leaving the Stalingrad area soon anyway. By the end of the day, the German southern pocket had been eliminated, but there were rumors that remnants of the 14 Panzer Division were moving to the north to join the northern pocket. All that was left for the 64th Army to do was to disarm the thousands of prisoners and to eliminate scattered holdouts in and around the city center. As for the thousands of wounded enemy troops hidden away in the ruins, well, they would mostly be left to take care of themselves. Reportedly, during the day, some 50,000 German and other Axis troops surrendered to units of the Red Army. Soon these new POWs would be formed into columns and marched out of the city to processing centers, mostly to the south around Beketovka.

These POW centers, for the most part, didn't really exist, so the starving prisoners were in fact just marching out into the winter wilderness. Total victory for the Red Army almost was complete.

Hitler was furious when he learned of the "dishonorable surrender to the enemy" by Field Marshal Paulus. In response, Hitler vowed to never create another Field Marshal during the war, and he kept this promise.

By nightfall, the 21st, 57th, and 64th Armies had stopped major combat operations within the central area of Stalingrad. But scattered fighting still took place in the ruins as groups of German troops refused to surrender. These holdouts were eliminated as quickly as possible. Maybe tomorrow, the weary Russian troops hoped, the fighting really would end.

1 February 1943	
64th Army	
Assigned – Don Front	
Lieutenant General M.S. Shumilov Commanding	
Rifle, Airborne and Cavalry Units	7 Rifle Corps (93, 96, 97 Rifle Brigades) 15, 36 GRDs 29, 38, 204, 422 Rifle Divisions 143 Rifle Brigade 20 Separate (Tank) Destroyer Brigade 45, 77, 115, 118 Fortified Regions 166, 172, 175, 177, 303 Separate Machinegun Battalions
Artillery Units	19 Artillery Division (70 Guards, 123 Artillery Regiment, 457, 1108, 1159 Gun Artillery Regiment, 5 High-Power Howitzer Artillery Regiment, 400 Separate High-Power Artillery Battalion) 1104, 1111, 1168 Gun Artillery Regiments 58, 184, 186, 493, 496, 500, 502, 536, 565, 762, 1188 Antitank Artillery Regiments 140 Army Mortar Regiment 18 Guards Mortar Brigade (rocket launchers) 1261 AA Regiment
Armored and Mechanized Units	90, 254 Tank Brigades 38 Separate Motorized Rifle Brigade
Engineer Units	175, 328, 329, 330 Separate Engineer Battalions
Reported/Estimated Combat Strength Estimated Ration Strength	47,305 men 65,804 men[398]

BSSA - 43, and BSA - 43

For clarification, the above estimated ration strength of 65,804 men are divided into the following general categories:

- Infantry units = 36,733
- Artillery = 9,863
- AA = 709
- Tank = 400
- Engineers = 500
- HQ and Support = 17,599

These totals do not include the estimated 16,147 wounded still in the hospital on 1 Feb 1943.

1 February 1943 - Day Twenty-Three of Operation Ring
Weather Report: Local fog, little cloud cover, clear night, 9°F (-12°C) day, 5°F (-15°C) night, icy.

The Luftwaffe airdropped 73 tons of supplies into the northern pocket where German troops still offered resistance. These 89 aircraft made this journey in the hope that some of these supplies would reach their intended target, the final group of cold, sick, starving troops of the 6th Army.

The remnants of the 14 Panzer Division attempted to move north toward the XI Corps pocket, from the Red Square area, but they had little chance of reaching it. The remaining five divisions of the XI Corps held out in the northern pocket scattered around the Dzerzhinsky Tractor Factory. Small groups of troops also continued to fight in the city center area.

Due to the restrictive nature of the fighting within the built-up area of Stalingrad at the end of January, the Don Front had too many armies in the front line. To solve this problem, Stavka ordered the 57th Army to transfer control of its surplus combat units to the 64th Army. This accounts for the newly increased combat and ration strength of the 64th Army. At the same time the 57th Army withdrew into the Stavka reserve, and eventually it was disbanded with its headquarters being used to form the new 68th Army. Likewise, the 21st Army was ordered to transfer its extra combat and support units to the 64th and 65th Armies. Because of these transfers, the total number of combat units assigned to the 64th Army increased to at least fifty-three formations (divisions, brigades, regiments, and battalions).[399] Tracking which units were transferred and when is difficult. Therefore, the estimated ration strength for the 64th Army during the early part of February quickly rose to over 85,000 men. These totals do not include the reported 16,147 wounded in the hospital on 1 Feb 1943. Once again, this increase in ration strength was mostly the result of the 64th Army taking over administrative control of scattered formations in and around the Stalingrad area. So, these totals could change drastically as separate units were added or subtracted from the Army to meet the needs of Stavka and the Don Front.

The 64th Army, like other armies in the southern part of Stalingrad, were busily engaged in gathering enemy prisoners, picking up abandoned weapons, and killing any enemy troops refusing to surrender. In essence, they were taking stock of the situation and policing the battlefield. Most soldiers could not believe the fighting really was over, and one must assume they were happy to have survived this great battle. Of note, the long-serving 66 and 154 NRBs left the 64th Army never to return.

In far off Moscow, the National Committee of Defense issued Secret Resolution #2812 on the collection of trophies (captured) property at the front, ensuring its storage. This resolution attempted to accelerate exports to the rear of domestic and trophy weapons, property, and scrap for reuse and repair. Chairman of the State Defense Committee, Joseph Stalin himself, signed this resolution. This high-level interest in scrap metal was due to it being a significant source of metal for factories producing new weapons and tanks.

Further north in Stalingrad, the fighting did continue. At 08:30 hours, the Don Front pounded the northern German pocket with massed artillery fire supported by attacks by the VVS. White flags of surrender began to appear in some areas; in others, resistance continued.

2 February - Day Twenty-Four of Operation Ring

Weather Report: Light clouds, light morning fog, frosty, clear, 10°F (-12°C) day, -4°F (-20°C) night.

The Luftwaffe airdropped 98 tons of supplies into the northern pocket. Only 40 supply aircraft made the trip on this day. With this delivery, the aerial effort to supply the 6th Army was essentially over. During the airlift, from 24 November to 2 February, the Luftwaffe delivered some 8,350 tons of supplies (this is equivalent to 14 supply trains at 600 tons each) and evacuated around 30,000 men.[400] During that time the Luftwaffe lost about 488 aircraft and 1,000 airmen. Note: These aircraft totals, the tons of supplies delivered, and the total number of men flown out still are being debated to this day.

At some point during the day, the 90 and 254 TBs transferred out of the 64th Army. The 90 TB went to the 62nd Army with 7 tanks, and the 254 TB went to the 65th Army with 6 tanks. After a further massive artillery bombardment and assault against the northern pocket by the Don Front, the Red Army had finally broken German resistance. Enemy troops began to surrender en masse. Finally, the guns went silent and quiet descended over the devastated city. The last organized groups within the northern pocket surrendered to the Red Army. The German 6th Army had been destroyed at last; Stalingrad was free.

"At 16 hours and 10 minutes a representative of the Supreme Command, Colonel General of Artillery NN Voronov and the commander of the Don Front, Colonel General Rokossovsky sent to the Supreme [Stalin] this report, 'Following your order, the Don Front at 16 hours 2 February 1943 we completed the rout and destruction of the encircled Stalingrad grouping opponent.' "[401] After seven months of combat in and around the city of Stalingrad, the battle finally was over.

Operation Ring had lasted a total of 24 days, not the 7 days as originally projected by General Voronov. Because of the drawn-out nature of the struggle, harsh weather conditions, inadequate resources, disease, and the desperate resistance offered by German forces, total Russian casualties were substantial. Adding to these problems, the prolonged battle also consumed huge amounts of supplies, which the Red Army was having a hard time replacing. Offensive operations were now over, but vast numbers of men, on both sides, still struggled to survive.

For the 64th Army, the accounting was just beginning. During Operation Ring's 24 days of combat, from 10 January to 2 February, the 64th Army alone suffered a total of 21,113 casualties: 14,845 injured in combat, 1,079 freezing, 233 shell-shocked, 180 burnt, and 4,776 sick. Out of theses total reported casualties, approximately 4,966 men were killed outright on the battlefield or died before reaching a hospital, leaving some 16,147 wounded to be admitted into the medical facilities of the 64th Army. Remember the 64th Army started in January with some 58,000 troops, so these losses amounted to about 37 percent of the starting total, which are huge losses for a 24-day offensive. This was also why the infantry formations wasted away to almost nothing as they were taking most of the losses. Additionally, medical staff reported that the average number of casualties during the offensive was around 600 men per day, but on some days the total reached as high as 1,400 casualties. Due to the surge in heavy fighting, the load on the existing medical facilities reached 6,000 to 8,000 men at any one time.[402] Considering the 64th Army started this battle with a maximum of 4,100 beds for wounded of all types, the medical facilities and personnel were simply overwhelmed. In trying to deal with the lack of beds, army medical personnel reorganized the distribution of wounded to all available medical units regardless of the type of injury. So battle wounded troops would now be sent to the one Infectious and two Evacuation Hospitals within the 64th Army. Needless to say, these hospitals were not prepared or equipped to treat the newly wounded. This makeshift effort expanded the number of beds to the maximum amount available, but even this was insufficient to deal with the flood of wounded. Medical units of all the other armies of the Don Front also suffered from an extreme overload of wounded and could not help. In the end, medical personnel of the 64th Army dealt with the crisis as best they could. The wounded of course had no choice; they suffered and died due to poor planning and inadequate medical resources. According to Russian sources, a high percentage of the wounded who were admitted to the hospital would die of complications, sometimes reaching 60-70 percent for some types of injuries.[403] Realistic data for these so-called "sanitary losses" are difficult to find, but they would greatly add to the total number of dead.[404]

To put the losses suffered by the 64th Army into perspective, let us look at the casualties for the entire Don Front during Operation Ring, including 64th Army losses. During the initial attack, covering the first six days of Operation Ring, 10-15 January, the Don Front's seven armies reported suffering the following losses:[405]

- 6,617 killed
- 18,893 injured
- 154 missing
- 718 other losses

For a total of 26,382 men and women, these figures amount to 4,397 casualties per day.

For the next four days of Operation Ring, during a 16-19 January lull in the fighting, the Don Front suffered the following estimated casualties: (Note: These are estimated losses due to the lack of hard data.)

- 2,800 killed
- 7,840 injured
- 112 missing
- 448 other losses

For a total of 11,200 casualties, these figures reflect 2,800 casualties per day.

For the next eleven days of Operation Ring, during the 20-30 January final push into the city, the Don Front reported suffering the following casualties:

- 6,661 killed
- 20,228 injured
- 148 missing
- 1,044 other losses

For a total of 28,031 casualties, these figures show a loss rate of 2,548 per day.

For the final three days of Operation Ring, during the destruction of the remaining two pockets of resistance, 31 January to 2 February, the Don Front suffered the following estimated (due to the lack of hard data) casualties:

- 800 killed
- 2,500 injured
- 18 missing
- 325 other losses

For a total of 3,643 casualties, at a rate of 1,214 per day.

Finally, the total reported and estimated casualties suffered by the entire Don Front, during Operation Ring, 10 Jan to 2 Feb, over a period of 24 days, were:

- 16,828 killed
- 49,461 injured
- 432 missing
- 2,535 other losses

This yields a grand total of 69,256 casualties for the Don Front, or a loss rate of 2,885 people per day. Numbers like these illustrate the level of fighting that took place during the final assault against Stalingrad.

Supporting these numbers, David Glantz and Jonathan House in their book *Endgame at Stalingrad, Book Two Vol 3,* p. 583 stated that "the [Don] front lost 48,000 men during the 21day period from January 10 through 30 January, for a loss rate of about 2,286 per day." They also state on page 582, the Don Front committed 281,158 troops to combat (during Operation Ring).

Note: The value of 48,000 men lost is less than what my research indicates, but I have included in my totals an additional three days of fighting, from 31 January to 2 February 1943. Along with this, I added, for the first time, accurate casualty figures for the 64th Army based upon a historical medical record.[406] So I feel the total losses for the Don Front of 69,256 men is a reasonable approximation.

So how does this information relate to the 64th Army? The 64th Army contained about 19 percent (53,748 men for the 64th Army) of the total number of troops committed by the Don Front at the start of Operation Ring (281,158 troops).

If we compare the reported losses for the 64th Army (21,113 casualties) to the overall reported and estimated Don Front casualties (69,256 casualties, my numbers), we find the 64th Army suffered about 30.5 percent of the total casualties during Operation Ring, which is huge. If we were to use the 48,000 casualties' value from Glantz/House, then the 64th Army would have lost about 45 percent of the total casualties during Operation Ring, which is not believable! Using either set of numbers, 48,000 or 69,256 yields an extremely high percentage of losses for the 64th Army. The 64th Army was conducting supporting attacks, away from the main battle area on the western side of the pocket, so these enormous proportions call into question the reported losses suffered by the other Don Front armies. The possibility remains the casualties for the other six armies of the Don Front were substantially underreported. (To explore this issue further, see Appendix 20)

Nevertheless, the reported and estimated losses for the Don Front still provide a good overall impression of the severity of the fighting. But the Axis units surrounded at Stalingrad suffered a great deal more. Reports said that after the battle, some 140,000 dead and frozen Axis troops were picked up from the Stalingrad battlefield and buried.[407] Other sources claim 146,300 corpses were collected, but who can say for sure which number is the real one? The best historians and researchers can do is estimate the losses based on whatever hard data is available, then reach a reasonable, believable value and continue from that point.

Also, on or about this 2 February date, the 64th Army issued a "Damage Inflicted" report of its accomplishments during the last phase of the Stalingrad fighting, between 10 January and 2 February 1943. This report claimed the Army took into captivity 34,906 soldiers; 1,491 officers; and 9 generals for a total of 36,406 captured enemy troops. They had also "destroyed" some 15,550 enemy soldiers and officers. If these numbers are accurate, then the 64th Army single-handedly killed or captured about 1/3 of the remaining enemy troops within the Stalingrad pocket at the start of Operation Ring.[408] Even though the 64th Army was way below strength and was suffering substantial losses itself, it stayed in the fight and finished the battle to the very last day.

Accounting for enemy troops captured during Operation Ring suffers from the same vague but believable values. The Russians reportedly took some 2,500 officers and 24 generals prisoner, and then most accounts go on to announce the capture of over 91,000 Axis soldiers and officers at Stalingrad.[409] Certainly the 91,000 total prisoners captured is a generally accepted value, but we may never know the true number. What we do know from the historical record is that most of these 91,000 prisoners were to suffer even further at the hands of their captors. Only 5,000 - 6,000 German troops returned home from Russia after the war, and an unknown number of other Axis forces returned to their native countries.

Once again, the total number of Axis troops captured at Stalingrad is a hotly debated issue. Numbers range from 91,000 to over 93,000 captured during Operation Ring. Then for the entire series of operations, from Uranus to Little Saturn, many more POWs were captured outside of the Stalingrad pocket. What became of these additional Axis prisoners of war? They number at least 109,000 Germans; 40,000 Romanians; 60,000 Italians; and about 40,000 Hungarian troops.[410] We know they fell into the hands of the advancing Red Army, but what happened to them after capture? How many of these forgotten POWs, numbering around 250,000 men, would return home after the end of the war? From this group, research indicates that of Italian POWs, it is thought only some 10,000 of them were repatriated after the war.[411] For the other prisoners captured outside Stalingrad, who knows how many survived?

But these totals are only for Axis troops; what about the *Hiwis*? Before, during, and after the Stalingrad battle, large numbers of Russian/Soviet military prisoners of war worked for the German and Axis armies. These so-call *Hiwis* (helpers) served mostly in noncombat positions behind the lines. Approximately 30,000 of these helpers worked in and around Stalingrad during November 1942. Most German divisions had several thousand *Hiwis* assigned to each of them. During Operation Uranus an estimated 20,300 *Hiwis* troops were surrounded at Stalingrad along with the German 6th and 4th Panzer Armies. Of these, approximately 10,000 *Hiwis* died during the fighting or were executed by the Soviets after capture. The remaining 10,000 were processed by NKVD forces and then sent to either penal or replacement units.[412][413] (see Appendix 21) We do know the Soviet authorities and NKVD troops spared no effort to find, identify and eliminate these traitors. The lucky ones lived to serve the Red Army again because spilling their blood on the battlefield would cleanse them of their crimes against the state. Also, hundreds of thousands of Soviet civilians were working for the Axis far behind the front lines. They were also called *Hiwis* by the Germans. Some of these civilians were supporting the Axis forces because they wanted to do so, others were forced to work, and still others just wanted to survive. Most were just trying to make the best of a bad situation.

Speaking about civilians, we also should not forget about the civilians who were caught up directly in the battle for Stalingrad. When the German summer offensive advanced toward the lower Don River, thousands of civilians fled toward Stalingrad and the perceived safety of the Volga River area. These refugees swelled the city's population to over 400,000. When fighting approached more closely, Soviet officials were reluctant to order or even talk about an evacuation of the city because Stalin had forbidden it for various reasons. In his book *Khrushchev Remembers*, Edward Crankshaw stated that Nikita Khrushchev, the political advisor and Military Council member to General Eremenko during the battle for Stalingrad, said that, "For anyone else [other than Stalin] to have suggested an evacuation would have been to invite some very unpleasant consequences." In other words, these unpleasant consequences would have been demotion, detention or death by an NKVD executioner. Accordingly, to begin a general evacuation of the non-combatants from the city was out of the question, so tens of thousands of men, women and children also became part of the battle. These deaths, like so many others, were the direct result of Stalin's total disregard for the lives of the people of the Soviet Union. Like most of the statistics presented here, we will never know the true totals, but it is thought that at the very least some 80,000 civilians died during the fighting from all causes: air attack, land combat, artillery barrages, starvation and disease. Along with these losses the Germans forcibly removed over 180,000 civilians from the Stalingrad area and sent them to the west. These so called *Fremdarbeiter* (Foreign Workers) or *Ostarbeiter* (Eastern Workers)[414] went to work in factories or farms, mines and labor camps all over occupied Europe. Many of these slaves also perished because of the war.[415] During the chaos of a great battle, who tallies civilian dead or missing anyway? A 100,000 here, 200,000 there, it is with victory or defeat that leaders assess a battle. Stalin and the Red Army had won their great victory on the Volga, but at what cost?

With the capture of Field Marshal Paulus and the destruction of both the southern and northern pockets of resistance within the city, Stavka acted quickly in issuing new directives to reassign troops around Stalingrad to other active fronts.

On 2 February 1943, at 14:10 hours, Stavka issued Directive #46038 to the commander of the Don Front, General Rokossovsky.

This directive in part ordered the transfer of the 21st and 64th Armies to Stavka control (the reserve of the Supreme Command). These two armies were to move, by rail, as soon as possible to an area behind the Briansk Front (an area northwest of Voronezh and southeast of Orel), to support a planned offensive in that area.

Directive #46038 must have set off an intense amount of activity within the Don Front. In turn, General Rokossovsky also must have sent back numerous communications to Stavka over the next 24 hours about the current situation in and around Stalingrad and specifically the condition and duties of the 64th Army. We do not know the contents of these communications, but undoubtedly General Rokossovsky and General Shumilov would have pointed out the 64th Army was exhausted after many months of battle, the army was also actively engaged in policing the battlefield and was fully involved in trying to manage some 90,000 enemy POWs. In response to these presumed messages from General Eremenko, Stavka soon changed its orders.

Also, on this day Stavka launched Operation Star, a wide-reaching winter offensive by the distant Voronezh Front. The right flank armies of the Voronezh Front, the 60th and 38th, were directed toward the Kursk and Oboyan areas. Its left flank armies, 40th, 69th and 3rd TA, were directed toward the Kharkov and Belgorod area. Additionally, another operation, Operation Gallop, using the Southwest and Southern Fronts, would cross the Donets River, turn south into the Don Bass area, and then compress German Army Group Don (formally Army Group B) against the Sea of Azov and destroy it there.

Generally speaking, with these two new offensives taking place almost simultaneously, Stavka clearly looked for additional forces to strengthen these attacks. The Stalingrad area was a good source of uncommitted formations. But when Directive #46038 was originally issued, Stavka most likely intended to use the 64th Army to support an attack by the Bryansk Front toward the city of Orel. But this plan was changed as the winter offensives unfolded. Eventually the 64th Army was used in the Kharkov and Belgorod area instead, but at a slightly later date.

Furthermore, on this day, the 173, 214, and 233 Rifle Divisions were assigned to the 64th Army. These types of assignments were strictly temporary administrative measures.

3 February

In a final act of desperation, 10 Luftwaffe aircraft dropped 7 tons of supplies into areas suspected of containing German troops. This was the last known aerial supply mission to Stalingrad.

Also during this day, the 21 Rifle Brigade of NKVD forces, searching within the ruins with a Reconnaissance Search Group (RPG), encountered enemy troops still holding out; the enemy troops were destroyed.[416]

4 February

Remarkably, with search and destroy operations still taking place inside Stalingrad, a mass rally was held within the ruins, where thousands of soldiers and officers from every military unit came together to celebrate the great victory. There was even talk of a five-day vacation for everyone. (This hasn't been verified by research.) Later that night, a gala evening was organized at the 64th Army HQ. Attending were the "Military Council [of] the 64th Army, together with the Stalingrad regional committee of the CPSU (b). Army Commander Gen. MS Shumilov and the secretary A. Chuyanov congratulated all with the victory. And after the first toast, [they] sang Russian songs of the soul. Everyone was happy, elated, really grand."[417] Similar types of celebrations took place in other armies and assorted headquarters all around Stalingrad.

Eventually, in the weeks that followed, "For military exploits in the battles of Stalingrad were awarded orders and medals to 18,896 fighters, commanders and political officers of the 64th Army."[418] Similar awards were given to all the other participating armies too. Stalin and Stavka were not reluctant to shower honors on the victors of Stalingrad!

Back in Moscow, Stavka changed its mind about the planned deployment of the 64th Army when they finally issued Directive #46046 at 04:00 hours, which amended Directive #46038. The Commander of the Don Front was to give up control of the 65th Army to the Supreme Command Reserve. The 65th Army with all parts, services and agencies, and the army rear area support troops would be sent from the Stalingrad area instead of the 64th Army. Apparently the 64th Army could not be spared from its work in the Stalingrad area, so it was given time to recover from seven months of nonstop combat. General Shumilov knew this was only a temporary reprieve from duty at the active front, but it gave him time to start rebuilding his army.

5 February

Then on 5 February, Stavka acted to control the forces in and around Stalingrad by issuing Directive #46053, which formed a Stalingrad operational group. Taking effect on 27 February, these units would later be renamed the Stalingrad Group of Forces.

DIRECTIVE Stavka number 46053

The commander of the Don Front

About the Organization Management of the Remaining Troops at Stalingrad

Cc: Chief of the General Staff of organizational management

February 5, 1943 17 h 50 min

GHQ orders:

1. In connection with the redeployment of the Don Front Headquarters, control of the remaining troops in the area of Stalingrad, are assign to the deputy commander of the Don Front Lieutenant General K.P. Trubnikov.

2. To provide management to the troops, Lieutenant General Trubnikov will create a small task force of the army staff officers and to ensure direct telephone and telegraph communication with the staffs of armies and General Headquarters.

3. The remaining 62, 64 and 66th Armies to include six infantry divisions:

62 Army - 27th and 39th Guards, 24, 45, 99 and 284 RDs.

64 Army - 15th and 36th Guards, 173, 204, 214 and 233 RDs.

66 Army - 13th and 66th Guards, 116, 226, 299 and 343 RDs.

The rest of the infantry formations and units of Stalingrad Operational Group are directly subordinate to Lieutenant General Trubnikov.

4. In the group commander responsibility:

a) For the accommodation units and formations, with a view to ensuring the conditions for them to rest and put themselves in order. They are to be deployed near the railroads.

b) For the organization of combat training of troops and improving unit cohesion between units within the army.

c) Guidance for the full order in the account personnel and assets in units and formations.

d) For the tidying up parts of all types of property.

e) For the complete demining of the surrounding area.

f) For the collection and removal of trophy assets.

g) For the organization supply troops.

5. The list of compounds and individual parts of the Stalingrad Operational Group, indicating their location and the nearest railway stations to submit to the General Staff by 8 February.

6. Sending units and formations according to Directives of the High Command and the General Staff, to ensure timely loading on railway transport.

To supply each unit supply food for 5-7 days and a 10-day supply of unloading, one and a half of ammunition and two of refueling, which, on the basis of these rules, establish in advance in each of the irreducible reserve. To forbid the units to take with them weapons, vehicles and assets not provided by the states and or authorized. Everything in excess to select, consider and transfer to the organizations for collecting trophies or front warehouses. The status of the units to be transferred, are to be sent by couriers to the General Staff.

7. A detailed report on the work carried out by break-in of armies, management in the accounting and property, sending all formations and units at Bids Directive and the General Staff to submit to the General Staff.[419]

With the Stalingrad area soon to be under the control of the newly formed operational group commanded by Lieutenant General K.P. Trubnikov, Stavka issued another new directive about the disbandment of the Don Front and the formation of the new Central Front. This was Stavka Directive #46056. General Rokossovsky left Stalingrad to command this new Central Front.

Also, Valeriy Zamulin makes note in his book, *The Forgotten Battle of the Kursk Salient*, p 584, that Major General N.A Vasil'ev became the deputy commander of the 64th Army on this day. Vasil'ev remained with the Army after Stalingrad and on 25 April 1943, he was appointed commander of the 24 Guards Rifle Corps. Vasil'ev was praised for his diligence in his staff work and would earn high marks during performance reviews before and after the battle for Kursk. This appointment of General Vasil'ev as deputy commander might have replaced Major General I. Laskin in this role. General Laskin remained chief of staff of the 64th Army until 15 May 1943.

6 February

With the end of organized combat at Stalingrad, the huge job of gathering enemy POWs proceeded. The 64th Army was designated as the main receiving army for these POWs. No less than sixteen POW camps were established in and around the Stalingrad area, with at least seven directly within the 64th Army's zone of control. Most of the smaller camps eventually were disbanded as the remaining, surviving prisoners were consolidated into fewer main camps.

Efforts to transport and to care for the prisoners were at first crude to say the least. Prisoners were simply formed into long columns and sent marching off mostly toward an unknown destination to the south. One group of 300 German officers was sent on a forced march due west out of Stalingrad for no apparent reason. After days of marching, they eventually crossed the Don River some 40 miles away. The survivors, along with other groups of German officers that joined them there, then spent several days west of the Don. Later on, this entire group was turned around and marched back to the city. Only after reaching Stalingrad again did a much-reduced column of officers turn south toward Beketovka. This was truly a death march through the frozen winter with little food, water or shelter.[420]

Little or no effort was taken to care for the thousands of sick or wounded Axis troops who were found around the city; basically, anyone that could not march was left to die where they were. But once the march started, if someone dropped out of the marching column due to exhaustion, sickness or wounds, the escorting guards would quickly finish them off. There was little mercy shown for the hated enemy. Normally little or no food, water or heated shelters were provided during the several days it took to reach the main camps in and around Beketovka. Thousands died or were killed along the way. For example, "Germans captured in northern Stalingrad went on a forced march – without food or water – through Tsaritsa Balka, Gorodishche, Gumrak and Beketovka. The march lasted five days. A solider from the 305 Infantry Division estimated that of the 1,200 men that began the march, only 120 reached the final destination. [In this case that represents a 90 percent loss rate] But overall, over 20 percent of the German prisoners of war who began marching to prisoner of war camps, [right after the end of the fighting] died in transit."[421] Ultimately eight POW hospitals[422] would be established in the Stalingrad area, and by then a limited but steady food distribution effort took place. Of course, by this point huge numbers of prisoners had already died. Truthfully, when these prisoners were sent off to the POW camps, there really were no camps at all. Everything was being improvised at the last moment. Prisoners were herded into burned out or damaged buildings with no windows, no heat and little or no food and water. A German solider related, "That evening we reached Beketovka Camp and were split up into houses without windows. As they [civilians and Russian troops] had taken our blankets, we had to sleep in the cold and damp on the stone floor. Soon there were 50,000 prisoners at the Beketovka Camp. Terrible things went on there because the men were starving. The camp had two Opel-Blitz Lorries [German trucks] which did nothing all day but remove the corpses to nearby gullies. There was a kitchen, but insufficient for so many prisoners, so one got something to eat once a week if one were lucky."[423]

One very new female Russian doctor, Yevgenia Mikhailovna, related her experience at the Beketovka Camps, "In February 1943, we, a group of graduates of the 1st Moscow Medical Institute, were summoned to the Lubyanka [Headquarters of the KGB]. We were informed at the Lubyanka, that we are in the Department of Prisoners of War and internees, and that our group is to fly to Stalingrad, where Soviet troops have captured many German soldiers. And although we have not received the diplomas yet, we were distributed as ordinary doctors to camps for prisoners of war." She went on to relate, "I will not describe the ruins of this city. Everything around was turned into a desert, noticeable by snow, for many miles. That's exactly how camp No. 108/20 looked like for the prisoners, where I was sent from the local department of the NKVD with three fellow students." The prisoners were located "[i]n large concrete tanks, where the cucumbers and cabbage were previously salted, the Germans sat. They were lucky, because at least they were hiding themselves, if not from the frost, then from the piercing icy wind. Others huddled under the canopies of the former potato (huts), some simply got into [groups] to cover at least their backs."[424] For her troubles, Dr. Yevgenia Mikhailovna eventually contracted typhus, from which she was very lucky to have survived. For many of her patients, things did not go as well. Eventually over 35,000 POWs died just in the camps around Beketovka, and even more perished in the other camps.

7 February
On this day another NKVD RPG unit encountered "a large number of Germans holding out within a basement." After a fight, the Germans were destroyed. In yet another incident, NKVD forces killed 180 enemy troops and captured 216 others. In further actions during this day, another 24 Germans were killed and 128 captured in and around the city.

8 February
All the armies in and around Stalingrad, including the 64th Army, received orders about policing the battlefield in their areas of control. Civilians and POWs also were used in this task. In essence, this involved cleaning up the battlefield and removing and/or accounting for all wreckage and remains found. Thousands of corpses from both sides, both humans and horses, were found and disposed of. Along with this effort, troops also were tasked in recovering weapons and equipment found lying about. All this abandoned equipment, both Russian and trophy assets from the enemy, eventually were stored in 26 warehouses at rail stations and on side tracks around Stalingrad. Additionally, the added hazard of unexploded ordinance was hidden beneath the snow everywhere. The NKVD helped with this tricky business by interrogating prisoners about the location of minefields, and they identified 59 different minefields. (This figure must have been a small percentage of the total number.) Clearing minefields and unexploded ordinance from the battlefield of course added to the number of dead and wounded German, Russian and civilians in and around the city.

This boring, gruesome and dangerous cleanup work went on for weeks. The 64th Army employed the 36 GRD, 173, 204, 214, 233 RDs and the 38 Rifle Brigade in this effort and by 2 March 1943 they had recovered 2,750 Red Army dead; 12,697 enemy dead; and numerous horses. In the spring, after the snow melt, further corpses, equipment and explosives were revealed. These too had to be dealt with if Stalingrad were ever to return to the land of the living.

Additionally, at this late date, the head of the NKVD, Lavrentiy Beria, finally issued orders covering the release of supplies, food, trucks, gasoline and other provisions for the care and treatment of POWs at Stalingrad. But it was one thing to order something to be done in Moscow, and another for these orders to be carried out in the field.

9 February

Further to the west, the new Russian winter offensive started to reach its objectives. The Russian 40th Army captured the city of Belgorod with units of the 69th Army and 3rd TA supporting from the south. At this point the Operation Star offensive was four days behind schedule. A bit further north, the Russian 60th Army captured Kursk; the front line was moving westward nicely, and Stalin was pleased.

10 February

On this day a report was issued on the state of health of the Stalingrad Army Group. In part this report stated that "in 45 hospitals and 19 army evacuation hospitals were 54,777 sick and wounded" (these were Russian troops). In addition, in March the army group had received a further 3,543 people, bringing the total of sick and wounded to 58,320 people. Of this amount, by 1 May the following changes had taken place:

- Evacuated to the rear: 21,596 people. (These would have been soldiers with serious wounds requiring long-term care.)
- Returned to their units: 27,607 soldiers.
- Transferred to the Stalingrad evacuation center hospital: 5,696 soldiers.
- Additional deaths during this period: 3,407 soldiers. (Note: This number represents the official 6 percent, sanitary death rate, for wounded soldiers that had reached medical support). During the war, the overall death rate per casualty, from battlefield to hospital, would have been much higher, possibly reaching as much as 25-30 percent.)

We also should not forget about the huge number of horses laboring within Red Army units at Stalingrad. A Memorandum from Deputy Commissar of Defense of the USSR E.A. Shtadenko stated that the lack of fodder in February caused a disruption in feeding of horses. This disruption of food supplies killed at least 385 horses outright and sent over 3,565 to veterinary hospitals. Eventually 3,389 horses were cured and sent back to military units. The harsh winter weather was the enemy of every living thing, and winter was far from over.

Also, starting on or about this date, combat units in and around Stalingrad began systematic training sessions. This transition from combat to training occurred slowly as these units were fully engaged in other work around the city.

On other fronts the fighting continued with the 40th and 69th Armies attacking directly toward the major city of Kharkov as German resistance mounted in and around this important city. Recalling that last year's battle around Kharkov, in May 1942, ended in disaster, the Red Army now attempted to correct that failure by finally recapturing the fourth largest city in the Soviet Union.

11 February

Some 375 miles to the west of Stalingrad, the battle for Kharkov continued as German forces were forced to give up ground and to fall back. Times had changed; the Red Army was now on the attack.

12 February

By this date some of the rifle divisions and/or combat units of the 64th Army already were reassigned to combat training. General Shumilov most likely started this type of training days before, but this was the first known reference to this effort by units of the 64th Army.

To increase the pressure on exhausted German forces, the 3rd Tank Army joined the fight for Kharkov. Slowly the defending German forces were pushed back toward the city.

13 February

Any available civilians around Stalingrad previously had been used to stack and then bury all the German and Russian dead found on the battlefield. But on or about this date, a 1,200-man unit of German POWs was laboring to clean up the city. This unit (allegedly) also had some 300 medical and 200 explosive experts assigned and well as 100 guards. One of the few survivors of this unit stated that most of these Germans soon died from typhus.[425]

Also, on this date, the Soviet command received reports listing 6,669 officers and 86,956 other ranks, (93,625 men in total) as POWs in and around Stalingrad in 16 camps. Of these, approximately 67,314 prisoners were in camps in the area controlled by the 64th Army.

Fearing that Kharkov might soon fall, Adolf Hitler ordered, "Kharkov is to be held at all costs."

14 February

On this day, the city of Rostov, at the mouth of the Don River, fell to advancing Russian troops. But before the city was lost, the 1st Panzer Army was able to slip out of the Caucasus region and over the Don River to the north, via bridges at Rostov. The 1st Panzer Army entered the Don Bass region where General von Manstein would soon be putting the Army to use in this new area of operations. Additionally, the German 17th Army also successfully retreated from the Caucasus to the Taman Peninsula and formed the large Kuban bridgehead. With the safety of the Crimea to their backs, the 17th Army was secure for a time.

15 February

Russian forces almost had surrounded the II SS Panzer Corps within Kharkov, the fourth largest city in the Soviet Union. With the battle raging on three sides, General Paul Hausser disobeyed direct orders from Hitler to hold the city and withdrew his II SS Panzer Corps out of the trap.

16 February

The 64th Army received a Secret Order #0804, checking on the results of the progress of combat training from 12 to 14 February 1943, for the 36, 15 GRD, 204, and the 173 RD. This long report noted, in part, the following deficiencies:

- *Classes are conducted slowly and insufficiently productive.*
- *Field manuals, parts 1 and 2, have not been studied.*
- *Weapons have not been cleaned up and put in order; in the 204 RD they have found many weapons have not been brushed and are rusty.*
- *State of discipline is overall satisfactory, but clothing is worn in a sloppy unbuttoned form with many people going without a belt and clothing and equipment is torn.*
- *Kitchen forces are often not equipped, with food cooked in the open air.*
- *Reports of diary of events, entries are produced irregularly, and very late.*
- *The studying of outstanding fighting episodes is only beginning, etc.*

A plan to correct these and other deficiencies was to be developed by 20 February and would cover "[t]he entire system of training and education and, along with it, putting in good condition all weapons and material and technical equipment to achieve [and] further enhance the ability of combat readiness units and [allow] them at any time to perform in battle." Evidently the High Command was seriously trying to bring the 64th Army back into fighting shape. An interesting side note, a related section to this order was not signed by General Shumilov but by Head of the NKVD of the Stalingrad area, the Commissioner of State Security Grade III Voronin. This section related to a specialist unit of NKVD Sappers containing some 180 men. This NKVD unit would be used to conduct checks of the areas cleared and demined by units of the 64th Army.

With the fall of Kharkov on this day to victorious Red Army units of the Voronezh Front, German forces were now in retreat all along the front line. Kharkov had fallen, but due to stiff German resistance Operation Star was seven days behind schedule and losses were mounting.

17 February

Adolf Hitler met with Field Marshal von Manstein at his HQs located at Zaporozhe, a city on the lower Dnieper River. During two days of talks, von Manstein outlined his plan for a powerful counterattack against the overextended Russian forces. Hitler finally agreed and gave von Manstein freedom of maneuver. Von Manstein was now free to take whatever action was needed to defeat the Russian winter offensive that was currently overrunning Army Group South.

Also on this date, the 64th Army reported that it controlled 12 rifle divisions and a large assortment of subunits totaling 94,503 troops on hand. Interestingly 48,226 POWs were included within the Army's ration strength, which raises the total number of people under the Army's control to 142,729.[426] These totals changed daily as troops, units and POWs were shifted around. The NKVD would soon take control of all POWs within the Stalingrad area.

18 February

The GKO (State Defense Committee) issued Decree #2911 on armies and divisions left at Stalingrad after the elimination of the Don front. This concerned the transfer of military units, individual units and institutions from the Don Front to the Stalingrad Army Group to be under the command of Lieutenant General Trubnikov.

Also, on 18 February, the GKO formed a special commission to control the transfer and refitting of military units and other forces in and around Stalingrad.[427] This commission was headed by:

- G.M. Malenkov, a member of the GKO (State Defense Committee).
- B.A. Shchadenko, Head of the Glavuprform, (Major Department of Formation and Recruitment of the Red Army).
- N.D. Yakovlev, Chief of the Main Artillery Directorate of the Red Army.
- N.I. Biryukov, Political Commissar.

Fellow researchers have debated about the exact membership of this commission. The four people listed above were the most likely members. Their duties also included assessing the readiness, composition, and the final assignment of these units.

19 February

In the Ukraine, not far from the Dnieper River, spearheads of the Russian 6th Army reached a point only 30 miles from von Manstein's HQ and a visiting Adolf Hitler. These advanced Russian units, having run out of supplies, soon were destroyed by German counterattacks. Later in the day, with his meeting completed, Hitler flew back to Germany. Von Manstein then launched his counterattack with the II SS Panzer Corps attacking from the north while the 1st and 4th Panzer Armies attacked from the south. At this point, the advancing Russian armies of the Voronezh and Southwestern Fronts were highly vulnerable to this type of flank attack.

20 February

A further NKVD RPG unit action on this day encountered the final remnants of German forces holding out within Stalingrad. They also state that "in separate regions," (outside Stalingrad) reports about enemy troops being killed or captured continued to come in until 1 March 1943.[428] In total, by March 1943, the 21 Rifle Brigade of NKVD forces had killed 2,418 enemy soldiers and captured 8,646, which were turned over to POW camps.[429]

Also on this date, the city of Stalingrad Defense Committee adopted a resolution "on the use of prisoners of war for the reconstruction of industrial enterprises and the city of Stalingrad," which the Office of the POW camps offered to enterprises of the city to allocate 45,300 people (POWs). However, it was not possible to implement these and other plans to use POWs to rebuild the city because of bad living conditions, insufficient food rations, and poor work organization. Ultimately it was found that "the average monthly output per payroll prisoner was only 61.74 rubles, [w]hile the cost of maintaining a prisoner was 200 rubles."[430]

At this time, the placement of major 64th Army units are displayed on Map #31, a local map of the Stalingrad area. As can be seen, the main army units were spread out in and around Stalingrad and Beketovka, but also near rail lines, per prior orders. This loose arrangement would change as soon as the 64th Army was transferred back to the active front.

Map #31 64th Army Deployment 20 February 1943

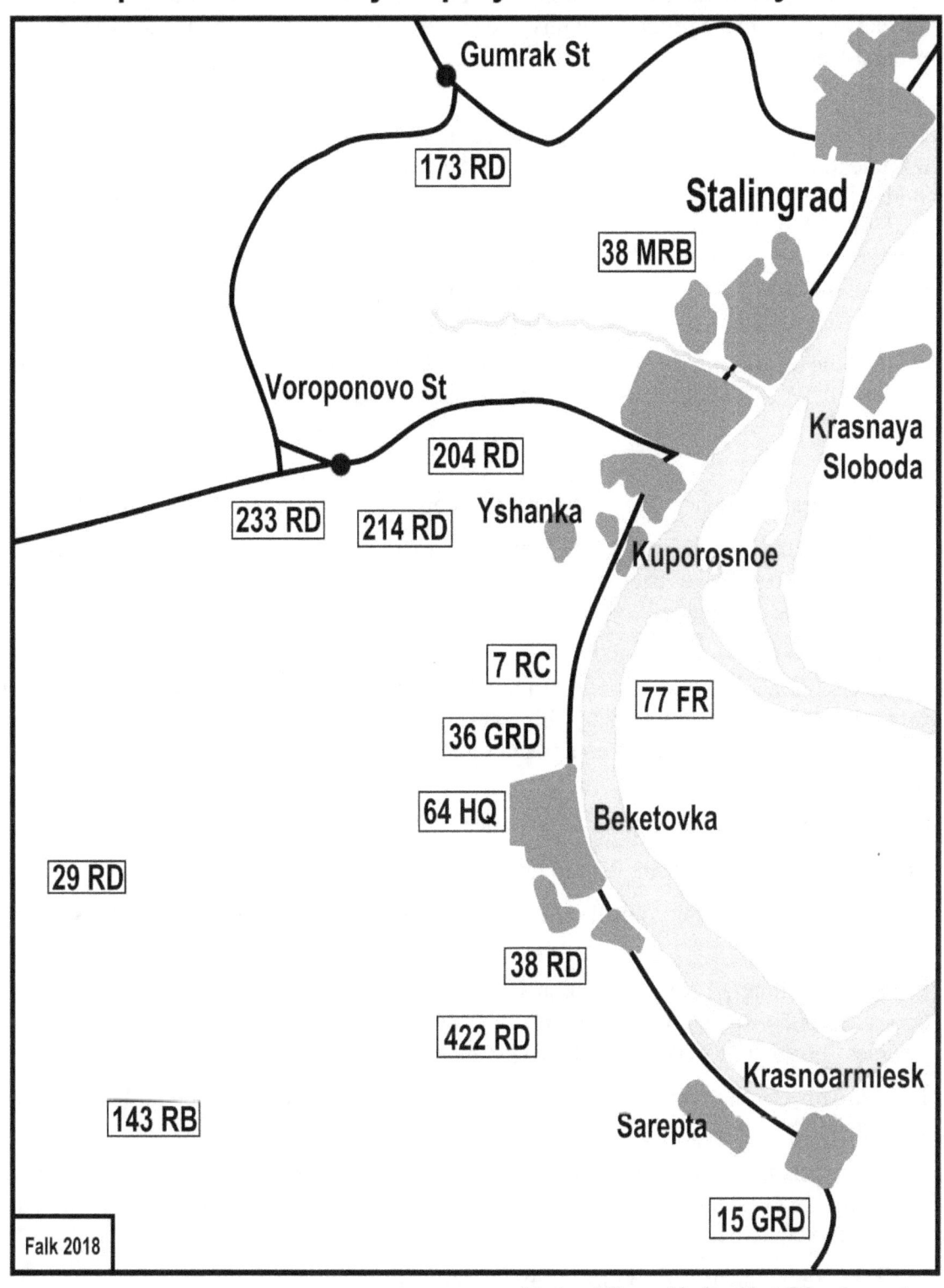

21 February

Heavy fighting continued in the Ukraine along the southern part of the front as the German counterattacks began to make progress against the flanks of the overextended Russian armies of the Southwest and Voronezh Fronts.

22 February

From Moscow, the People's Commissariat of Defense of the USSR ordered the elimination of landmines, unexploded shells and bombs in the Stalingrad region. This effort was headed by combat units, but also included POWs, civilians, Air Defense Units, and some 31 battalions of engineers, sappers and NKVD Specialists.[431] This cleanup work already had started, but now it was official. By May this effort had removed some 2.5 million landmines, bombs, grenades and other types of ordinance from the city area. Eventually this dangerous work continued for many more months or even years before finishing.

Stavka realized the serious nature of the German counterattack in the Ukraine and ordered the 69th Army and the 3rd TA to shift their attack southward against advancing German forces west of Kharkov.

23 February

Fresh Russian forces were sent to the active front to stop the unexpected German offensive.

While at Stalingrad, the NKVD head of the Stalingrad area of Commissioner of State Security, Voronin, issued Memo #5086 about demining areas in and around Stalingrad. Apparently, the work was not proceeding fast enough.

24 February

Acting on the GKO Decree #2911, Stavka formally transferred the armies and divisions remaining at Stalingrad to the Stalingrad Army Group. This transfer was originally planned for 27 February but was moved up a few days. This group comprised three combined arms infantry armies, 62nd, 64th and 66th, with the following forces:

- Rear service agencies containing 40,320 men; 6,230 horses; and 1,908 vehicles.
- 111,235 infantry troops, with 21,146 horses and 3,659 vehicles.
- Rear units and institutions outside the army's control with 12,822 men; 1,882 horses; and 867 vehicles.
- Within the Stalingrad Army Group, just the combat units contained 226,232 men; 30,290 horses; 11,005 vehicles; 1,532 tractors; 119 tanks (some not operational); 30 other armored vehicles; 11 pontoon parks; and 61 rocket launchers.

The Stalingrad Army Group represented a sizeable force that could be employed at the front with a bit more preparation.

Clarifying documents from 20 - 24 of February, show that the following rifle divisions were assigned to the 64th Army: 15 GRD, 36 GRD, 173 RD, 204 RD, 214 RD, and the 233 RD. They contained the following men and equipment:

- 15 GRD - 4,889 men; 1,033 horses; 149 vehicles; 12 tractors and 139 mortars (50 -120mm); 60 guns (45 -122mm).
- 36 GRD - 3,483 men; 248 horses; 161 vehicles; 27 tractors and 59 mortars; 65 guns.

- 173 RD - 3,870 men; 705 horses; 118 vehicles; 14 tractors and 129 mortars; 77 guns.
- 204 RD - 2,612 men; 294 horses; 128 vehicles; 17 tractors and 75 mortars; 34 guns.
- 214 RD - 3,416 men; 625 horses; 114 vehicles; 8 tractors and 169 mortars; 79 guns.
- 233 RD - 3,507 men; 509 horses; 160 vehicles; no tractors and 45 mortars; 54 guns.

Along with these major combat units, the rear area logistical units of the 64th Army contained an additional 26,685 men; 2,584 horses; 523 vehicles and 49 tractors. At this time the 64th Army also had administrative control over some 160 sub units. These units had belonged to other armies and were transferred to the 64th Army's control during the final days of combat in and around Stalingrad.[432]

Other units of interest that until recently were assigned to the 64th Army were:

- 7 Rifle Corps (93, 96, 97 rifle brigades) - 8,383 men; 1,378 horses; 284 vehicles; 4 tractors, and 306 mortars; 86 guns.
- 143 Rifle Brigade - 1,944 men; 254 horses; 139 vehicles; 2 tractors, and 49 mortars; 31 guns.

In early 1943, under a new TO&E (Table of Organization and Equipment), a full-strength rifle division should have contained between 9-10,000 men. So, these veteran Stalingrad divisions were greatly depleted, containing only 20 to 40 percent of authorized strength. They needed time to rebuild both in men and equipment to reach full strength once again.

Speaking about rear area organization, the head of the NKVD of the Stalingrad area of Commissioner of State Security, Voronin, once again issued a report, this time #5088, about security at POW camps around Stalingrad. He noted that 16 days after Beria issued orders about the care and treatment of POWs, only about 20-25 percent of the required food had arrived. Of the vehicles that were to be delivered, only about 50 percent of the trucks and cars had arrived, and of these, only about half were in running condition. He also noted that despite specific orders, more than 10,000 sick and wounded prisoners of war were in the camps and not in hospitals. Due to all these issues, and more, the death rate had increased for POWs to some 700 men per day. Additionally, he noted that "the command of the 64th Army made a complete selection of personal belongings of prisoners of war (watches, rings, lighters, soap, etc.). All of these things were put into the warehouse and enrolled in the Army Foundation." To remedy these problems, he made eight specific recommendations to the command of the Don Front:

1. Follow the instructions on the importation of the necessary amount of food into the camp warehouses.
2. Transfer to the prison camps the following serviceable vehicles: 100 cargo trucks and 15 passenger cars.
3. Release 5 full tanks of gasoline.
4. To ensure the refurbishment of the transferred vehicles, the 64 Army should transfer control of one of their body shops, in good repair, to the camps.
5. To provide transportation between the camps, convey 450 horses with carts to the camps.
6. Deliver the following tools to the POW camps: 1,000 shovels, 1,000 axes, 500 crowbars and 100 saws.

7. Instruct the Medical Administration of the South and Don Fronts to organize the treatment of wounded and sick prisoners of war.

8. Have the 64th Army commander (Shumilov) return to the POW camps all personal items taken from prisoners of war.

No one knows how many of these recommendations were acted upon. Probably few of them were carried out, and then only in part. In general, harsh winter conditions, the poor overall supply situation, and the needs of the Red Army insured few resources were diverted to the care of POWs.

Back at the active front, forces of the Southwest Front commanded by General Vatutin were ordered onto the defensive. Powerful German attacks continued as the Red Army forces were driven back, losing further ground.

25 February
The NKVD took total control of all the remaining POW camps from the Red Army in and around the Stalingrad area. They listed in 13 different camps 142,861 prisoners.[433]

Clearly some of these prisoners were captured during Operation Uranus in November and during the continuing advancement of the Red Army westward in December, January, and February. Nevertheless, the long-term outlook was grim for these prisoners of war.

"Chances of survival in the Soviet camp system were rank-dependent. Over 95 percent of noncommissioned officers and enlisted men perished; 55 percent of junior officers died; but only 5 percent of senior officers expired." By the spring of 1943, 55,228 prisoners captured at Stalingrad had already died.[434] Cold, starvation, and the lack of medical care took their toll, with out of control typhus being the biggest killer.

Accounts vary on just how many German prisoners of war captured at Stalingrad actually survived the war and returned to Germany. These estimates range from 5,000 to 6,000 and similar totals could be presented for the other nationalities like Italian, Romanian and Hungarians troops who also fought along the Don or Volga Rivers. All of the Axis armies that fought at Stalingrad paid a terrible human price both during the early victories of 1942 and even more in later defeats of 1943. Eventually, the German 6th Army, 4th Panzer Army, as well as the Romanian 3rd and 4th Armies were swept away by the Red Army. Then in December and January, the Italian 8th Army and the Hungarian 2nd Army also disappeared from the Axis order of battle as the advancing Red Army pushed the invaders back. Armies on both sides lost huge numbers of men and even larger amounts of equipment, all without a truly war-winning result. At the start of February 1943, German forces once again were advancing eastward and were approaching the Donets River. They even were headed for the recently-liberated city of Kharkov. For the men of the 64th Army, what looked like an end to the fighting a month before had turned out to be only a shifting of the battle to a new combat zone.

Also, on this day the special commission from the GKO (State Defense Committee) reported back to Stalin in Moscow. This telegram, signed by Malenkov, Shchadenko, Yakovlev, and Biryukov, submitted for Stalin's approval numerous proposals concerning the future deployment of forces within the Stalingrad Army Group. In part they proposed relocating the 64th Army of General Shumilov. The Army, consisting of the 15 GRD, 36 GRD, 204 RD, 29 RD, 38 RD and 422 RD, was to be moved to the area of Valuiki (east of Belgorod) where it would be rebuilt. Reinforcements for this army would be prepared by the *Glavupraform*, a major department for Formations and Recruitment of the Red Army.[435]

This proposal not only changed the rifle divisions that were assigned to the 64th Army, but it also stated the city of Valuiki as a replenishment area. This area was under the control of the Voronezh Front, soon to be commanded once again by General Nikolai Fyodorovich Vatutin, who would take over command on 28 March 1943. This GKO proposal also alerted General Shumilov that his army would be headed back to the front very soon.

26 February
Advancing German Panzer units surrounded and destroyed retreating Russian formations all along the front as they fled toward the east. Stavka recognized that existing Russian forces were unable to stop these German attacks.

27 February
Stavka ordered the 1st Guards Army and the 6th Army of the Southwestern Front back over to the east side of the Donets River to hold that section of the line against fierce German attacks.

28 February
Stavka ordered the 3rd TA to swing southward in an attempt to free surrounded units of the 6th Army. Despite the urgent need, the 3rd TA required several days to organize this desperate attack.

On this date, the 64th Army had a reported combat strength of 31,638 men with an estimated support strength of 12,000, for a total ration strength of 43,638 men.[436]

With the changing situation at the active front, the 64th Army's time in the rear area was over. Stavka needed to bring up fresh forces. The 64th Army was headed to the Voronezh Front area to help stabilize the front and to prevent the Germans from gaining more territory. The first of two orders/directives concerning the transfer of the 64th Army was the NKO Order #36799 on 28 February 1943 at 05:30 hours. This order provided detailed instructions about the transfer of the 64th Army by rail to the Voronezh Front. The order outlined which units were being moved, how many trains were allocated, and which entraining (loading) and detraining (unloading) stations would be used. (see Appendix 22) Stavka directive #46063 soon followed:

Stavka Directive #46063

Authorized Stavka, to the command of the Voronezh Front and the 64th Army

About reassignment of the Army

28 February 1943 17 h 00 min GHQ orders:

1. The 64th Army of six infantry divisions (15 GR. 36 GR. 29, 38, 204 and 422 RD); 156 and 1111 Gun Artillery Regiment; 186, 493, 496, and 500 Antitank Artillery Regiment; 838 Artillery Observation Battalion; 27 Guards Tank Brigade; 224 and 245 Tank Regiments (the two TRs are coming from Gorky) *relocate to the area of Valuiki, at the disposal of the commander of the Voronezh Front.*

2. Redeployment to proceed by rail. Start sending on March 1, with the transport of the army into the new area to finish by March 15.

3. Resupply all infantry divisions with personnel, horses, arms, and in the new areas of deployment, bringing the number of men in each infantry division, including the Guards, to 8,000.

Reply upon Execution. GHQ STALIN

This directive officially removed the 173 RD, 214 RD, and the 233 RD from the 64th Army and substituted the 29 RD, 38 RD, and the 422 RD. The reason for this change of rifle divisions assigned to the army is unknown. These three newly assigned rifle divisions had the following strength in mid-February:

- 29 RD - 3,083 men; 446 horses; 56 vehicles; 5 tractors and 32 mortars; 27 guns.
- 38 RD - 3,534 men; 659 horses; 114 vehicles; 13 tractors and 72 mortars; 29 guns.
- 422 RD - 4,307 men; 451 horses; 117 vehicles; 11 tractors and 79 mortars; 50 guns.

In total, the final six rifle divisions had combined strength of 21,908 men; 3,131 horses; 725 vehicles; 85 tractors; 456 mortars; and 265 guns. This was a powerful force but far below authorized strength. Hopefully this new deployment would allow enough time to bring these units back to full fighting strength. Also, the special commission from the GKO stated that 50-60 percent of the weapons and equipment (within the Stalingrad Group of forces) needed repair and that it was impossible to carry out these repairs under the conditions at Stalingrad. Hopefully, these repairs could be undertaken at the new front.

Directive #46063 also was hopelessly optimistic. To comply with this order, the 64th Army needed to:

- Gather the required combat units, medical and support formations and then provide them with supplies for the journey (food, water, fodder, wood, fuel and ammunition).
- Assemble the different sub-formations at the designated entraining stations at different times.
- Load troops and equipment as quickly as possible if the required number of trains and types of cars appeared at the correct stations.
- Set off on the long journey, but only if the rail lines were clear and the steam engines were filled with fuel and water. Only then would the 64th Army begin its slow trip along overloaded rail lines toward the front.

To complicate matters, due to logistical issues, units of the 64th Army were sent via four different routes from Stalingrad and Gorki to their destinations in the Valuiki area. (see Map #32, Rail map showing the four routes from Stalingrad and Gorky to the Valuiki area).

1. The first route covered 730 miles: Stalingrad – Povorino – Gryazi –Yelets – Kastornoe – to the Valuiki area. This route was taken by the 64th Army Headquarters, 15 GRD, 38 RD, 29 RD and 27 GTB and most if not all of the first group of supply, medical and rear support units.
2. The second route covered 869 miles and was taken by the 36 GRD, 204 RD and the 422 RD. From Stalingrad they crossed to the east side of the frozen Volga on foot and entrained at Zaplavyoe. Then they travelled via rail to Verhn Baskuncak, then turned north to Urbakh and then west crossing back to the west side of the Volga to Saratov. Finally they travelled via Rtishchevo – Balashov – Povorino – Liski to arrive in the Valuiki area.
3. The third route covered the remaining units of the army at Stalingrad, mostly artillery and AT units. They took a shorter route that covered about 400 miles: Stalingrad – Povorino – Liski - Valuiki.
4. Additionally, the two newly assigned Tank Regiments, 245, 230 TRs, travelled all the way from Gorky, some 720 miles, to reach the Valuiki area.

After traveling for many days, these slow-moving troop trains would eventually arrive at a series of vaguely defined and dispersed locations for detraining. Finally, as these army sub-formations detrained, they would have to march through the snow to assembly areas in order to reform their combat units.

All this was to be accomplished in only 15 days during the middle of winter. Furthermore, this transfer was intended to help block a powerful German counteroffensive that was advancing into the same general region where the 64th Army would detrain. In a situation like this, a slight miscalculation on the part of Stavka or the Voronezh Front headquarters could be disastrous for the 64th Army.

Regardless of the difficulties in carrying out this order, the 64th Army was duty-bound to start its long journey back to the active front, leaving behind Stalingrad and all its terrors. With a new battleground calling them forward, the Army gathered its strength and prepared to leave. However, its leaders and men were no longer part of a new and inexperienced army, but rather members of an army that had faced a great trial by fire and survived with honor! Many future battles lay ahead for the 64th Combined Arms Infantry Army, but that account is yet another story.

Conclusion

A final thought about the battle for Stalingrad: we will never know the true human cost of this battle; the best we can do is to provide an approximation. David M. Glantz and Jonathan M. House, in their book *Endgame at Stalingrad Vol. 3, book 2* stated, "It has been estimated that during the entire Stalingrad Campaign [July 42 to January 43], Soviet forces suffered about 1,700,000 casualties in the defense of the city, with the Axis forces suffering just over 1,000,000 casualties trying to capture it." Of course, any civilian losses would only add to these totals. Sadly, with this level of carnage, the exact number of casualties is irrelevant. In essence, everyone involved in this battle lost. In the end, all this death and destruction was ultimately about the possession of a single city on the banks of the Volga River: Stalingrad.

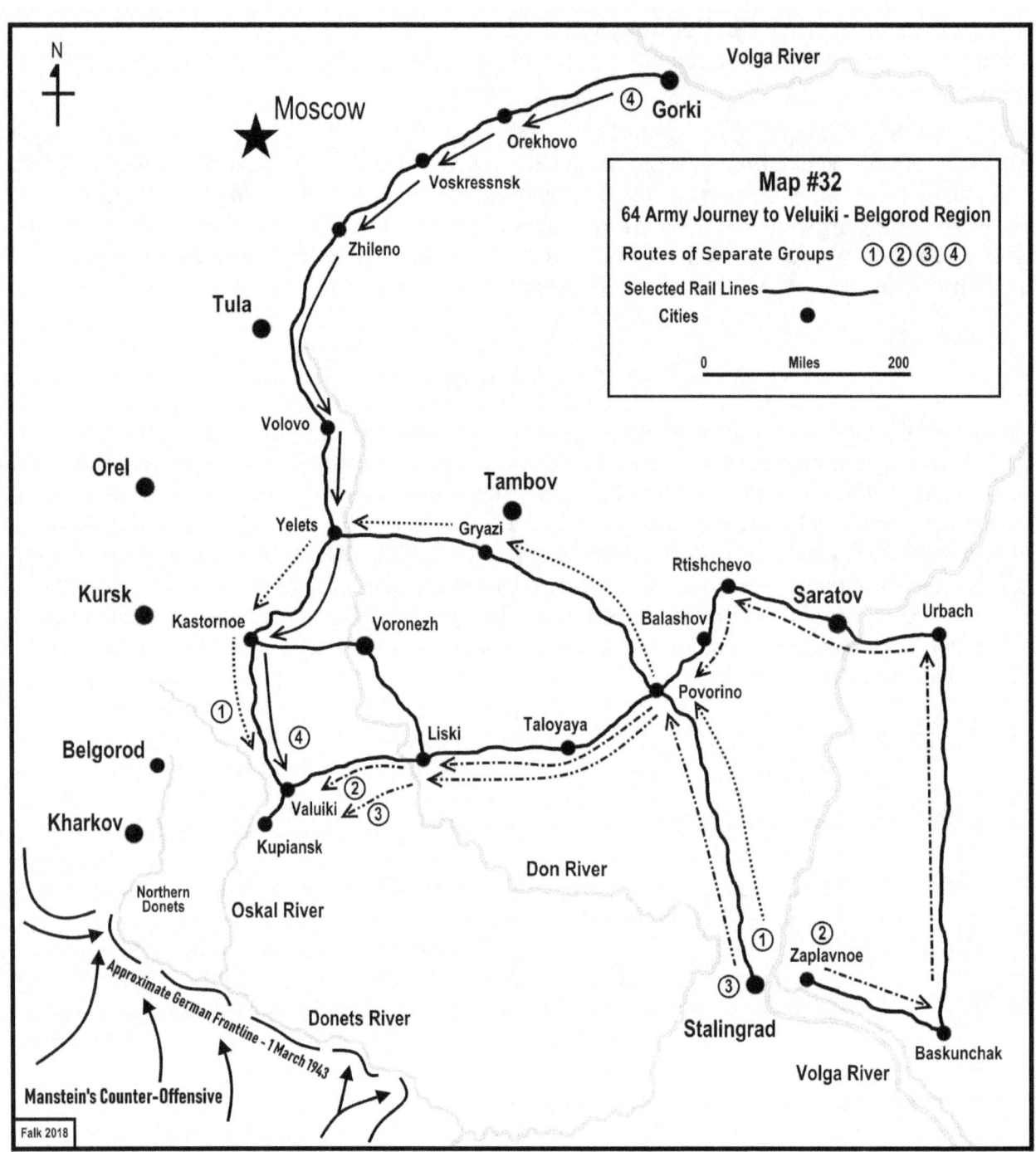

Appendix

Appendix 1

BSSA (*Boevoi Sostav Sovetskoi Armii* – Combat Composition of the Red Army). BSSA tables list all designated units of the active fronts. Data was compiled the first day of each month; therefore, it provides a snapshot of information once a month for the entire Red Army. Further monthly references from this source will be listed as BSSA - 41, 42, 43. The links below provide access to these invaluable records:

http://www.soldat.ru/files/f/boevojsostavsa1941.pdf
http://www.soldat.ru/files/f/boevojsostavsa1942.pdf
http://www.soldat.ru/files/f/boevojsostavsa1943.pdf
http://www.soldat.ru/files/f/boevojsostavsa1944.pdf
http://www.soldat.ru/files/f/boevojsostavsa1945.pdf

BSA (*Boyevoy Sostav Obshchevojskovykh Armiy* - Composition of the Combined Arms Combat Armies). BSA data was compiled three times each month, so it gives a bit more detail than BSSA data. Further monthly references from this source will be listed as BSA - 41, 42, 43. These historical BSA files are difficult to find, and I know of no public source. BSSA and BSA records do not always agree with each other, which forces historians to reconcile the discrepancies. In many cases, a third source, historical records from the TsAMO archive (Central Archives of the Ministry of Defense) also were consulted to clarify and to expand the known units that were assigned to the 64th Army.

Appendix 2

When I lack solid, reported information from Soviet/Russian archives about combat or ration strengths, I will estimate these strengths for the 64th Army. I will include any hard data that I can find and approximate the unknown values for units assigned. My intention in listing these combat and ration strengths is to give the reader an idea of the forces involved and the losses suffered. These values are not perfect. They are my estimates, based upon known and estimated starting strengths and then known and estimated losses during battle. These values are subjected to change if better quality data is someday found.

Appendix 3

Besides Major General M.M. Ivanov as the first 64th Army commander from December 1941 - March 1942; Major General of Artillery Lakov I. Broud was the commander from March - May 1942. Member of the Military Council divisional commissioner K. K. Abramov was assigned from December 1941 - May 1942, and as the Chief of a Staff Colonel N. N. Lozhkin was assigned to the 64th Army from December 1941-May 1942.

Appendix 4

Due to more combat units being assigned to the 64th Army for longer periods of time, I have increased the Army's supporting troops to 2,000 men and women. Large numbers of women were included in Soviet Military Forces during the war and their numbers increased as losses at the front grew. In all, 490,235 women were conscripted into military service with many thousands more serving as volunteers in various roles. Overall, women accounted for 41 percent of all doctors and 43 percent of front-line medical staff within the Red Army. They too suffered their share of casualties.

Appendix 5

This information is based upon personal correspondence with researcher and author Charles C. Sharp. Note: Two other armies were renamed the same day. In Directive #170331, the 41st Army was renamed as 3rd Reserve Army, and in Directive #170332, the 27th Army was renamed as 2nd Reserve Army. However, next month, on 1 June 1942, BSSA data only displays the 1st Reserve Army as being formed. Other reserve armies only started to be recorded on the July 42 BSSA data list. The reason for these inconsistencies is unknown.

Appendix 6

On 10 July the new 64th Army was listed as having a strength of 72,800 people. Note: There is some debate whether this figure relates to ration strength (total number of people assigned to the Army) or combat strength (people assigned to combat units). My opinion is the 72,800 value is the combat strength of the Army. If this is correct, the ration strength, which includes support troops, might add another 5-10,000 people to the Army's total. In this case, I have estimated the support troops totaled about 7,000 people, which brings the ration strength of the Army to 79,800 men and women.

Appendix 7

Stavka Directives #994101 transferred the following units:

- 205 RD - from Khabarovsk
- 96 RD - from Kuibyshevka
- 204 RD - from Cheremhovo
- 422 RD - from Rozengartovki
- 87 RD - from Spassky
- 208 RD - from Slavyanka
- 126 RD - from Razdolnaya Putsilovki
- 98 RD - from Khorolya
- 250 RB - from Birobidzhan
- 248 RB - from Zakandvorovki, Primorye
- 253 RB - from Shkotovo

Stavka Directive #994102 transferred:

- 321 RD - from District Borzi
- 399 RD - from District Borzi
- 229 RB - from District Borzi

Appendix 8

Information obtained from the Axis History Forum Contributor, *Der Alte Fritz*, https://forum.axishistory.com// and RKKA in World War 2 Forum Contributors, Art (Artyom A.), and Alex (amvas) http://www.armchairgeneral.com/rkkaww2/.

Appendix 9

This distance was established using period Russian RR maps via the following route:

- Stalinogorsk - Ryazhsk, 151 km (93.82 miles)
- Ryazhsk - Michurinsk, 89 km (55.3 miles)
- Michurinsk - Tambov, 67 km (41.63 miles)
- Tambov - Balashov, 200 km (124.27 miles)
- Balashov - Povorino, 71 km (44.11 miles)
- Povorino - Stalingrad, 344 km (213.75 miles)

for a total of 922 km or 572.88 miles.

Appendix 10

The seven interdiction targets listed are: 1-All types of moving columns and march movements. 2- Troop concentrations. 3- Rail interdiction targets (including rolling stock and permanent installations). 4- Road interdiction targets. 5- Waterway interdiction targets. 6- Man-made structures (bridges, viaducts, etc.). 7- Port installations.

Appendix 11

Major General Mikhail Stepanovich Shumilov, who took over command of the 64th Army at the end of July, would continue as its commander until 16 April 1943 when the 64th Army was awarded guards status, becoming the 7th Guards Army. On 31 December 1942, General Shumilov was promoted to Lieutenant General. Then in October 1943, Shumilov was promoted to Colonel General and would remain in command of the 7th Guards Army until the end of the war.

Appendix 12

The breakdown of the 64th Army's Estimated Combat Strength – 52,890 men on 7 August 1942:

- 38 RD - 10,000
- 126 RD - 10,000
- 204 RD - 6,000
- 214 RD - 4,500
- 66 NRB - 2,600
- 118 FR - 2,000
- Three Zhytomyr Cadet Rifle Regiments - 6,000 (2,000 each)
- 1st Ordzhonikidzenskih Infantry Training School Regiment - 2,000
- Krasnodar Machine-Gun and Mortar School Regiment - 2,000
- 13 TC (6 Guards, 13, 56, 133, 254 TB) - 4,500
- 28 Armored Train Battalion - 400
- 1111 Cannon Artillery Regiment - 480
- 1251 Tank Destroyer Regiment - 480
- 140 Mortar Regiment - 470
- 76 Guards Mortar Regiment (Katyusha rocket launchers) - 800
- 1363 Sapper Battalion - 160

Group Chuikov Estimated Combat Strength – 23,800 men:

- 29 RD - 10,000
- 138 RD - 4,000
- 157 RD - 1,200
- 208 RD - 3,000
- 154 NRB - 3,000
- 255th Cavalry Regiment - 1,000
- 18 Guards Mortar Regiments (Katyusha rocket launchers) - 800
- 19 Guards Mortar Regiments (Katyusha rocket launchers) - 800

Appendix 13

I used information taken from BSA data to assign these three units to the 64th Army at this time:

- The ration strength of the 204 RD would have been in the 2-3,000 range. I picked 2,500 as an average.
- The 208 RD almost certainly consisted of only a headquarters, so 200 men would be an approximate strength.
- The strength of the 77 Fortified Region would have been similar to that of the 118 FR, so I picked 1,400 for its value.

Appendix 14

Another aspect of this episode is that General Chuikov might have been dismissed from his command by General Eremenko on 14 October. Stalin then intervened to keep Chuikov on as the commander of the 62nd Army and that is why Eremenko and his deputy, General Popov, showed up in the city. Read more details of this possibility in *Stalingrad, How the Red Army Survived the German Onslaught* by Michael K. Jones.

Appendix 15

During WW II most trains were powered by a single steam engine because of control issues when linking more engines together. With two engines, the separate crews had to coordinate their efforts very closely, or the engines would not pull together as one unit. In other words, if the crews were not careful, one engine would start to pull or push the other; at that point, the underperforming engine would become a load and not actually help move the train. Modern trains circumvent this problem with the use of interlinking communication systems and computers. These systems allow a single crewmember to smoothly control multiple engines, even if they are dispersed throughout the train.

Appendix 16

The TsAMO Archive doc lists 43,263 people in infantry units. Other sources list 13,252 people in non-combat units. Taking these totals into consideration, I added my strength estimate for the following:

- Artillery units - 9,288 men
- Tanks and armored trains - 1,250
- Engineers - 550

for a total ration strength of 67,603 and a combat strength of 54,351 men.

Appendix 17

The only other major battle in history, that I can recall, in which the same double encirclement occurred is during the siege of Alesia in 52 BC. Julius Caesar and the Roman Army were laying siege to the city of Alesia in Gaul (modern France) when a very large relieving Gallic force arrived and besieged the Roman forces in turn. Ultimately, Caesar was able to defeat the outside Gallic force and then successfully complete the siege of Alesia.

Appendix 18

This estimate of ration strength totaling 65,243 men breaks down into the following general categories:

- Infantry units - 44,980
- Artillery - 3,748
- AA - 373
- Tank (+ 38 Motorized Brigade) - 2,710
- Armored Train - 242
- Engineers - 690
- HQ and Support - 12,500

These numbers also are based on BSSA data.

Appendix 19

The killed verses wounded ratio during six days of combat for the 64th Army was 2.91. The killed verses wounded ratio during 11 days of combat for the adjacent 57th Army was 4.34. The average between the two comes to 3.62. I then doubled the six-day killed value for the 64th Army to develop an estimated value of 784 killed over 11 days. I then used the average ratio of 3.62 to estimate the number of wounded over the same period, 784 x 3.62 = 2,838 wounded. The missing and other losses are my pure estimates based upon earlier reported values in these categories.

Appendix 20

If we explore this issue further and speculate somewhat with numbers, we might reveal a hidden truth.

The 64th Army had a reported loss of 21,113 men. So, what if this reported loss represented a more plausible 18 percent of the Don Front's total casualties during Operation Ring? If true, that would push the actual Don Front losses to around 120,000 men or about 40 percent of the starting force. These are significant losses. But, supporting this speculation, Glantz and House have stated that during Operation Uranus (November 42), Soviet forces lost 169,925 casualties or 55 percent of the starting force of 307,500. These numbers are from *Endgame at Stalingrad, Book Two Vol 3,* p. 582. These values make the hypothetical 40 percent loss rate for Operation Ring much more believable. Obviously further research needs to be undertaken to clarify this issue.

Appendix 21

Manfred Kehrig provides data on *Hiwis* and attached personnel serving with German troops at Stalingrad on 16 November. In turn, David Glantz writes in the *Journal of Slavic Studies* 21: 377-471, 2008 on page 468-469 which displays two tables that show in mid-November 1942, the 4th Panzer Army, VI Corps and the 29 Motorized Division and the 6th Army in total contained large numbers of *Hiwis* and other attached personnel. It must be noted these two sources add values for *Hiwis* and other attached personnel together, producing inflated totals for *Hiwis*. However, additional information provided by researchers on the internet via the RKKA in WW II Forum, http://www.armchairgeneral.com/rkkaww2/ and the Axis History Forum, https://forum.axishistory.com//, offers updated information which corrects these minor errors and provides what I believe to be more realistic and accurate values. In the end, I found that approximately 30,000 *Hiwis* were serving with Axis combat units in and around Stalingrad during November 42. Of these, about 20,300 were trapped within the pocket. Finally, reports reveal that 10,271 were detained by Soviet forces during and after the battle and sent to replacement or penal units. So approximately 10,000 of the surrounded *Hiwis* died during the fighting or were executed by the Soviets after capture.

Appendix 22

COMMANDING OF STALINGRAD GROUP OF FORCES, BY FORCES OF VORONEZH FRONT, 64 ARMY ABOUT THE REASSIGNMENT OF THE ARMY

On February 28, 1943. 05 h of 30 min

The People Commissioner of Defense ordered:

1. *To direct 64 Army at the disposal of the Voronezh front. In army to include 15 GRD, 36 GRD, 204, 29, 38 and 422 RD, 27 Guards Tank Brigade, 1111 and 156 Cannon Artillery Regiment, 186, 493, 496 and 500 Antitank Artillery Regiment and 838 Artillery Observation Battalion.*

2. *Sending 64 Army and reinforcing means to produce by the following order:*

    a) Administration of army (HQ) - 20 echelons (trains), loading at Station Beketovskaya, starting at 06.00 hours on 2.3. 1943, loading at the rate of 2 trains/day, trains № 25201 – 25221.
    b) 15th GRD, 10 echelons, loading – Station Sarepta, at 06.00 hours on 2.3.1943, rate – 3 trains/day, trains № 25221 – 25230.
    c) 36 GRD, 7 echelons, loading – Station Zaplavnaya, at 06.00 hours on 5.3.194, rate – 3, trains № 25281 – 25287.
    d) 29 RD, 6 echelons, loading – Station Voroponovo, at 12.00 hours on 8.3.1943, rate – 3, trains № 25241 – 25246.
    e) 38 RD, 7 echelons, loading – Station Elshanka, at 18.00 hours on 5.3.1943, rate - 3, trains № 25233 – 25239.

f) 204 RD, 6 echelons, loading – Station Zaplavnaya, at 06.00 hours on 3.3. 1943, rate - 3, trains № 25274 – 25279.

g) 422 RD, 6 echelons, loading – Station Zaplavnaya, at 12.00 hours on 7.3.1943, rate - 3, trains № 25291 – 25296.

h) 1111 Cannon Artillery Regiment, 2 echelons, loading – Station Gumrak, at 06.00 hours on 2.3.1943, rate - 2, trains № 25251 – 25252.

i) 156 Cannon Artillery Regiment, 2 echelons, loading – Station Gumrak, at 06.00 hours on 2.3.1943, rate - 2, trains № 25254 – 25255.

j) 186 AT Regiment, 1 echelon, loading – Station Gumrak, at 06.00 hours on 7.3.1943, train № 25259.

k) 500 AT Regiment, 1 echelon, loading – Station Gumrak, at 06.00 hours on 6.3.1943, Train № 25257.

l) 493 AT Regiment, 1 echelon, loading – Station Voroponovo, at 06.00 hours on 8.3.1943, train № 25261.

m) 496 AT Regiment, 1 echelon, loading – Station Elshanka, at 06.00 on 2.3.1943, train № 25263.

n) 838 Artillery Observation Battalion, 1 echelon, loading – Station Sarepta, at 06. 00 on 2.3.1943, train № 25265.

o) 27 Guards Tank Brigade, 3 echelons, loading – Station Gumrak, at 06.00 on 7.3.1943, rate - 3. train № 25248 - 25250.

Region of unloading for all parts - Valuyki. (which is ease of Belgorod)

3. From the general composition of the administration 64 of army to send 20 echelons, where to immerse the parts of control, connection, of engineer support and strictly necessary the part of army rear. Transportation of the remaining part of the army apparatus and establishments it will be realized tentatively in the month.
4. Those sent to supply: by ammunition - 1 fire unit, by fuel - 1 servicing, by food and forage - on the route of 7 days and, furthermore, 3- days unloading reserve.
5. Komsomol of Voronezh Front to organize meeting, reception, accommodation of those arriving and to enroll them to all forms of allowance.
6. Commander of the troops of the Stalingrad Group of Forces prior to dispatch of the 64th Army to finish sending the 157th and 169th Divisions and the 11th Artillery division. Stop sending the 68th and 69th GRD. The loaded echelons of these divisions are to be returned to the loading station and unloaded. The empty cars to transport parts of 157th, 169th divisions and 11th artillery division.
7. On the progress of sending, arriving to report to the General Staff every day by 18.00. Deputy chief of the General Staff.

This order clearly lays out which units are to be included within the 64th Army as it leaves Stalingrad. It also assigns how many trains will be used and the loading rate.

Section 6 is interesting because the General Staff wants to send the 157th and 169th Divisions and the 11th Artillery division from Stalingrad before the 64th

Army. To accomplish this, it ordered trains that were already loaded with units of the 68th and 69th Guard Rifle Divisions (both are newly formed and not part of the 64th Army) back to the stations to unload, then load these empty rail cars with units of the 157th and 169th Rifle Divisions and the 11th Artillery Division.

Notes

Abbreviations

TsAMO – Tsentral'nyi arkhiv Ministerstva Oborony (Central Archives of the Ministry of Defense)
BSSA – Boevoi Sostav Sovetskoi Armii (Combat Composition of the Red Army)
BSA – Boyevoy Sostav Obshchevojskovykh Armiy (Composition of the Combined Arms Combat Armies)

1. Sharp, Charles C. *Soviet Order of Battle World War II, Volume IX, Red Tide*, George F. Nafziger, 1996, p. 2-3.
2. Dunn, Walter S. JR. *Hitler's Nemesis*, Praeger, 1994, p. 35-37. (JR 1994).
3. BSSA, part 1 (June-December 1941). Moscow, 1963.
4. Based upon the Historical Document, *Material on the Staffing of the Army, for 7 December 1942*, Retrieved from TsAMO Archive ID# 131307644.
5. Grinevsky, V.V., A.D. Ovsyannikov., I.M. Ryzhov., & A.N. Yanchinsky. *The heroic sixty-fourth*, Ninshe-Volzhskoye Publishing House, Volgograd, 1981.
6. *Journal of Combat Operations Staff, 24th Army*. (n.d.) Retrieved from the Russian web site Memory of the People https://pamyatnaroda.ru/.
7. Sharp, Charles C. *Red Volunteers*, 1996, Vol.11. p. 1-2.
8. Grinevsky, V.V., A.D. Ovsyannikov., I.M. Ryzhov., & A.N. Yanchinsky. *The heroic sixty-fourth*, Ninshe-Volzhskoye Publishing House, Volgograd, 1981.
9. Sharp, Charles C. *Red Volunteers*,1996, – The three unknown rifle brigades will most likely be found within this group of ten, the 4, 15, 38, 52, 60, 86, 105, 106, 107, 109. Unfortunately, it has proven to be impossible to find any records that reveal which three. This best guess information is based on my personal correspondence with Mr. Sharp.
10. BSA – 1941, this data shows these 7 RBs assigned to the 24th Army during December.
11. Based upon the Historical Document, *Material on the Staffing of the Army, for 7 December 1942*, Retrieved from Pamyat Naroda web site, TsAMO Archive, ID# 131307644.
12. Based upon the Historical Document, *Operational Summary of the Staff 24 A, for 1 January 1942*, Retrieved from TsAMO Archive, ID# 135231597.
13. Based upon the Historical Document, *Operational Summary of the Staff 24 A, for 3 January 1942*, Retrieved from TsAMO Archive, ID# 135231641.
14. *Stavka VGK Documents and Materials 1942*, INSTITUTE OF THE MILITARY HISTORY OF THE MINISTRY OF DEFENSE OF THE RUSSIAN FEDERATION, Moscow, 1996.
15. *Journal of Combat Operations Staff, 24th Army*, (n.d.) Retrieved from the Russian web site Memory of the People https://pamyatnaroda.ru/ (ID# Unknown).
16. Based upon the Historical Document, *Material on the Staffing of the Army, for 7 December 1942*, Retrieved from TsAMO Archive, ID# 131307644.
17. *Journal of Combat Operations Staff, 24th Army*, (n.d.) Retrieved from the Russian web site Memory of the People https://pamyatnaroda.ru/ (ID# Unknown). Note: the proceeding web site is very difficult to use.
18. *Journal of Combat Operations Staff, 24th Army*, (n.d.) Retrieved from the Russian web site Memory of the People https://pamyatnaroda.ru/ (ID# Unknown). Note: the proceeding web site is very difficult to use.
19. 1942 Stavka/VGK Directives list, from various sources on the internet.
20. Jukes, Geoffrey. *Hitler's Stalingrad decisions*, University of California Press 1985. p 27.
21. *Journal of Combat Operations Staff, 24 Army*, (n.d.) Retrieved from the Russian web site Memory of the People https://pamyatnaroda.ru/ (ID# Unknown).
22. Overy, Richard. *Russia's War*, London, Penguin Putnam Inc. 1997, p. 146.
23. *Journal of Combat Operations Staff, 24 Army*, (n.d.) From Pamyat Naroda web site, TsAMO Archive, (ID# Unknown).
24. Grinevsky, V.V., A.D. Ovsyannikov., I.M. Ryzhov., & A.N. Yanchinsky. *The heroic sixty-fourth*, Ninshe-Volzhskoye Publishing House, Volgograd, 1981. p. 2.
25. Kerr, Walter. *The Secret of Stalingrad*, 1978, Macdonald and Jane's, p. 32-33.
26. The medical information retrieved found the web site Soldat.ru/hospital.html.
27. *Journal of Combat Operations Staff, 24th Army*, (n.d.) Retrieved from Pamyat Naroda web site, TsAMO Archive, (ID# Unknown). Other listings within this section of text are from the same source.
28. Sharp, Charles C. *Vol. 10 Red Swarm*, George F. Nafziger, 1996.

[29] Glantz, David M. - *Zhukov's Greatest Defeat*. University Press Kansas, 1999.
[30] Chuikov, V. I. - *Beginning of the Road*, Macgibbon & Kee 1963, p. 14.
[31] Ibid., 14.
[32] Ziemke, Earl F. *Moscow to Stalingrad: Decision in the East*, Center of Military History United States Army, 1987, p. 282.
[33] Ibid., 321.
[34] Based upon the Historical Document, *Material on the Staffing of the Army, for 7 December 1942*, Retrieved from TsAMO Archive. ID# 131307644.
[35] *Journal of Combat Operations Staff, 24th Army*, (n.d.) Retrieved from Pamyat Naroda web site, TsAMO Archive, (ID# Unknown.) .
[36] From BSSA – 1942, part 2.
[37] Kerr, Walter. *The Secret of Stalingrad*, Macdonald and Jane's, 1978, p. 30-33.
[38] Grinevsky, V.V., A.D. Ovsyannikov., I.M. Ryzhov., & A.N. Yanchinsky. *Heroic Sixty Fourth*, Ninshe-Volzhskoye Publishing House, Volgograd, 1981.
[39] *Stavka VGK Documents and Materials 1942*, INSTITUTE OF THE MILITARY HISTORY OF THE MINISTRY OF DEFENSE OF THE RUSSIAN FEDERATION, Moscow, 1996.
[40] Isaev, Aleksey. *Stalingrad: There is No Land for Us Beyond the Volga*, Prologue Chapter, Moscow: Iauza Eksmo, 2008.
[41] Chuikov, V. I. *The Beginning of the Road*, Macgibbon & Kee, 1963 London, p. 14.
[42] Bergstrom, Christer. *Stalingrad - the Air Battle*, Midland, 2007, p. 54.
[43] Gordon L. Rottman. *Soviet Rifleman 1941-45*. Osprey Publishing Ltd, 2007 p. 14-17.
[44] *Stavka VGK Documents and Materials 1942*, INSTITUTE OF THE MILITARY HISTORY OF THE MINISTRY OF DEFENSE OF THE RUSSIAN FEDERATION, Moscow, 1996.
[45] Davie, H.G.W. The Journal of Slavic Military Studies, *The Influence of Railways on Military Operations in the Russo-German War 1041-1945*, Taylor & Francis Group, Philadelphia, Vol 30, Number 2, April-June 2017.
[46] Based upon the Historical Document, *Card for registering the numerical and combat strength of units that are not part of the army 10 July 1942*. Retrieved from TsAMO Archive. ID #135232288. It should be noted this document listed a Shtat manning level of 12,807 men, which is slightly above the 04/200 Shtat of 12,725 men. The reason for this variation in strength is unknown.
[47] Sharp, Charles C. *Soviet Order of Battle WW2 Vol 10 Red Swarm*, George F. Nafziger, 1996.
[48] Grinevsky, V.V., A.D. Ovsyannikov., I.M. Ryzhov., & A.N. Yanchinsky. *Heroic Sixty Fourth*, Ninshe-Volzhskoye Publishing House, Volgograd, 1981, p. 10.
[49] *Battle of Stalingrad Preparatory Phase Documents Document #9*. Voyenno-Istoricheskiy Zhurnal article #8 Aug 82, Retrieved from: http://www.dtic.mil/dtic/tr/fulltext/u2/a362749.pdf.
[50] Zaloga, Steven J., Leland S. Ness. *Red Army Handbook 1939-1945*, Sutton Publishing, 1998, p. 23 - 35. This info is also taken from Shtat 04/200, which is the one most likely to apply in this case, but the overall numbers of equipment could differ.
[51] *Battle of Stalingrad Preparatory Phase Documents. Document #10*. Voyenno-Istoricheskiy Zhurnal article #8 Aug 82 .
[52] Based upon Historical Documents, *A summary of the movement of trains 64 A. 19-22 July 1942.*, Retrieved from TsAMO Archive. ID #131306260, 131306276, & 131306289.
[53] Chuikov, V. I. *The Beginning of the Road*, Macgibbon & Kee, 1963 London, p. 16.
[54] Zheltov, Lyubov Valentinovna. Retrieved from the reference section: *Military health care in the Battle of Stalingrad: health maintenance organizations and the activities of the medical service of the Red Army*. Thesis and dissertation Ph.D. 1999 Volgograd. (http://www.dissercat.com/content/voennoe-zdravookhranenie-v-stalingradskoi-bitveorganizatsiya-meditsinskogo-obespecheniya-i-).
[55] Gudkova, Galina D. *They Will Live*, Moscow, 1986, Chapter one.
[56] *Stavka VGK Documents and Materials 1942*, INSTITUTE OF THE MILITARY HISTORY OF THE MINISTRY OF DEFENSE OF THE RUSSIAN FEDERATION, Moscow, 1996.
[57] Based upon the Historical Document, *Combat order of the headquarters of 64 A, 10 July 1942*. Retrieved from TsAMO Archive, ID #131305247.
[58] *Stavka VGK Documents and Materials 1942*, INSTITUTE OF THE MILITARY HISTORY OF THE MINISTRY OF DEFENSE OF THE RUSSIAN FEDERATION, Moscow, 1996.
[59] Chuikov, V. I. *The Beginning of the Road*, Macgibbon & Kee, 1963 London, p. 16.

60. Ibid.,
61. Based upon the Historical Documents, *A summary of the movement of trains arriving at 64 A, 19 July 1942, and A summary of the movement of echelons of 64 A 22 July 42*. Retrieved from TsAMO Archive, ID # 131306260 & #131306289.
62. Chuikov, V. I. *The Beginning of the Road*, Macgibbon & Kee, 1963 London, p. 16.
63. Deichmann, General der Flieger Paul., Edited by Dr. Alfred Price. *Spearhead for Blitzkrieg - Luftwaffe Operations in Support of the Army 1939-1945*. Ivy Books, New York 1996, p. 117.
64. Hayward, Joel S. A. *Stopped at Stalingrad - The Luftwaffe and Hitler's Defeat in the East 1942-1943*. University Press Kansas, 1998, p. 137.
65. Bergstrom, Christer. *Stalingrad - the Air Battle*, Midland, 2007, p. 55.
66. Hayward, Joel S. A., *Stopped at Stalingrad - The Luftwaffe and Hitler's Defeat in the East 1942-1943*. University Press Kansas, 1998, p. 183.
67. Grinevsky, V.V., A.D. Ovsyannikov., I.M. Ryzhov., & A.N. Yanchinsky. *Heroic Sixty Fourth*, Ninshe-Volzhskoye Publishing House, Volgograd, 1981, p. 7-11.
68. Paterson, James Hamilton. *Marked for Death the First War in the Air*, Pegasus Books, New York, London, 2016, p. 116.
69. Vladimirovich, Kovalev Ivan. *Transport in the Great Patriotic War 1941-1945*. Chapter Seven, Transport in the Battle of Stalingrad, Moscow: Nauka, 1981, militera.lib.ru/h/kovalev_iv/index.html.
70. Ibid.,
71. Chuikov, V. I. *The Beginning of the Road*, Macgibbon & Kee, 1963 London, p. 15.
72. Vladimirovich, Kovalev Ivan. *Transport in the Great Patriotic War 1941-1945*. Chapter Seven, Transport in the Battle of Stalingrad, Moscow: Nauka, 1981, militera.lib.ru/h/kovalev_iv/index.html.
73. Deichmann, General der Flieger Paul., Edited by Dr. Alfred Price. *Spearhead for Blitzkrieg - Luftwaffe Operations in Support of the Army 1939-1945*. Ivy Books, New York 1996, p. 132.
74. Vladimirovich, Kovalev Ivan. *Transport in the Great Patriotic War 1941-1945*. Chapter Seven, Transport in the Battle of Stalingrad, Moscow: Nauka, 1981, militera.lib.ru/h/kovalev_iv/index.html.
75. Based upon the Historical Document, *Extract from the operational Directive of the Stalingrad Front, 17 July 1942*. Retrieved from TsAMO Archive, ID #131305223.
76. These transfers were noted by an *Operational summary of the headquarters of 64 A on 17 July 42*. Retrieved from TsAMO Archive, ID #131309232.
77. Rotundo, Louis., *Battle for Stalingrad, 1943 Soviet General Staff Study,* Pergamon-Brassey's 1989, p. 45.
78. *Battle of Stalingrad: Preparatory Phase Documents*, Voyenno-Istoricheskiy Zhurnal #8 August 1982, signed to press 26 July 1982, p. 27-31. TsAMO, folio 48, inv. 1, file 69, sheet 140.
79. Mark, Jason. *Panzer Krieg Vol 1, German Armoured Operations at Stalingrad*, Leaping Horseman Books, Sydney Australia, 2017, p. 216-222.
80. Kerr, Walter. *The Secret of Stalingrad*, Macdonald and Jane's London, 1978, p. 75.
81. Chuikov, V. I. *The Beginning of the Road*, Macgibbon & Kee, 1963 London, p. 16-17.
82. Ibid., 17.
83. Mikhailovich, Samsonov Alexander. *The Battle of Stalingrad*, Moscow: Nauka, 1989, Chapter Three - In the great bend of the Don and on the outskirts of the Volga. p. 48.
84. Grinevsky, V.V., A.D. Ovsyannikov., I.M. Ryzhov., & A.N. Yanchinsky. *Heroic Sixty Fourth,* Ninshe-Volzhskoye Publishing House, Volgograd, 1981, p. 12.
85. Sharp, Charles C. *Soviet Order of Battle WW2 Vol 1*, George F. Nafziger 1996, p. 81.
86. Sharp, Charles C. *Soviet Order of Battle WW2 Vol 2*, George F. Nafziger 1996, p. 72.
87. Popov, P.P., A.V. Kozlov., B.G. Usik. *Turning Point*, Leaping Horseman Books, 2008, p. 54-57.
88. Chuikov, V. I. *The Beginning of the Road*, Macgibbon & Kee, 1963 London, p. 19-20.
89. Isael, Alexey V. *Сталинград За Волгой для нас земли нет* (Stalingrad, there is no land for us beyond the Volga), Moscow, 2008, p. 10.
90. *Sailors in the Battle of Stalingrad,* (An article by Doctor of Historical Sciences, professor, participant of the Battle of Stalingrad, Major VI Tomarov's Guards. From the collection of materials of the scientific-practical conference on November 19, 2002, "60 years of the Battle of Stalingrad in the Great Patriotic War", from. 96-100, Moscow, Book and Business, 2003) Retrieved from http://samsv.narod.ru/Br/Sbr/omsbr154/h3.html.
91. Chuikov, V. I. *The Beginning of the Road*, Macgibbon & Kee, 1963 London, p. 27.

[92] Isael, Alexey V. *Сталинград За Волгой для нас земли нет* (Stalingrad, there is no land for us beyond the Volga), Moscow, 2008, p. 15.

[93] Chuikov, V. I. *The Beginning of the Road*, Macgibbon & Kee, 1963 London, p. 20.

[94] Grinevsky, V.V., A.D. Ovsyannikov., I.M. Ryzhov., & A.N. Yanchinsky. *Heroic Sixty Fourth*, Ninshe-Volzhskoye Publishing House, Volgograd, 1981, p. 3.

[95] Glantz, David M. *Atlas of Operation Blue, June – November 1942*, David M. Glantz, 1998, Map 90.

[96] Rottman, Gordon L. *Soviet Field Fortifications 1941-45*, Osprey Publishing, 2007, p. 29-30.

[97] Chuikov, V. I. *The Beginning of the Road*, Macgibbon & Kee, 1963 London, p. 27.

[98] Ibid., 28.

[99] *Journal of Combat Operations Staff, 64th Army,* (n.d.) Retrieved from the Russian web site Memory of the People https://pamyatnaroda.ru/.

[100] Chuikov, V. I. *The Beginning of the Road*, Macgibbon & Kee, 1963 London, p. 29-32.

[101] Sharp, Charles C. *Soviet Order of Battle WW2 Vol 9,* George F. Nafziger 1996, p. 19.

[102] V. I. Chuikov. *The Beginning of the Road*, Macgibbon & Kee, 1963 London, p. 31.

[103] Falk, Dann. *The Military Geographic Significance of the Volgograd Region*, California State University Chico, Thesis 1992, p. 7-8.

[104] Chuikov, V. I. *The Beginning of the Road*, Macgibbon & Kee, 1963 London, p. 36. It should be noted this change of command might have occurred on or about 3 August.

[105] Isael, Alexey V. *Сталинград За Волгой для нас земли нет* (Stalingrad, there is no land for us beyond the Volga), Moscow, 2008, p. 52.

[106] Chuikov, V. I. *The Beginning of the Road*, Macgibbon & Kee, 1963 London, p. 36-37.

[107] Ibid., 37.

[108] Glantz, David M., Jonathan M. House. *To the Gates of Stalingrad Vol. 1,* University Press Kansas, 2009, p. 246.

[109] Based upon the Historical Document, *Report on the loss of artillery materiel to the military formations and units of the 64 A. 25-31 July 1942*. Retrieved from TsAMO Archive. ID # 131308196.

[110] Mark, Jason. *Panzer Krieg Vol 1, German Armoured Operations at Stalingrad*, Leaping Horseman Books, Sydney Australia, 2017, p 229-230.

[111] Based upon the Historical Document, *Material on the Staffing of the Army, for 7 December 1942*, Retrieved from TsAMO Archive. ID# 131307644.

[112] Glantz, David M. *Atlas of Operation Blue June-Nov*, David M. Glantz, 1998. Map 100.

[113] Glantz, David M., Jonathan M. House. *To the Gates of Stalingrad Vol. 1*, University Press Kansas 2009 p. 279.

[114] Marchand, Jean-Luc. *Order of Battle Soviet Army in WW2 Vol 6*, 2010, p. 90.

[115] Chuikov, V. I. *The Beginning of the Road,* Macgibbon & Kee, 1963 London, p. 44.

[116] Ibid., 46.

[117] Haritonovich, Gregory. Account retrieved from the web site: http://www.pvesti.ru/ru/10651/Poisk/260/K-68годовщине-победы-в-Сталинградской-битве.htm (68th anniversary-victory-in-Stalingrad-battle) (Accessed 2017).

[118] Lemelsen, Joachim. *29 Division*, Podzun-Verlag, 1960, p. 194.

[119] Glantz, David M., Jonathan M. House. *To the Gates of Stalingrad Vol. 1*, University Press Kansas, 2009, p. 279.

[120] Chuikov, V. I. *The Beginning of the Road*, Macgibbon & Kee, 1963 London, p. 47.

[121] Bergstrom, Christer. Andrey Mikhailov, *Black Cross Red Star*, Pacific Military History, 2001, p. 249.

[122] Chuikov, V. I. *The Beginning of the Road*, Macgibbon & Kee, 1963 London, p. 47.

[123] Haritonovich, Gregory. Account retrieved from the web site: http://www.pvesti.ru/ru/10651/Poisk/260/K-68-годовщине-победы-в-Сталинградской-битве.htm (68th anniversary-victory-in-Stalingrad-battle) (Accessed 2017).

[124] Chuikov, V. I. *Battle of the Century*, Moscow 1975, South Group chapter.

[125] Chuikov, V. I. *The Beginning of the Road*, Macgibbon & Kee, 1963 London, p. 47.

[126] Ibid., 49.

[127] Bergstrom, Christer. *Stalingrad - the Air Battle*, Midland, 2007, p. 62.

[128] Gudkova, Galina D. *They Will Live*, Moscow, 1986, Chapter three.

[129] Valentinovna, Zheltov, Lyubov. Retrieved from the reference section: *Military health care in the Battle of Stalingrad: health maintenance organizations and the activities of the medical service of the Red Army.* Thesis and dissertation Ph.D. 1999 Volgograd. http://www.dissercat.com/content/voennoe-zdravookhranenie-v-stalingradskoi-bitveorganizatsiya-meditsinskogo-obespecheniya-i-.

[130] Chuikov, V. I. *The Beginning of the Road*, Macgibbon & Kee, 1963 London, p. 49-50.

[131] Bergstrom, Christer. *Stalingrad - the Air Battle*, Midland, 2007, p. 62.

[132] Glantz, David M., Jonathan M. House. *To the Gates of Stalingrad Vol. 1*, University Press Kansas, 2009, p. 279.

[133] Popov, P.P., A.V. Kozlov., B.G. Usik. *Turning Point*, Leaping Horseman Books, 2008, p. 40.

[134] Ibid., p. 69-72.

[135] Wilfried, Kopenhagen. *Armored trains of the Soviet Union, 1917-1945*, Schiffer Military History, 1996, p.16-17.

[136] Opalev, M.N. *Unknown Facts about Red Army Armored Artillery Train Units During the Battle of Stalingrad,* (n.d.) Retrieved from the web site: http://www.volsu.ru/upload/medialibrary/2d1/9_ypzxlkhirrgo.pdf.

[137] Chuikov, V. I. *The Beginning of the Road*, Macgibbon & Kee, 1963 London, p. 50-51.

[138] Ibid., 51.

[139] Ibid., 52.

[140] Ibid., 53.

[141] Glantz, David M., Jonathan M. House. *To the Gates of Stalingrad Vol. 1*, University Press Kansas, 2009, p. 286.

[142] Popov, P.P., A.V. Kozlov., B.G. Usik. *Turning Point*, Leaping Horseman Books, 2008, p. 76.

[143] Gudkova, Galina D. *They Will Live*, Moscow, 1986, Chapter three.

[144] Glantz, David M., Jonathan M. House, *To the Gates of Stalingrad Vol. 1*, University Press Kansas 2009, p. 287.

[145] Bergstrom, Christer. *Stalingrad - the Air Battle,* Midland, 2007, p 64.

[146] Glantz, David M., Jonathan M. House. *To the Gates of Stalingrad Vol. 1*, University Press Kansas, 2009, p. 287-288.

[147] Bergstrom, Christer., Andrey Mikhailov, *Black Cross Red Star Vol 2*, Pacifica Military History, 2001, p. 63.

[148] Popov, P.P., A.V. Kozlov., B.G. Usik. *Turning Point*, Leaping Horseman Books, 2008, p. 70-71.

[149] *Stavka Documents and Materials 1942* - Military History Institute of the Russian Defense Ministry – Moscow 1996.

[150] The movements of this and other German and Romanian units are based on surviving OKH Large Ost Maps (East Maps). Retrieved from the web site: http://www.wwii-photos-maps.com/index.html.

[151] Glantz, David M., Jonathan M. House. *To the Gates of Stalingrad Vol. 1*, University Press Kansas 2009, p. 299.

[152] Ibid., 557.

[153] *Stavka Documents and Materials 1942* - Military History Institute of the Russian Defense Ministry – Moscow, 1996.

[154] *Battle of Stalingrad. Chronicle, Facts, People.* (in Russian) Team of Authors, Olma-Press 2002, Vol 1, p. 373, The 214 RD shows up assigned to the 4th Tank Army on this date.

[155] Kerr, Walter. *The Secret of Stalingrad*, Macdonald and Jane's, London, 1978, p. 123.

[156] Retrieved from the web site http://www.divizia126.narod.ru/journal.htm (Accessed 2017).

[157] Chuikov, V. I. *The Beginning of the Road*, Macgibbon & Kee, 1963 London, p. 43.

[158] Ibid., 54.

[159] *Staff situation report 208 RD, Operational Report #1, 17 August 1942.* Retrieved from TsAMO Archive. (ID# Unknown) The 1,678 number was almost certainly the divisions combat strength. So, its ration strength would have been several thousand more.

[160] Chuikov, V. I. *The Beginning of the Road*, Macgibbon & Kee, 1963 London, p. 56.

[161] Schroter, Heinz. *Stalingrad the Battle that Changed the World,* E. P. Dutton & Company Inc. New York 1958, p. 30-31.

[162] Retrieved from the web site http://www.divizia126.narod.ru/journal.htm (Accessed 2017).

[163] Bergstrom, Christer. *Stalingrad - the Air Battle*, Midland, 2007, p. 72.

[164] The 65th anniversary of the Victory in the GREAT PATRIOTIC WAR 1941-1945. *126 Infantry Division* - Volgograd regional Public organization "SEARCH" 2009.

[165] Bergstrom, Christer. *Stalingrad - the Air Battle*, Midland, 2007, p. 73.
[166] The 65th anniversary of the Victory in the GREAT PATRIOTIC WAR 1941-1945. *126 Infantry Division* - Volgograd regional Public organization "SEARCH" 2009.
[167] *Stavka Documents and Materials 1942* - Military History Institute of the Russian Defense Ministry - Moscow 1996.
[168] Doerr, Hans. *The March on Stalingrad*, (in German) Darmstadt, 1955, p. 140.
[169] Glantz, David M., Jonathan M. House. *To the Gates of Stalingrad Vol. 1*, University Press Kansas, 2009, p. 342.
[170] Mark, Jason D. *Death of the Leaping Horseman*, Leaping Horseman Books, Sydney Australia, 2003, p. 62.
[171] Bergstrom, Christer. *Stalingrad - the Air Battle,* Midland, 2007, p. 73.
[172] Chuikov, V. I. *The Beginning of the Road*, Macgibbon & Kee, 1963 London, p. 59.
[173] Jason D. Mark. *Death of the Leaping Horseman*, Leaping Horseman Books, Sydney Australia, 2003, p. 66-67.
[174] Glantz, David M., Jonathan M. House. *To the Gates of Stalingrad Vol. 1*, University Press Kansas, 2009, p. 363.
[175] Isael, Alexey V. *Сталинград За Волгой для нас земли нет* (Stalingrad, there is no land for us beyond the Volga), Moscow, 2008. Retrieved from http://militera.lib.ru/h/isaev_av8/05.html.
[176] Laskin, Andreevich, Ivan. *The Volga and the Kuban*, Moscow, Military Publishing, 1986, Chapter two.
[177] Retrieved from web site – poiskfebs.narod.ru/memorial/kniga_stalingrad.doc, (n.d.) (Accessed 2017).
[178] Colonel Vladimir Evseevich Sorokin ended up in concentration camp number 277. Retrieved from http://samsv.narod.ru/Div/Sd/sd126/default.html.
[179] Retrieved from the web site http://www.divizia126.narod.ru/journal.htm (n.d.) (Accessed 2017).
[180] Sharp, Charles C. *Soviet Order of Battle in WW2, Volume X*, p. 83.
[181] Retrieved from web site - http://samsv.narod.ru/Br/Sbr/omsbr154/default.html (n.d.) (Accessed 2017).
[182] Bergstrom, Christer. *Stalingrad - the Air Battle*, Midland, 2007, p. 74.
[183] Retrieved from web site – poiskfebs.narod.ru/memorial/kniga_stalingrad.doc (n.d.) (Accessed 2017).
[184] Valentinovna, Zheltov Lyubov. Retrieved from the reference section: *Military health care in the Battle of Stalingrad: health maintenance organizations and the activities of the medical service of the Red Army*. Thesis and dissertation Ph.D. 1999 Volgograd. http://www.dissercat.com/content/voennoe-zdravookhranenie-v-stalingradskoi-bitveorganizatsiya-meditsinskogo-obespecheniya-i- .
[185] Based upon the Historical Document, *Material on the Staffing of the Army, for 7 December 1942*, Retrieved from TsAMO Archive, ID# 131307644.
[186] Chuikov, V. I. *The Beginning of the Road*, Macgibbon & Kee, 1963 London, p. 63-64.
[187] Laskin, Ivan Andreevich. *On the way to the break*, Moscow: Military Publishing, 1977, p. 236.
[188] Mark, Jason D. *Death of the Leaping Horseman*, Sydney Australia, 2003, p. 96-108.
[189] Glantz, David M., Jonathan M. House. *To the Gates of Stalingrad Vol. 1,* University Press Kansas, 2009, p. 375.
[190] Gudkova, Galina D. *They Will Live*, Moscow, 1986, Chapter six.
[191] Glantz, David M., Jonathan M. House. *To the Gates of Stalingrad Vol. 1*, University Press Kansas 2009, p. 31.
[192] Mark, Jason D. *Death of the Leaping Horseman*, Leaping Horseman Books, Sydney Australia, 2003, p. 114.
[193] Ibid., 112.
[194] Chuikov, V. I. *The Beginning of the Road*, Macgibbon & Kee, 1963 London, p. 66.
[195] Ibid.,
[196] Retrieved from the two web sites - http://www.litmir.me/br/?b–194705&p=64 & http://www.soldat.ru/spravka/voen_uch/ .(n.d.). (Accessed 2017)
[197] Popov, P.P., A.V. Kozlov., B.G. Usik. *Turning Point*, Leaping Horseman Books, 2008 p. 60.
[198] Mark, Jason D. *Death of the Leaping Horseman,* Leaping Horseman Books, Sydney Australia, 2003 p. 119-123.
[199] Glantz, David M., Jonathan M. House. *To the Gates of Stalingrad Vol. 1*, University Press Kansas, 2009, p. 47.
[200] Hardesty, Von., Ilya Crinberg. *Red Phoenix Rising*, University Press Kansas, 2012, p. 122.
[201] *Battle of Stalingrad. Chronicle, Facts, People*. (in Russian) Team of Authors, Olma-Press 2002, Vol 1, p. 501 and related German OKH maps.
[202] Mark, Jason D. *Death of the Leaping Horseman*, Leaping Horseman Books, Sydney Australia, 2003 p. 130.

203. Ibid., 104-130.
204. Ziemke, Earl F., Magna E. Bauer. *Moscow to Stalingrad: Decision in the East,* Army Historical Series, 1987, p. 462.
205. Daines, Vladimir. 2008, Information Retrieved from the book *Penal Battalions and Blocking detachments of the Red Army,* on the following web site, http://www.e-reading.link/bookreader.php/1000831/Daynes_Vladimir__Shtrafbaty_i_zagradotryady_Krasnoy_Armii.html.
206. Sharp, Charles C. *Soviet Order of Battle in WW2, Volume VII,* 1995, p. 54-64.
207. It should be noted that Blocking Detachments were also used in 1941.
208. Glantz, David M., Jonathan M. House. *To the Gates of Stalingrad Vol. 1*, University Press Kansas, 2009, p. 197.
209. Ibid., 82.
210. Ibid., 80.
211. *Battle of Stalingrad. Chronicle, Facts, People.* (in Russian) Team of Authors, Olma-Press 2002, Vol 1, p. 525.
212. Glantz, David M., Jonathan M. House. *To the Gates of Stalingrad Vol. 1*, University Press Kansas, 2009, p. 82.
213. *Stavka Documents and Materials 1942* - Military History Institute of the Russian Defense Ministry – Moscow 1996.
214. Chuikov, V. I. *The Beginning of the Road*, Macgibbon & Kee, 1963 p. 72-76.
215. Glantz, David M., Jonathan M. House. *To the Gates of Stalingrad Vol. 1*, University Press Kansas, 2009, p. 84.
216. Chuikov, V. I. *The Beginning of the Road*, Macgibbon & Kee, 1963 London, p. 70.
217. Glantz, David M., Jonathan M. House, *To the Gates of Stalingrad Vol. 2*, University Press Kansas, 2009, p. 91.
218. Ziemke, Earl F., Magna E. Bauer. *Moscow to Stalingrad: Decision in the East*, Army Historical Series, 1987, p. 392-393.
219. Retrieved from web site - http://samsv.narod.ru/Arm/a64/h1.html (Accessed 2018).
220. Data from: *Battle of Stalingrad. Chronicle, Facts, People.* (in Russian) Team of Authors, Olma-Press 2002, Vol 1, p. 549.
221. Chuikov, V. I. *The Beginning of the Road*, Macgibbon & Kee, 1963 London, p. 72.
222. Glantz, David M., Jonathan M. House. *To the Gates of Stalingrad Vol. 1*, University Press Kansas, 2009, p. 91.
223. Schroter, Heinz. *Stalingrad the Battle that Changed the World*, E. P. Dutton & Company Inc. New York, 1958, p. 42-44.
224. Glantz, David M., Jonathan M. House. *To the Gates of Stalingrad Vol. 1*, University Press Kansas, 2009, p. 93-96.
225. Falk, Dann. Master's Thesis, *The Military Geographic Significance of the Volgograd Region*, California State University Chico, 1992.
226. Chuikov, V. I. *The Beginning of the Road*, Macgibbon & Kee, 1963 London p. 87.
227. Glantz, David M., Jonathan M. House. *To the Gates of Stalingrad Vol. 1*, University Press Kansas, 2009, p. 116.
228. Popov, P.P., A.V. Kozlov., B.G. Usik. *Turning Point*, Leaping Horseman Books, 2008 p. 57.
229. Chuikov, V. I. *The Beginning of the Road*, Macgibbon & Kee, 1963 London, p. 104-105.
230. Glantz, David M., Jonathan M. House, *To the Gates of Stalingrad Vol. 1*, University Press Kansas, 2009, p. 168-174.
231. Isael, Alexey V. *Сталинград За Волгой для нас земли нет* (Stalingrad, there is no land for us beyond the Volga), Moscow, 2008, table #10.
232. Glantz, David M., Jonathan M. House. *To the Gates of Stalingrad Vol. 1,* University Press Kansas, 2009, p. 168-181.
233. Popov, P.P., A.V. Kozlov, B.G. Usik. *Turning Point*, Leaping Horseman Books, 2008, p. 60.
234. Erickson, John. *The Road to Stalingrad*, Weidenfeld and Nicolson, London, 1975, p. 412-413.
235. Ziemke, Earl F., Magna E. Bauer. *Moscow to Stalingrad: Decision in the East,* Army Historical Series, 1987, p. 396.
236. Ibid., 416.

237 Glantz, David M., Jonathan M. House. *To the Gates of Stalingrad Vol. 2*, University Press Kansas, 2009, p. 286.
238 Ziemke, Earl F., Magna E. Bauer. *Moscow to Stalingrad*, p. 416.
239 Ilyich, Loktionov Ivan. *Volga flotilla in the World War II*, Moscow, Military Publishing, 1974, p. 24.
240 Chuikov, V. I. *The Beginning of the Road*, Macgibbon & Kee, 1963 London, p. 168-169.
241 Glantz, David M., Jonathan M. House *Armageddon in Stalingrad Vol. 2*, University Press Kansas, 2009, p. 359.
242 Maclean, French L. *Stalingrad, The Death of the German Sixth Army on the Volga, 1942-1943 Vol 1*.
243 Glantz, David M., Jonathan M. House. *Armageddon in Stalingrad Vol. 2*, University Press Kansas, 2009, p. 369.
244 *Battle of Stalingrad. Chronicle, Facts, People Vol 1.* Team of Authors, Olma-Press, 2002, 8 October.
245 Taylor, Brian. *Barbarossa to Berlin Vol 1,* Spellmount, Staplehurst, 2003, p. 302.
246 Retrieved from web site: http://www.soldat.ru/doc/nko/1942.html for NKO Directives. (Accessed 2018)
247 Taylor, Brian. *Barbarossa to Berlin, Vol 1,* Spellmount, Staplehurst, 2003, p. 303.
248 Zaitsev, Vassili. *Notes of a Russian Sniper*, Frontline Books, London, 2009, p. 106-107.
249 Ziemke, Earl f., Magna E. Bauer. *Moscow to Stalingrad: Decision in the East*, Center of Military History United States Army, Washington D.C,1987, p. 449-451.
250 Joly, Anton. *Stalingrad Battle Atlas, Oct 14 –Nov 18*, p. 18.
251 *Battle of Stalingrad. Chronicle, Facts, People.* Team of Authors, Olma-Press, 2002, Vol 1,15 October.
252 Bellamy, Chris. *Absolute War*, Alfred A. Knopf, New York, 2007, p. 248.
253 Chuikov, V. I. *The Beginning of the Road*, Macgibbon & Kee, 1963 London, p. 185.
254 Glantz, David M., Jonathan M. House. *Armageddon in Stalingrad Vol. 2,* University Press Kansas, 2009, p. 400.
255 Chuikov, V. I. *The Beginning of the Road*, Macgibbon & Kee, 1963 London, p. 187.
256 Dinardo, R.L. *Mechanized Juggernaut or Military Anachronism*, Stackpole Books, 2008, p. 61-64.
257 Turbiville, Graham H. Jr. Paper *Soviet Operational Logistics 1939-1990*, p. 298.
258 Klemin, A. S. Voyenno-Istoricheskiy Zhurnal #3 March 1985, *Military Transport During the years of the Great Patriotic War*. p. 85.
259 Dunn, Walter S. Jr. *Soviet Blitzkrieg*, Stackpole Books, 2000, p. 36.
260 Ibid., 29-30.
261 This estimate is based upon personal correspondence with Der Alte Fritz, via the https://forum.axishistory.com// forum. *German Railways in the East*.
262 Information obtained from personal correspondence from Axis History Forum contributor Der Alte Fritz, via the axishistory.com forum.
263 Rotundo, Louis. *Battle for Stalingrad - The 1943 Soviet General Staff Study*, p. 258-267.
264 Glantz, David M., Jonathan M. House. *Armageddon in Stalingrad Vol. 2*, University Press Kansas, 2009, p. 463-466.
265 Opalev, M.N. *Unknown Facts about Red Army Armored Artillery Train Units During the Battle of Stalingrad,* (n.d.) Retrieved from the web site: http://www.volsu.ru/upload/medialibrary/2d1/9_ypzxlkhirrgo.pdf
266 Chuikov, V. I. *The Beginning of the Road*, Macgibbon & Kee, 1963 London, p. 189.
267 Zvonnitsa, M.G. *Stalingrad Epic,* 2000, Units of the NKVD of the USSR: The Department of Special Departments and the 3rd branch of the 2nd Special Division - military counterintelligence and censorship Book. Retrieved from the web site: http://militera.lib.ru/docs/da/stalingrad1/index.html.
268 Glantz, David M., Jonathan M. House. *Armageddon in Stalingrad Vol. 2*, University Press Kansas, 2009, p. 434.
269 Ibid., 449.
270 Zvonnitsa, M.G. *Stalingrad Epic,* 2000, Units of the NKVD of the USSR: the Department of Special Departments and the 3rd branch of the 2nd Special Division - military counterintelligence and censorship Book Retrieved from the web site: http://militera.lib.ru/docs/da/stalingrad1/index.html (M.G. 2000).
271 Morozov, Ivan. *From Stalingrad To Prague,* Nizhne-Volzhskoye Publishing House, 1976, Volgograd, p. 57.
272 Glantz, David M. *Combat Documents Vol 1, 3 Sept to Nov 18 1942*, David M. Glantz, 2007, p. 109.
273 Maclean, French L. *Stalingrad, The Death of the German Sixth Army on the Volga, 1942-1943 Vol 1*. Note: multiple reports/references, all during October, were taken from Volume 1 of this two Volume set.
274 *Battle of Stalingrad. Chronicle, Facts, People Vol 1*, Team of Authors, Olma-Press, 2002, 8 October.

275 Based upon the Historical Document, *Material on the Staffing of the Army, for 1 November 1942,* Retrieved from TsAMO Archive, ID# 131308268.
276 Maclean, French L. *Stalingrad, The Death of the German Sixth Army on the Volga, 1942-1943 Vol 2,* Schiffer, 2013. Note: multiple, daily weather reports along with references for the month starting in November 1942 to February 1943, are taken from Volume 2 of this set. This information is very hard to find and Mr. Maclean provides the most comprehensive source that I have found.
277 Maclean, French L. *Stalingrad, The Death of the German Sixth Army on the Volga, 1942-1943 Vol 2,* Schiffer, 2013.
278 *Battle of Stalingrad. Chronicle, Facts, People Vol 1.* (in Russian) Team of Authors, Olma-Press 2002, Multiple daily reports & references, throughout the November chapter, were taken from this book.
279 Glantz, David M. *Combat Documents Vol 1, 3 Sept to Nov 18 1942,* David M. Glantz, 2007, p. 118.
280 Ibid., p. 120.
281 Glantz, David M., Jonathan M. House. *Armageddon in Stalingrad Vol. 2,* University Press Kansas, 2009, p. 639.
282 Hellbeck, Jochen. *Stalingrad - The City that Defeated the Third Reich,* PublicAffairs, New York, 2015, p. 90-124.
283 Fialkovskii, Leonid. *Stalingrad Apocalypse. Tank Brigade in hell,* Moscow, Yauza, Eksmo, 2011, Chapter Two, 13 August 1942. http://militera.lib.ru/db/fialkovskiy_li01/index.html (254 Tank Brigade).
284 Glantz, David M., Jonathan M. House. *Armageddon in Stalingrad Vol. 2,* University Press Kansas, 2009, p. 702.
285 Chuikov, V. I. *The Beginning of the Road,* Macgibbon & Kee, 1963, London, p. 214-215.
286 Eremenko, Andrey Ivanovich. *Stalingrad: Memoirs of the commander of the front,* Moscow: Military Publishing, 1961. Chapter XIII.
287 Maclean, French L. *Stalingrad, The Death of the German Sixth Army on the Volga, 1942-1943Vol 2.* Schiffer, 2013, p. 40.
288 Glantz, David M. *Combat Documents Vol 1, 3 Sept to Nov 18 1942,* David M. Glantz, 2007 p. 126.
289 Eremenko, Andrey Ivanovich. *Stalingrad: Memoirs of the commander of the front,* Moscow: Military Publishing, 1961, Chapter XIII.
290 Ibid.,
291 *BSA Data, 1942 section 3 and 4.*
292 Zhukov, Georgy. *Reminiscences and reflections,* Vol 2, Progress Publishers Moscow, 1985, p. 115-118.
293 Glantz, David M. *Combat Documents Vol 1, 3 Sept to Nov 18 1942,* David M. Glantz, 2007, p. 130-131.
294 Zhukov, Georgy. *Reminiscences and Reflections,* Vol 2, Progress Publishers Moscow, 1985, p. 119-121.
295 Glantz, David M. *Combat Documents Vol 1, 3 Sept to Nov 18 1942,* David M. Glantz, 2007, p. 135.
296 Mark, Jason D. *Death of the Leaping Horseman,* Leaping Horseman Books, Sydney Australia, 2003, p. 369-373.
297 Glantz, David M. *Combat Documents Vol 1, 3 Sept to Nov 18 1942,* David M. Glantz, 2007, p. 139.
298 Kerr, Walter. *The Secret of Stalingrad,* Macdonald and Jane's, London, 1978, p. 189.
299 Maclean, French L. *Stalingrad, The Death of the German Sixth Army on the Volga, 1942-1943 Vol 2,* Schiffer, 2013, p. 60.
300 Ibid., 65.
301 Mark, Jason D. *Death of the Leaping Horseman,* Leaping Horseman Books, Sydney Australia, 2003, p. 373.
302 Joly, Anton. *Stalingrad Battle Atlas, Oct 14 –Nov 18,* Stal Data Publications, 2014, p. 123.
303 Melvin, Major General Mungo. *Manstein- Hitler's Greatest General,* Thomas Dunne Books St. Martin's Press New York, 2010, p. 290.
304 Maclean, French L. *Stalingrad, The Death of the German Sixth Army on the Volga, 1942-1943 Vol 2.* Schiffer, 2013, p. 80.
305 Glantz, David M. Jonathan M. House. *Endgame at Stalingrad Vol. 3,* University Press Kansas, 2014, p. 272.
306 Ibid., 291.
307 Ibid., 330.
308 Ibid., 338.
309 Ibid., 359.

[310] Kehrig, Manfred. *Stalingrad: Analyse und Dokumentation einer Schlacht*, Deutsche Verlags-Anstalt, Stuttgart, 1974, p.633-637. Note: Totals for the tonnage delivered to the Stalingrad pocket vary from source to source; therefore, I will be using numbers from Kehrig exclusively unless a better source is found. I will then quote that source.

[311] Manstein, Field Marshal Eric von. *Lost Victories*, Presidio Press, 1982, p. 307 note.

[312] Maclean, French L. *Stalingrad, The Death of the German Sixth Army on the Volga, 1942-1943 Vol 2*. Schiffer, 2013, p. 90.

[313] Tarrant, V.E. *Stalingrad*, Leo Cooper, Lindon, 1992, p. 149.

[314] The 24 of November is also von Manstein's 55 birthday.

[315] Glantz, David M., Jonathan M. House. *Endgame at Stalingrad Vol. 3*, University Press Kansas, 2014, p. 388-389.

[316] Maclean, French L. *Stalingrad, The Death of the German Sixth Army on the Volga, 1942 - 1943 Vol 2*, Schiffer, 2013, p. 90.

[317] Raus, Erhard. *Panzer Operations*, Da Capo Press, 2003, p. 137-144.

[318] Melvin, Major General Mungo. *Manstein- Hitler's Greatest General,* Thomas Dunne Books St. Martin's. Press New York, 2010, p. 299.

[319] Glantz, David M., Jonathan M. House. *Endgame at Stalingrad Vol. 3*, University Press Kansas, 2014, p. 499-501.

[320] Glantz, David M., Jonathan M. House. *Endgame at Stalingrad Vol. 3,* University Press Kansas, 2014, p. 513.

[321] Based upon the Historical Documents, *Material on the Staffing of the Army, for early December 1942*, Retrieved from TsAMO Archive, ID# 131308101, 131308120, 131307644.

[322] Maclean, French L. *Stalingrad, The Death of the German Sixth Army on the Volga, 1942-1943 Vol 2*, Schiffer, 2013, p. 116.

[323] *Battle of Stalingrad. Chronicle, Facts, People.* (Сталинградская битва. Хроника, факты, люди) Team of Authors, Olma-Press, Moscow, 2002, book 2, p146-147. Daily reports throughout this chapter are taken from this book which gives the reader a Day by Day view of operations for the 57th & 64th Armies. Needless to say, this two-volume book is absolutely vital for anyone studying the battle of Stalingrad from the Soviet perspective.

[324] Maclean, French L. *Stalingrad, The Death of the German Sixth Army on the Volga, 1942-1943 Vol 2*, Schiffer, 2013, p. 128.

[325] *Battle of Stalingrad. Chronicle, Facts, People.* (Сталинградская битва. Хроника, факты, люди) Team of Authors, Olma-Press, Moscow, 2002, book 2, p. 146-147.

[326] Information about the number of Hiwis trapped within the pocket was explored on the following thread *Hiwi's at Stalingrad*. Retrieved from the following web site: http://www.armchairgeneral.com/forums/showthread.php?t=169613&highlight=hiwis.

[327] Glantz, David M. *Companion to Endgame at Stalingrad*, University Press Kansas, 2014, p. 272.

[328] Wijers, Hans. *Winter Storm*, Stackpole Military, 2012, p. 189.

[329] Laskin, Ivan A. *On the path to the Turning Point*. Moscow, Voenizdat, 1977, Chapter 5.

[330] Raus, General Erhard. *Panzers on the Eastern Front*, Greenhill Books, 2006, p. 156 & 159.

[331] Jukes, Geoffrey. *Hitler's Stalingrad Decisions*, University of California Press, 1985, p. 124.

[332] Glantz, David M., Jonathan M. House. *Endgame at Stalingrad Vol. 3, book 2*, University Press Kansas, 2014, p. 30-31.

[333] Glantz, David M. *Companion to Endgame at Stalingrad*, University Press Kansas, 2014, p. 278.

[334] Wijers, Hans. *Winter Storm*, Stackpole Military, 2012, p. 225.

[335] Glantz, David M., Jonathan M. House. *Endgame at Stalingrad Vol. 3, book 2*, University Press Kansas, 2014, p. 205.

[336] Wijers, Hans. *Winter Storm*, Stackpole Military, 2012, p. 233-239.

[337] Busch, Reinhold. *Survivors of Stalingrad*, Frontline Books, London, 2012, various stories.

[338] Glantz, David M., Jonathan M. House. *Endgame at Stalingrad Vol. 3, book 2,* University Press Kansas, 2014, p. 371-375.

[339] Wijers, Hans. *Winter Storm*, Stackpole Military, 2012, p. 252-254.

[340] Jukes, Geoffrey. *Hitler's Stalingrad Decisions*, University of California Press, 1985, p. 125.

[341] Wijers, Hans. *Winter Storm*, Stackpole Military, 2012, p. 253.

[342] Ibid., 255.

[343] Hoyt, Edwin P. *The Battle for Stalingrad 199 days*, A Tom Doherty Associates Book, New York, 1993, p. 228-229.
[344] Ibid., 229.
[345] Beevor, Anthony. *Stalingrad The Fateful Siege: 1942-1943*, Viking, 1998, p. 313.
[346] Forczyk, Robert. *Red Christmas, The Tatsinskaya Airfield Raid 1942*, Osprey Publishing, 2012.
[347] Maclean, French L. *Stalingrad, The Death of the German Sixth Army on the Volga, 1942-1943 Vol 2*, Schiffer, 2013, p. 205.
[348] Jukes, Geoffrey. *Hitler's Stalingrad Decisions*, University of California Press, 1985, p. 127.
[349] Glantz, David M. *Companion to Endgame at Stalingrad*, University Press Kansas, 2014, p. 412.
[350] Sidorov, S.G. *Prisoner at Stalingrad*, 2003, National History (To the 60th Anniversary Victorious conclusion of the battle of Stalingrad).
[351] Telegin, Konstantin F. *Wars Uncounted Milestones*, Moscow,1988, Chapter 2.
[352] Beevor, Anthony. *Stalingrad the Fateful Siege: 1942-1943*, Viking, 1998, p. 318.
[353] Based upon the Historical Documents, *Material on the Staffing of the Army, for early January 1943*, Retrieved from TsAMO Archive, ID# 133373126, 133373128.
[354] Maclean, French L. *Stalingrad, The Death of the German Sixth Army on the Volga, 1942-1943 Vol 2*.
[355] Kehrig, Manfred. *Stalingrad: Analyse und Dokumentation einer Schlacht*, DVA Stuttgart, 1974, p. 633-637.
[356] Rokossovsky, K. *A Soldiers Duty*, Progress Publishers, Moscow 1985, p. 158-159.
[357] Opalev, M.N. *Unknown pages of participation of the United ARTILLERY armored trains of Red Army in the Battle of Stalingrad*, Journal of Volgograd State University, Issue number 2/2011.
[358] Sella, Ammon. *The Value of Human Life in Soviet Warfare*, Routledge, 2015, p. 35-37.
[359] Voitenko, Professor M.F. *Characteristics of the medical support for the 64th Army during the destruction of the encircled enemy troops at Stalingrad*. Voenno-Meditsinskii-Zhurnal, Vol 1987/Nov, p. 28-31.
[360] Georgievskii, A.S., O.S. Lobastov., F.A. Ivan'Kovich. *Organization of the medical services to the Soviet troops during the defense of Stalingrad (*on the 45 anniversary of the battle) 1987/Nov, p. 24.
[361] Maclean, French L. *Stalingrad, The Death of the German Sixth Army on the Volga, 1942-1943 Vol 2*, Schiffer, 2013, p. 258.
[362] The full text of the ultimatum Retrieved from the following web site, under the Highlights section. http://www.stalingrad.net/russian-hq/the-russian-ultimatum/rusultimatum.html,.
[363] Tarrant, V.E. *Stalingrad*, Leo Cooper, London, 1992, p. 197.
[364] Based upon the Historical Document, *Material on the Staffing of the Army, for 9 January 1943*, Retrieved from TsAMO Archive, ID# 133373144.
[365] Glantz, David M. *Companion to Endgame at Stalingrad*, University Press Kansas, 2014, Table 35, p. 502.
[366] Glantz, David M., Jonathan M. House, *Endgame at Stalingrad Vol. 3, book 2*, University Press Kansas, 2014, Table 12 and 13, p. 431.
[367] Tarrant, V.E. *Stalingrad*, Leo Cooper, London, 1992, p. 207. This data originally comes from Istoriya Velikoi Otechestvennoi voiny Sovetskovo Soyuza 1941-1945, Vol 3. P. 62, Moscow 1965.
[368] Aleksey, Isaev. *Stalingrad: There is No Land for Us Beyond the Volga*, Iauza Eksmo, Moscow, 2008.
[369] Glantz, David M., Jonathan M. House. *Endgame at Stalingrad Vol. 3, book 2*, University Press Kansas, 2014, p. 443, 452.
[370] Maclean, French L. *Stalingrad, The Death of the German Sixth Army on the Volga, 1942-1943 Vol 2*. p. 280.
[371] Aleksey, Isaev. *Stalingrad: There is No Land for Us Beyond the Volga*, Iauza Eksmo, Moscow, 2008.
[372] Voitenko, M.F. *Characteristics of the medical support for the 64th army during the destruction of the encircled enemy troops at Stalingrad*. Voenno-Meditsinskii Zhurnal (Military Medical Journal), Vol-1987/Nov, p. 28-31.
[373] Maclean, French L. *Stalingrad, The Death of the German Sixth Army on the Volga, 1942-1943 Vol 2*, Schiffer, 2013, p. 283.
[374] Glantz, David M. *Companion to Endgame at Stalingrad*, University Press Kansas, 2014, p. 537-538.
[375] Bastable, Jonathan. *Voices from Stalingrad*, D&C, 2006, p. 242.
[376] Glantz, David M., Jonathan M. House. *Endgame at Stalingrad Vol. 3, book 2*, University Press Kansas, 2014, p. 510.
[377] Ziemke, Earl F., Magna E. Bauer. *Moscow to Stalingrad: Decision in the East*, Army Historical Series, 1987, p.78.
[378] Beevor, Anthony. *Stalingrad The Fateful Siege: 1942-1943*, Viking, 1998, p. 412.

[379] Joly, Anton. *Stalingrad Battle Atlas, Oct 14 –Nov 18*, Stal Data Publications, 2014, p. 24.
[380] Aleksey, Isaev. *Stalingrad: There is No Land for Us Beyond the Volga*, Iauza Eksmo, Moscow, 2008, table 28.
[381] Glantz, David M. *Companion to Endgame at Stalingrad*, University Press Kansas, 2014, p. 565.
[382] Ibid., 566.
[383] Tarrant, V.E. *Stalingrad*, Leo Cooper, London, 1992, p. 215.
[384] Ibid., 217.
[385] Sidorov, S.G. *Prisoner at Stalingrad*, 2003, National History (To the 60th Anniversary Victorious conclusion of the battle of Stalingrad).
[386] Laskin, Ivan A. *On the path to the Turning Point*. Moscow, Voenizdat, 1977, Part 2 - Chapter 5.
[387] Glantz, David M., Jonathan M. House. *Endgame at Stalingrad Vol. 3, book 2*, University Press Kansas, 2014, p. 521.
[388] Maclean, French L. *Stalingrad, The Death of the German Sixth Army on the Volga, 1942-1943 Vol 2*, Schiffer, 2013, p. 322.
[389] Aleksey, Isaev. *Stalingrad: There is No Land for Us Beyond the Volga*, Iauza Eksmo, Moscow 2008, p. 417.
[390] Laskin, Ivan A. *On the path to the Turning Point*. Moscow, Voenizdat, 1977 Part 2 Chapter 5, 1977.
[391] Glantz, David M. *Companion to Endgame at Stalingrad*, University Press Kansas, 2014, p. 584.
[392] Abdulin, Mansur. *Red Road from Stalingrad*, Stackpole Books, 2004.
[393] Zhagal, Viktor Makarovich. *Clearing the path of the infantry*, Military Publishing, Moscow, 1985, Chapter: Tornado on the Volga.
[394] Tarrant, V.E. *Stalingrad*, Leo Cooper, London, 1992, p. 220.
[395] Adam, Wilhelm. Otto Ruhle. *With Paulus at Stalingrad*, Pen and Sword, 2015, p. 214-219.
[396] Laskin, Ivan A. *On the path to the Turning Point*. Part 2 - Chapter 5, 1977.
[397] From an article called *SHUMILOV Mikhail Stepanovich Hero of the Soviet Union, the colonel general.* (n.d.) Retrieved from the web site: http://wwii-soldat.narod.ru/MARSHALS/ARTICLES/shumilov.htm.
[398] This is based on BSSA data & Historical Documents, *Strength Reports* from, Retrieved from TsAMO Archive ID #133373193.
[399] Marchand, Jean-Luc. *Order of Battle Soviet Army World War 2 Vol IX*, Nafziger Collection, Ohio, 2010, p. 99. Also, these totals are based upon BSA, BSSA data.
[400] Glantz, David M., Jonathan M. House. *Endgame at Stalingrad Vol. 3, book 2*, University Press Kansas, 2014, p. 501.
[401] Laskin, Ivan A. *On the path to the Turning Point*. Moscow, Voenizdat, 1977 Part 2 - Chapter 5.
[402] Voitenko, Professor M.F. *Characteristics of the medical support for the 64th Army during the destruction of the encircled enemy troops at Stalingrad.* Voenno-Meditsinskii-Zhurnal, vol 1987/Nov, p. 28-31.
[403] Elansky, Nikolai N. *Military Field Surgery*, Features of military surgery section, 1942. Retrieved from the web site: http://www.rkka.ru/docs/med.htm#g26.
[404] Historical Document, *Report card of DonF activities from 10.1 to 22.1.43.*, lists 16,947 casualties. Retrieved from TsAMO Archive, ID#100210761. This is far less than what is reported from medical sources, 21,113 casualties by Professor M.F. Voitenko.
[405] These values are based on my estimates of losses for the 64th Army, and the book - *Stalingrad: There is no land for us Behind the Volga*, Isaev Aleksey. Moscow, 2008. Table 27 & 28.
[406] Voitenko, Professor M.F. *Characteristics of the medical support for the 64th Army during the destruction of the encircled enemy troops at Stalingrad.* Voenno-Meditsinskii-Zhurnal, Vol 1987/Nov, p. 28-31.
[407] Laskin, Ivan A. *On the path to the Turning Point*. Moscow, Voenizdat, 1977 Part 2 - Chapter 5.
[408] Historical Document, *Information on the damage inflicted by the troops of 64 A for 10 Jan to 2 Feb 43*, Retrieved from TsAMO Archive, ID # 133373248.
[409] Aleksey, Isaev. *Stalingrad: There is No Land for Us Beyond the Volga*, Iauza Eksmo, Moscow, 2008.
[410] Glantz, David M. *Companion to Endgame at Stalingrad*, University Press Kansas, 2014 Appendix 20 D, Table 47, p. 805.
[411] Corti, Eugenio. *Few Returned, Twenty-eight days on the Russian Front, Winter 1942-1943*, University of Missouri Press, 1997, p. 244-245.
[412] Kehrig, Manfred. *Stalingrad Analyse und Dokumentation Einer Schlacht*. 1974, DVA Stuttgart, Chart 5, p. 662-663.

[413] Glantz, David M. *Journal of Slavic Studies 21: 377-471,* 2008, p. 468-469.
[414] Moorhouse, Roger. *Berlin at War*, Basic Books, 2012, p. 118, 124.
[415] Glantz, David M., Jonathan M. House. *Endgame at Stalingrad Vol. 3, book 2*, University Press Kansas, 2014, p. 583-584.
[416] Glantz, David M. *Companion to Endgame at Stalingrad*, University Press Kansas, 2014 Appendix 18 W, p. 601-603.
[417] Laskin, Ivan A. *On the path to the Turning Point*. Moscow, Voenizdat, Part 2, 1977, Chapter 5.
[418] Ibid.,
[419] Retrieved from the following web site: militera.lib.ru. *Collection of documents, Stavka VGK, Documents and materials. 1941-1945.* 1996-1999.
[420] Holl, Adelbert. *After Stalingrad, Seven Years as a Soviet Prisoner of War,* Pen & Sword, 2016, p. 10-29.
[421] Maclean, French L. *Stalingrad, The Death of the German Sixth Army on the Volga, 1942-1943 vol 2*, p. 334.
[422] *Hospitals- Part 6, Hospitals for prisoners of war.* Retrieved from the following web site: http://www.teatrskazka.com/Raznoe/Perechni_voisk/Perechen_28_06.html, (n.d.). They were POW hospitals #1, 2, 3, 4, 5, 6, 8, 9. All but #2 & #6 were disbanded during the war.
[423] Busch, Reinhold. *Survivors of Stalingrad, Eyewitness Accounts from the Sixth Army, 1942-43*. Frontline Books, 2012, p. 59.
[424] Cherkashin, Nikolay. *Why did the captive Germans pray for a Russian doctor?* (n.d.). Retrieved from the web site: http://padalkoy-d.livejournal.com/152925.html
[425] Maclean, French L. *Stalingrad, The Death of the German Sixth Army on the Volga, 1942-1943 vol 2,* Schiffer, 2013, p. 333-336.
[426] Based upon the Historical Document, *Information on the number of 64 A and the parts that are on allowance, for 17 February 1943,* Retrieved from TsAMO Archive ID # 133373235.
[427] Zagorulko, Maxim., Boris Usyk. From the paper *Stalingradskaya Gruppa Voisk* (Stalingrad Group of Forces, February – May 1943) Volgograd State University & State Museum-panorama "The Battle of Stalingrad". Retrieved from http://www.marshals-victory.senat.org/marshalvictory/Stalingrad/stalingrad-grupp.html
[428] Glantz, David M. *Companion to Endgame at Stalingrad,* University Press Kansas, 2014, Appendix 18 W, p. 603. It's not clear what is meant by "separate regions", but it is evidently referring to Axis solders outside Stalingrad proper that were trying to escape back to friendly territory.
[429] Glantz, David M., Jonathan M. House. *Endgame at Stalingrad vol. 3, book 2*, University Press Kansas, 2014, p. 573.
[430] Sidorov, S.G. *Prisoner at Stalingrad*, 2003, National History (To the 60th Anniversary Victorious conclusion of the battle of Stalingrad).
[431] Zagorulko, Maxim., Boris Usyk. From the paper *Stalingradskaya Gruppa Voisk* (Stalingrad Group of Forces, February - May 1943) Volgograd State University & State Museum-panorama "The Battle of Stalingrad". Retrieved from http://www.marshals-victory.senat.org/marshalvictory/Stalingrad/stalingrad-grupp.html
[432] Maxim, Zagorulko., Boris Usyk. From the paper *Stalingradskaya Gruppa Voisk* (Stalingrad Group of Forces, February - May 1943) Volgograd State University & State Museum-panorama "The Battle of Stalingrad". Document #20, 20.1 from February and March 1943, p. 44, 47-52.
[433] Maclean, French L. *Stalingrad, The Death of the German Sixth Army on the Volga, 1942-1943 vol 2,* Schiffer, 2013, p. 334-335.
[434] Ibid., 335.
[435] Zagorulko, Maxim., Boris Usyk. From the paper *Stalingradskaya Gruppa Voisk* (Stalingrad Group of Forces, February – May 1943) Volgograd State University & State Museum-panorama "The Battle of Stalingrad". Document #22. P. 68-69. Retrieved from http://www.marshals-victory.senat.org/marshalvictory/Stalingrad/stalingrad-grupp.html
[436] Historical Document, *Account card on numerical and combat composition of units, 64 A, 28 Feb 43,* Retrieved from TsAMO Archive, ID #133373251.

Selected Bibliography

Abdulin, Mansur. *Red Road from Stalingrad.* Stackpole Books, 2008.

Adam, Wilhelm, and Otto Ruhle. *With Paulus at Stalingrad.* Pen and Sword, 2015.

Aleksey, Isaev, *Stalingrad There is No Land for Us Beyond the Volga.* Moscow, Iauza Eksmo, 2008.

Bastable, Jonathan. *Voices from Stalingrad.* D&C, 2006.

Beevor, Anthony. *Stalingrad The Fateful Siege 1942-1943.* Viking, 1998.

Bellamy, Chris. *Absolute War.* New York, Alfred A. Knopf, 2007.

Bergstrom, Christer, and Andrey Mikhailov. *Black Cross Red Star vol 2.* Pacifica, Pacifica Military History, 2001.

Bergstrom, Christer. *Stalingrad - the Air Battle.* Hinckley Midland, 2007.

Boevoi sostav Sovetskoi armii, chast'1 (июнь - декабрь 1941 года) – Combat Composition of the Red Army, part 1 (June-December 1941). (BSSA) Moscow, Military History Department, 1963.

Busch, Reinhold. *Survivors of Stalingrad.* London, Frontline Books, 2012.

Chuikov, V. I. *Battle of the Century.* Moscow, militera.lib.ru, 1975.

—. *The Beginning of the Road.* London, Macgibbon & Kee, 1963.

Corti, Eugenio. *Few Returned, Twenty-eight days on the Russian Front, Winter 1942-1943.* University of Missouri Press, 1997.

Davie, H.G.W. *The Influence of Railways on Military Operations in the Russo-German War 1041-1945. The Journal of Slavic Military Studies, Taylor & Francis Group*, vol 30, Number 2, April-June 2017, p. 321-346.

Davie, Hugh G. W. *German Railways in the East.* 2018, https://forum.axishistory.com//forum. (Accessed 2013).

Deichmann, General der Flieger Paul., Edited by Dr. Alfred Price. *Spearhead for Blitzkrieg - Luftwaffe Operations in Support of the Army 1939-1945.* New York, Ivy Books, 1996.

Dinardo, R.L. *Mechanized Juggernaut or Military Anachronism.* Mechanicsburg, Stackpole Books, 2008.

Doerr, Hans. *The March on Stalingrad. (in German).* Darmstadt E.S., Mittler & Sohn GmbH, 1955.

Dunn, JR, Walter S. *Hitler's Nemesis.* London, Praeger, 1994.

—. *Soviet Blitzkrieg.* Mechanicsburg, Stackpole Books, 2008.

Elansky, Nikolai N. *Military Field Surgery, Features of military surgery section, 1942.* (n.d.) http://www.rkka.ru/docs/med.htm#g26.

Falk, Dann, *The Military Geographic Significance of the Volgograd Region*. Master's Thesis. Chico, California State University Chico (CSUC), 1992.

Fialkovskii, Leonid. *Stalingrad Apocalypse. Tank Brigade in Hell,* Chapter Two, 13 August 1942. Moscow, Yauza, Eksmo, 2011.

Forczyk, Robert. *Red Christmas, The Tatsinskaya Airfield Raid 1942.* Oxford, Osprey Publishing, 2012.

Georgievskii, A.S., O.S. Lobastov, F.A. Ivan Kovich. *Organization of medical support for Soviet troops in the Stalingrad offensive operation.* (on the 45 anniversary of the battle). *Voenno Meditsinskii-Zhurnal*, 1987/Nov 11.

Glantz, David M. *Atlas of Operation Blue, June - November 1942.* David M. Glantz, 1998.

—. *Combat Documents, The Struggle for Stalingrad City, vol 1, 3 Sept to Nov 18 1942.* David M. Glantz, 2007.

—. *Companion to Endgame at Stalingrad.* Kansas University Press Kansas, 2014.

Glantz, David M., and Jonathan M. House, *To the Gates of Stalingrad Vol. 1.* University Press Kansas, 2009.

—. *Armageddon in Stalingrad Vol. 2.* University Press Kansas, 2009.

—. *Endgame at Stalingrad Vol 3 book 1.* University Press Kansas, 2014.

—. *Endgame at Stalingrad Vol 3 book 2.* University Press Kansas, 2014.

Grinevsky, V.V., A.D. Ovsyannikov., I.M. Ryzhov., & A.N. Yanchinsky. *Heroic Sixty Fourth.* Volgograd, Ninshe-Volzhskoye Publishing House, 1981.

Gudkova, Galina D. *They Will Live,* Chapter six. Moscow, Mol. Guard, 1986.

Hardesty, Von., and Ilya Crinberg. *Red Phoenix Rising.* University Press Kansas, 2012.

Hayward, Joel S. A. *Stopped at Stalingrad - The Luftwaffe and Hitler's Defeat in the East 1942-1943.* Lawrence, University Press Kansas, 1998.

Hellbeck, Jochen. *Stalingrad – The City that Defeated the Third Reich.* New York, PublicAffairs, 2015.

Holl, Adelbert. *After Stalingrad, Seven Years as a Soviet Prisoner of War.* Pen & Sword, 2016.

Hoyt, Edwin P. *The Battle for Stalingrad 199 days.* New York, A Tom Doherty Associates Book, 1993.

Ilyich, Loktionov Ivan. *Volga Flotilla in the World War II.* Moscow, Military Publishing, 1974.

Ivanovich, Eremenko Andrey. *Stalingrad: Memoirs of the Commander of the Front, Chapter XIII.* Moscow, Military Publishing, 1961.

Joly, Anton. *Stalingrad Battle Atlas, Oct 14 –Nov 18.* Stal Data Publications, 2014.

Jones, Michael K. *Stalingrad, How the Red Army Survived the German Onslaught.* Philadelphia, Casemate, 2007.

Jukes, Geoffrey. *Hitler's Stalingrad Decisions.* Berkeley CA, University of California Press, 1985.

Kehrig, Manfred. *Stalingrad Analyse und Dokumentation Einer Schlacht.* Stuttgart, Deutsche Verlag Anstalt, 1974.

Kerr, Walter. *The Secret of Stalingrad.* London, Macdonald and Jane's, 1978.

Klemin, A. S. *Military Transport During the years of the Great Patriotic War, #3 March. Voyenno-Istoricheskiy Zhurnal*, 1985, p. 78-87.

Kopenhagen, Wilfried. *Armored Trains of the Soviet Union, 1917-1945.* Atglen, Schiffer Military History, 1996.

Laskin, Ivan A. *On the Path to the Turning Point,* Part 2. Moscow, Voenizdat, 1977.

Lemelsen, Joachim. *29 Division.* Bad Nauheim, Podzun-Verlag, 1960.

Maclean, Colonel French L. *Stalingrad The Death of the German Sixth Army on the Volga 1942-1943 vol 1-2.* Atglen PA, Schiffer, 2013.

Manstein, Field Marshal Eric von. *Lost Victories.* Novato, Presidio Press, 1982.

Mark, Jason D. *Death of the Leaping Horseman.* Sydney, Leaping Horseman Books, 2003.

Mark, Jason D. *Panzer Krieg vol 1, German Armoured Operations at Stalingrad.* Sydney Australia, Leaping Horseman Books, 2017.

Melvin, Major General Mungo. *Manstein - Hitler's Greatest General.* New York, Thomas Dunne Books St. Martin's Press, 2010.

Moorhouse, Roger. *Berlin at War.* Basic Books, 2012.

Morozov, Ivan. *From Stalingrad To Prague.* Volgograd, Nizhne-Volzhskoye Publishing House, 1976.

Opalev, M.N. U*nknown pages of participation of the United Artillery armored trains of Red Army in the Battle of Stalingrad.* Journal of Volgograd State University, 2011: Issue number 2/2011.

Overy, Richard. *Russia's War.* London, Penguin Putnam Inc, 1997.

Popov, P.P., A.V. Kozlov, and B.G. Usik. *Turning Point.* Sydney, Leaping Horseman Books, 2008.

Pamyat Naroda, TsAMO Archive - Tsentral'nyi arkhiv Ministerstva Oborony (Central Archives of the Ministry of Defense). (n.d.) https://pamyat-naroda.ru/ (Accessed 2015-2018).

Raus, General Erhard. *Panzer Operations.* Cambridge, Da Capo Press, 2003.

—. *Panzers on the Eastern Front.* London, Greenhill Books, 2006.

Rokossovsky, K. *A Soldiers Duty.* Moscow, Progress Publishers, 1985.

Rotundo, Louis. (Ed) *Battle for Stalingrad - The 1943 Soviet General Staff Study.* Exeter, Pergamon Brassey's, 1989.

Schroter, Heinz. *Stalingrad the Battle that Changed the World.* New York, E. P. Dutton & Company Inc, 1958.

Sella, Ammon. *The Value of Human Life in Soviet Warfare.* Routledge, 2015.

Sharp, Charles C. *Soviet Order of Battle WW2* vol 1-12. West Chester: George F. Nafziger, 1995-1998.

Sidorov, S.G. *Prisoner at Stalingrad.* National History - To the 60th Anniversary Victorious Conclusion of the Battle of Stalingrad, 2003.

Tarrant, V.E. *Stalingrad.* London, Leo Cooper, 1992.

Taylor, Brian. *Barbarossa to Berlin,* vol 1 & 2. Staplehurst, Spellmount, 2003.

Telegin, Konstantin F. *Wars Uncounted Milestones, Chapter 2.* Moscow, 1988.

Voitenko, Professor M.F. *Characteristics of the medical support for the 64th Army during the destruction of the encircled enemy troops at Stalingrad.* Voenno-Meditsinskii-Zhurnal, 1987/Nov: p. 28-31.

Wijers, Hans. *Winter Storm.* Mechanicsburg, Stackpole Military, 2012.

Zagorulko, Maxim., and Boris Usyk. *Stalingradskaya Gruppa Voisk. (Stalingrad Group of Forces,* February – May 1943*).* n.d. http://www.marshals-victory.senat.org/marshalvictory/Stalingrad/stalingradgrupp.html (Accessed 2017).

Zaitsev, Vassili. *Notes of a Russian Sniper.* Yorkshire, Frontline Books, 2009.

Zaloga, Steven J., Leland S. Ness. *Red Army Handbook 1939-1945.* Phoenix Mill, Sutton Publishing, 1998.

Zhagal, Viktor Makarovich. *Clearing the path of the infantry* - Chapter: Tornado on the Volga. Moscow, Military Publishing, 1985.

Zhilin V.A., V.A. Gredzhev, O.V. Saksonov. Chernogor V. Yu., Shirokov V.L. *Battle of Stalingrad. Chronicle, Facts, People. (Сталинградская битва. Хроника, факты, люди)* Team of Authors. Moscow, Olma-Press, 2002.

Zhukov, Georgy. *Reminiscences and Reflections, vol 2.* Moscow, Progress Publishers, 1974.

Ziemke, Earl F., Magna E. Bauer. *Moscow to Stalingrad Decision in the East.* Army Historical Series, 1987.

Zvonnitsa, M.G. *Stalingrad Epic.* Units of the NKVD of the USSR: The Department of Special Departments and the 3rd branch of the 2nd Special Division - military counterintelligence and censorship. 2000. Retrieved from http://militera.lib.ru/docs/da/sta (Accessed 2016).

Websites

English language:

Soviet Army (RKKA) in World War II. http://www.armchairgeneral.com/rkkaww2/.
Axis and Allied forces in World War II. https://forum.axishistory.com//.
Maps and Aerial Photos from World War II. http://www.wwii-photos-maps.com/index.html
Site devoted to the Battle for Stalingrad. www.staldata.com.
Site devoted to the Battle for Stalingrad. www.stalingrad.net.

Russian language:

Raw materials about RKKA in the World War II and before. http://www.soldat.ru/.
Soviet/Russian Military Books and Memoirs. http://militera.lib.ru/.
Military Maps of the Second World War and info about the Red Army. http://www.samsv.narod.ru/.

Index

1 Ru ID, 88, 89
112 RD, 37, 47, 50, 52, 62, 63, 67, 72, 74, 75
118 FR, 72, 75, 87, 93, 94, 95, 106, 133, 139, 144, 167
126 RD, 43, 67, 75, 90, 94, 95, 96, 99, 100, 103, 104, 106, 107, 133, 138, 139, 143, 162, 165, 166, 168, 171, 175
13 Guards Rifle Division, 144, 160
13 Tank Brigade, 90, 184
13 Tank Corps, 83, 84, 87, 89, 90, 99, 133, 135, 143, 172, 184, 190
131 RD, 37, 47, 50, 51, 52, 56, 57, 139
133 TB, 90, 99, 133, 135, 138
137 Tank Brigades, 57, 62
137 TB, 62, 63, 67, 68, 71, 72, 75, 86
138 RD, 79, 80, 94, 95, 99, 133, 138, 139, 142, 143, 144, 151, 152, 155, 168
15 Guards, 77, 214, 217, 219
15 Guards Rifle Division, 99, 135, 171, 248, 251, 253, 255
154 Naval Rifle Brigade, 62, 216, 217, 219
154 NRB, 62, 63, 65, 68, 71, 74, 75, 76, 79, 80, 88, 93, 99, 106, 133, 139, 142, 168, 189
157 RD, 79, 80, 87, 88, 89, 94, 95, 101, 106, 133, 135, 138, 139, 151, 184, 185, 194, 215
166 Separate Tank Regiment, 190, 214
16th Air Army, 86, 106, 131, 135
173 Rifle Division, 242, 243, 248, 251, 252, 255
17th Army, 27, 42, 64, 247
18 RD, 25, 37, 47, 50, 52, 53, 56
195 RD, 32, 37
1st Guards Army, 84, 92, 100, 178, 254
1st Panzer Army, 27, 42, 64, 66, 247
1st Reserve Army, 25, 30, 31, 32, 33, 34, 35, 36, 37, 38, 41, 42, 43, 45, 46
1st Tank Army, 57, 75, 84
2 Ru ID, 103, 104
20 Ru ID, 73, 80, 89, 103, 132, 151, 153, 156, 169, 170, 184, 185, 187, 190, 217, 230
204 RD, 72, 75, 87, 89, 96, 99, 133, 135, 139, 151, 169, 180, 182, 183, 184, 185, 215, 230, 231, 248, 251, 252, 253, 255

208 RD, 80, 81, 82, 88, 89, 90, 95, 99, 100, 106, 107, 139
214 RD, 37, 46, 47, 50, 51, 52, 54, 62, 67, 68, 71, 74, 75, 79, 87, 93, 94, 251, 252, 255
229 RD, 36, 37, 41, 47, 49, 50, 51, 52, 53, 54, 62, 63, 65, 66, 67, 68, 72, 74, 75, 80, 86
23 TC, 75
24 Panzer Division, 67, 68, 72, 92, 99, 100, 102, 132, 133, 135, 136, 176
24th Army, 14, 16, 19, 20, 21, 25, 26, 27, 29, 30, 33, 49, 100, 145, 146, 228
254 Tank Brigades, 83, 89
28 Armored Train Battalion, 86, 87, 89, 90, 151
29 Motorized Division, 81, 148, 170
29 Motorized Infantry Division, 58, 81
29 RD, 27, 33, 37, 47, 50, 51, 52, 62, 71, 75, 79, 80, 83, 88, 89, 93, 96, 99, 103, 131, 133, 135, 139, 142, 144, 169, 185, 187, 194, 231, 253, 255
292 RD, 47, 49
297 Infantry Division, 92, 93, 97, 151, 152, 153, 156, 161, 165, 180, 182, 184, 187, 189, 190, 194, 206, 211, 213, 214, 215, 216, 217, 218, 219, 227
2nd Guards Army, 195, 199, 200, 201, 202, 203
321 RD, 75
36 Guards Rifle Division, 133, 135, 139, 142, 145, 146, 147, 161, 162, 165, 166, 168, 171, 175, 184, 185, 187, 195, 206, 230, 231, 246, 251, 253, 255
371 Infantry Division, 93, 99, 151, 152, 153, 154, 160, 161, 162, 165, 166, 168, 175, 176, 185, 188, 198, 210, 214, 215, 219
38 Mortised Rifle Brigade, 83, 230, 231
38 RD, 77, 88, 89, 90, 99, 133, 135, 139, 144, 147, 183, 184, 197, 198, 204, 253, 255
3rd TankArmy, 241
4 Ru ID, 89
40 Tank Brigade, 62
40 TB, 62, 75
422 RD, 89, 93, 99, 144, 145, 146, 151, 156, 160, 161, 163, 165, 166, 195, 196, 197, 198, 253, 254, 255
4th Panzer Army, 27, 38, 39, 42, 65, 66, 74, 76, 79, 80, 83, 86, 88, 92, 93, 94, 96, 97, 99, 101, 102,

103, 104, 106, 132, 135, 142, 143, 151, 156, 165, 179, 180, 182, 185
4th Tank Army, 37, 56, 76, 84, 92, 94, 99
51st Army, 60, 74, 76, 78, 79, 171, 178, 179, 182, 183, 184, 185, 187, 188, 190, 198, 199, 200, 202, 205
57th Army, 34, 77, 84, 87, 89, 90, 96, 97, 99, 102, 129, 135, 142, 144, 154, 155, 171, 178, 179, 182, 184, 187, 189, 190, 195, 196, 197, 198, 199, 202, 204, 206, 207, 211, 214, 215, 216, 217, 218, 219, 220, 221, 223, 224, 225, 226, 227, 229, 232, 233, 236
62nd Army, 43, 47, 54, 57, 60, 63, 64, 65, 66, 67, 71, 72, 74, 75, 76, 80, 83, 86, 87, 88, 89, 90, 92, 93, 94, 97, 99, 103, 129, 132, 133, 134, 136, 137, 138, 139, 140, 142, 144, 145, 146, 147, 148, 151, 152, 155, 156, 157, 160, 162, 165, 166, 168, 172, 174, 177, 185, 194, 207, 213, 227, 231
64th Army, 14, 43, 45, 46, 47, 48, 49, 50, 52, 54, 56, 57, 58, 60, 62, 63, 64, 65, 66, 67, 68, 71, 72, 74, 75, 76, 77, 79, 80, 83, 86, 87, 88, 89, 90, 92, 93, 94, 96, 97, 99, 100, 102, 103, 104, 106, 107, 129, 131, 132, 133, 135, 136, 137, 138, 139, 140, 142, 143, 144, 145, 146, 147, 148, 151, 152, 153, 154, 155, 156, 158, 159, 160, 161, 162, 163, 165, 166, 167, 168, 169, 170, 171, 172, 173, 174, 175, 176, 178, 179, 182, 183, 184, 185, 187, 188, 189, 190, 192, 193, 194, 195, 196, 197, 198, 199, 200, 201, 202, 206, 207, 208, 209, 210, 211, 212, 213, 214, 215, 216, 217, 219, 220, 221, 222, 223, 224, 226, 227, 228, 229, 230, 231, 232, 233, 236, 237, 238, 239, 240, 241, 242, 244, 245, 246, 247, 248, 249, 252, 253, 254, 255, 256
66 NRB, 62, 63, 67, 68, 71, 72, 74, 75, 79, 87, 93, 94, 95, 106, 133, 135, 139, 142, 144, 151, 152, 153
66th Army, 49, 100, 145, 146
6th Army, 27, 34, 39, 41, 42, 63, 64, 65, 66, 67, 71, 75, 76, 88, 89, 92, 93, 94, 96, 97, 99, 100, 101, 102, 103, 106, 131, 132, 142, 143, 151, 155, 156, 157, 161, 162, 165, 168, 170, 172, 176, 182, 183, 185, 186, 187, 188, 189, 193, 194, 198, 199, 201, 202, 203, 204, 205, 206, 211, 213, 214, 218, 220, 222, 223, 224, 225, 226, 227, 228, 229, 230, 231, 232, 233, 237, 249, 253, 254
6th Reserve Army, 36
7 RC, 165
7 Rifle Corps, 153, 154, 155, 162, 165, 166, 187, 188, 197, 209, 216, 219, 220, 221, 231, 252
71 Infantry Division, 68, 72, 73, 79, 132, 176
74 Km Station, 87, 88, 90, 92, 94, 99
77 FR, 99, 139, 167, 200
77 Infantry Division, 74
7th Reserve Army, 36, 43, 63
81 Cavalry Division, 188
8th Air Army, 76, 83, 86, 89, 99, 106, 131, 135
8th Reserve Army, 36, 100
90 Tank Brigade, 214, 219
93 RD, 37
93 Rifle Brigade, 209, 225
94 Infantry Division, 92, 135
96 Rifle Brigade, 168, 231
97 Rifle Brigade, 162, 187, 219, 231
9th Army, 34, 80
9th Reserve Army, 47, 100
Abganerovo, 68, 83, 86, 87, 88, 89, 90, 92, 93, 95, 96, 97, 99, 100, 104
Adolf Hitler, 154, 155, 231, 247, 248, 249
Army Group A, 64, 65, 66, 92, 190, 194, 199, 204, 206, 208, 213, 228
Army Group B, 64, 65, 165, 176, 178, 184, 186, 194, 199, 200, 242
Army Group Don, 182, 186, 187, 194, 223, 224, 226, 227, 242
Balashov, 52, 54, 56, 255
Baskunchak, 53, 55
Beketovka, 88, 101, 133, 135, 142, 144, 151, 154, 155, 161, 165, 169, 176, 213, 227, 231, 232, 233, 244, 245, 249
Belgorod, 189, 241, 242, 246, 253
Blocking Detachments, 136, 137, 167
Cadet Rifle Regiments, 57, 62, 87
Chir River, 57, 60, 63, 67, 68, 72, 73, 74, 75, 76, 156, 178, 183, 185, 189, 194, 195, 199, 203, 205
Detraining, 46, 52, 54, 58, 60, 62, 81, 82, 255, 256
Directive #07391, 25
Directive #1, 155

Directive #1035042, 46, 47
Directive #1035043, 47
Directive #1035044, 47, 49
Directive #1035046, 47
Directive #1035055, 49
Directive #151541, 23
Directive #151549, 23
Directive #170168, 27
Directive #170333, 30
Directive #170446, 38
Directive #170495, 49
Directive #170520, 60
Directive #170535, 72
Directive #170554, 83, 86
Directive #170562, 90
Directive #170566, 93
Directive #170588, 100
Directive #170603, 137
Directive #170669, 155
Directive #170720, 206
Directive #171718, 202
Directive #46038, 241, 242
Directive #46046, 242
Directive #46053, 242
Directive #46056, 244
Directive #46063, 254
Directive #575, 50
Directive #994012, 34
Directive #994101, 43
Directive #994102, 43
Directive #994103, 43, 45, 46
Directive #994111, 49
Directive #994170, 100
Directive #994171, 100
Directive #994200, 138
Directive #994209, 147
Don Front, 147, 161, 163, 170, 178, 184, 188, 189, 193, 195, 199, 207, 209, 210, 211, 212, 213, 214, 215, 217, 218, 219, 221, 223, 224, 227, 232, 233, 236, 237, 238, 239, 241, 242, 243, 244, 249, 252
Don River, 14, 27, 30, 38, 40, 41, 42, 43, 46, 49, 54, 57, 58, 60, 62, 63, 64, 65, 66, 67, 70, 74, 75, 76, 77, 79, 80, 86, 88, 89, 92, 93, 94, 100, 143, 156, 171, 178, 183, 184, 189, 190, 194, 196, 199, 201, 208, 241, 244, 247
Elkhi, 135, 154, 180, 182, 183, 184, 185, 194, 213, 220
Field Marshal Paulus, 231, 241
General Chuikov, 34, 35, 38, 41, 46, 49, 50, 52, 56, 58, 60, 62, 63, 64, 66, 67, 68, 75, 78, 80, 81, 82, 87, 88, 94, 101, 102, 130, 132, 138, 142, 144, 147, 155, 160, 177, 185, 227
General Eremenko, 86, 90, 93, 100, 101, 103, 106, 131, 132, 137, 138, 144, 147, 148, 152, 153, 155, 156, 161, 163, 170, 171, 175, 178, 211, 241
General Gordov, 63, 64, 71, 76, 77, 79, 80, 84, 86, 90
General Hoth, 64, 66, 102, 103
General Kolpakchi, 43, 65, 71, 80
General Laskin, 227, 231, 232
General Lopatin, 80, 92, 137
General Paulus, 64, 65, 99, 106, 155, 176, 178, 183, 184, 185, 186, 214, 221, 223, 224, 225, 227, 229
General Richthofen, 54, 86, 100, 102
General Shumilov, 72, 75, 76, 78, 80, 86, 87, 88, 89, 90, 92, 93, 99, 100, 101, 102, 103, 104, 106, 129, 132, 133, 134, 138, 143, 146, 147, 151, 152, 161, 162, 163, 165, 166, 171, 177, 187, 194, 198, 201, 207, 212, 217, 225, 230, 232, 233, 241, 242, 247, 248, 253, 254
General Timoshenko, 64
General Warner Kempf, vii, 96, 102, 118, 119
General Zhukov, 18, 19, 143, 145, 170, 171, 172, 175
General. Eremenko, 84
GKO, 106, 170, 249, 251, 254, 255
GKO Decree #2911, 249, 251
GKO Order #00248, 106
Glavupraform, 253
Golodnyi Island, 152
Gorky, 254, 255
Group Chuikov, 79, 80, 87, 88, 89, 90, 92, 93, 94, 95, 96
Hitler, 12, 13, 16, 27, 28, 30, 35, 39, 40, 42, 43, 64, 65, 66, 150, 154, 176, 182, 183, 184, 186, 188, 193, 199, 202, 206, 208, 213, 214, 223, 224, 225, 228, 229, 231, 233, 234, 247, 248, 249

Hitler Directive #41, 30
Hitler Operations Order #1, 155
Hitler Operations Order #2, 206
Hiwis, 137, 186, 196, 201, 240, 264
Hungarian 2 Army, 174
Italian 8 Army, 174
IV Corps - Romanian, 178, 182, 229
Joseph Stalin, 12, 226
Kalach, 67, 93, 156, 171, 177, 178, 183, 184, 185, 189
Kharkov, 27, 34, 35, 38, 58, 177, 241, 242, 246, 247, 248, 251, 253
Khrushchev, 93, 138, 153, 171, 241
Kotelnikovo, 74, 79, 80, 81, 82, 156, 171, 188, 189, 190, 195, 205
Kuporosnoe, 139, 143, 144, 146, 148, 162, 165, 175, 176, 214, 226
Lieutenant General Vasilii I. Chuikov, 34
Logistical, 27, 31, 34, 38, 44, 45, 46, 50, 56, 64, 138, 147, 148, 156, 157, 158, 159, 160, 172, 186, 194, 195, 199, 202, 208, 210, 252, 255
Logistics, 227
Luftwaffe, 12, 13, 39, 42, 54, 55, 56, 58, 68, 75, 81, 83, 86, 88, 90, 94, 95, 96, 99, 100, 102, 103, 104, 106, 129, 131, 132, 142, 144, 147, 154, 160, 165, 168, 178, 185, 186, 187, 188, 189, 190, 191, 193, 194, 195, 196, 197, 198, 199, 200, 201, 202, 203, 205, 206, 207, 210, 211, 213, 214, 216, 217, 218, 219, 220, 221, 222, 223, 224, 225, 227, 228, 229, 230, 231, 236, 237, 242
LVII Corps, 189, 195, 198
Major General M.S. Shumilov, 71, 129, 150, 167, 192, 207
Manstein Directive #1, 194
Manstein Gag Order #23079, 226
Moscow, 13, 14, 16, 18, 19, 20, 21, 22, 23, 25, 27, 31, 32, 33, 34, 38, 39, 41, 42, 43, 45, 47, 48, 49, 62, 64, 86, 100, 143, 170, 174, 175, 179, 202, 221, 236, 242, 245, 246, 251, 253
Nixhne-Chirskaia, 74
NKO, 23, 45, 153, 254
NKO Order #36799, 254
NKVD, 25, 136, 137, 160, 167, 207, 211, 240, 241, 242, 245, 246, 248, 249, 251, 252, 253
NKVD Memo #5086, 251
OKH, 27, 154, 155, 165, 168, 169, 170, 174, 193, 201, 213
OKW, 154, 161, 165, 174
Operation Barbarossa, 12, 13, 18
Operation Blue, 27, 28, 30, 34, 35, 38, 40, 42, 43, 64, 65, 92, 94, 194
Operation Little Saturn, 199, 200, 201, 203, 204, 208, 211
Operation Ring, 193, 195, 199, 200, 202, 204, 206, 208, 209, 210, 211, 212, 213, 214, 215, 220, 228, 237, 238, 239, 240
Operation Saturn, 190, 193, 194, 195, 197, 199
Operation Thunderclap, 194, 202
Operation Typhoon, 14, 16, 18
Operation Uranus, 143, 153, 166, 167, 170, 172, 176, 177, 179, 184, 186, 189, 190, 191, 192, 193, 213, 220, 240, 253
Operation Winter Tempest, 193, 194, 198, 201, 203, 205, 208
Order #170699, 197
Order #227, 72
Order #307, 153
Parkha River, 20, 22
Penal Battalions, 137
Peschanka, 162, 163, 197, 220, 221, 224
Povorino, 52, 54, 55, 255
POW, 96, 137, 207, 220, 227, 234, 244, 249, 252, 253
PPG - Mobile Field Hospital, 31
Razputitsa, 33
Red Airforce, 55
Romanian 3rd Army, 178, 182, 183
Romanian 4th Army, 132, 179, 182
Rtishchevo, 52, 53, 54, 255
Ryazan, 23, 25, 32, 33, 37, 38, 47, 50, 53
Saratov, 36, 53, 100, 255
Sarpinskii Island, 167
Secret Order #0804, 248
Shumilov Order #0186, 166
Southeastern Front, 84, 90, 93, 94, 99, 103, 104, 131, 134, 135, 137, 138, 145
Southwestern Front, 170, 185, 189, 195, 254
StalGRES power plant, 169

Stalin, 12, 13, 14, 18, 21, 27, 28, 30, 31, 34, 35, 38, 39, 40, 41, 42, 43, 47, 49, 58, 60, 63, 64, 66, 72, 83, 86, 88, 93, 100, 101, 102, 136, 137, 138, 143, 148, 150, 151, 152, 153, 154, 156, 157, 161, 167, 172, 174, 175, 179, 182, 186, 188, 190, 191, 193, 194, 195, 197, 198, 199, 201, 202, 208, 214, 221, 222, 227, 236, 237, 241, 242, 246, 253

Stalin Directive #227, 136

Stalingrad, 27, 30, 36, 37, 40, 42, 43, 44, 46, 47, 49, 50, 52, 53, 54, 55, 56, 57, 58, 60, 62, 63, 64, 65, 66, 67, 71, 72, 74, 76, 77, 79, 80, 82, 83, 84, 86, 88, 90, 92, 93, 94, 96, 97, 99, 100, 101, 102, 103, 104, 105, 106, 107, 129, 130, 131, 132, 133, 134, 135, 136, 137, 138, 139, 140, 142, 143, 144, 145, 146, 147, 148, 150, 151, 152, 153, 154, 155, 156, 157, 159, 160, 161, 162, 163, 165, 166, 167, 168, 170, 171, 172, 174, 175, 176, 177, 178, 182, 183, 184, 185, 186, 187, 188, 189, 190, 191, 192, 193, 194, 195, 197, 198, 199, 200, 201, 202, 203, 205, 206, 207, 208, 209, 210, 211, 212, 213, 214, 216, 217, 218, 219, 220, 221, 223, 224, 226, 227, 228, 229, 230, 232, 233, 234, 236, 237, 239, 240, 241, 242, 243, 244, 245, 246, 247, 248, 249, 251, 252, 253, 255, 256, 263

Stalingrad Front, 49, 52, 56, 57, 58, 60, 63, 64, 67, 71, 72, 74, 76, 77, 80, 82, 83, 84, 86, 88, 90, 92, 93, 107, 135, 137, 145, 146, 147, 152, 153, 154, 155, 157, 167, 170, 171, 174, 175, 178, 182, 183, 187, 189, 193, 195, 197, 199, 206, 209, 212

Stavka, 14, 16, 18, 19, 23, 25, 26, 27, 29, 30, 31, 33, 34, 36, 38, 39, 40, 41, 42, 43, 44, 45, 46, 49, 52, 57, 58, 60, 62, 63, 67, 72, 80, 83, 86, 90, 92, 93, 94, 100, 103, 132, 136, 137, 143, 145, 147, 152, 153, 154, 155, 167, 179, 182, 184, 186, 187, 189, 190, 191, 193, 194, 195, 197, 198, 199, 200, 201, 202, 206, 208, 214, 219, 221, 228, 229, 236, 241, 242, 243, 244, 251, 254, 256

Stavka Order #170708, 200

SW Front Order #4, 131

Taman Peninsula, 27, 65, 247

Tambov, 27, 36, 37, 41, 52, 53, 54

Tatsinskaya Airfield, 205, 206

Tinguta Station, 87, 99, 100, 106

Troop Trains, 50, 52, 55, 56, 58, 60, 82, 255

Tsaritsa River, 143, 224, 227, 228, 244

Tsimlianskaia, 76

Tula, 16, 23, 24, 25, 27, 29, 31, 32, 33, 36, 37, 38, 42, 43, 47, 49, 50, 52, 54, 57

Univermag Department Store, 231

Urbakh, 53, 55, 255

Valuiki, 253, 254, 255

VGK, 19, 45, 49, 52

VI Ru Army Corps, 74, 80

Volga River, 27, 40, 64, 66, 96, 99, 143, 148, 159, 162, 165, 167, 176, 213, 241, 256

Voronezh, 27, 36, 38, 39, 40, 41, 42, 43, 54, 56, 64, 194, 227, 241, 248, 249, 251, 254, 256

VOSO, 45, 46, 49

VVS, 13, 29, 55, 65, 76, 82, 83, 86, 90, 96, 97, 103, 106, 131, 132, 135, 142, 177, 178, 179, 189, 191, 199, 200, 214, 215, 216, 217, 218, 219, 220, 221, 222, 223, 224, 225, 227, 228, 229, 230, 231, 237

XI Corps, 178, 183, 184, 187, 188, 227, 231, 236

XLVIII Corps, 189

XXXXVIII Panzer Corps, 96, 102, 103, 139

Yelets, 54, 255

Zhutovo Station, 50, 81

www.ingramcontent.com/pod-product-compliance
Lightning Source LLC
Chambersburg PA
CBHW082112230426
43671CB00015B/2671